Dedication

This book is dedicated to all who place their lives on the line for the benefit and safety of others. In the post 9-11 world, we have developed a renewed appreciation for those who give the ultimate gift of life to others through service and self-sacrifice, and our world is a better place as a direct result of the contributions of these heroes.

We also wish to dedicate this book to those who devote their lives to providing a fire-safe environment for others through competent automatic sprinkler system design. Fire protection designers, technicians, and engineers—as well as architects, code officials, authorities having jurisdiction, and fire science instructors—save the lives of persons whose names they will never know. The design of automatic sprinkler systems is a technical profession where the lasting value of one's work has the potential to provide life-saving service long after the design is complete and long after the designer has departed from this earth. In the conduct of one's life, no earthly reward is as great as the knowledge that one's life has served to save and improve the lives of others on this planet.

Metrication

The authors of the *Designer's Guide to Automatic Sprinkler Systems* have used SI units wherever practical. In some instances, however, U.S. customary units have been retained. For example, when equations or design methodologies have input variables or constants that have been developed from data originally in U.S. customary units, those units are retained.

Contents

Preface

This book is a byproduct of the confluence of two influential organizations representing the fire protection profession, the Society of Fire Protection Engineers and the National Fire Protection Association. Each of the individuals who dedicated well over a year of their lives to this book believes in the goals of these organizations and works hard to fulfill their missions. The positive effects of all who serve SFPE and NFPA and contribute to their missions are felt in all corners of the world, and all SFPE and NFPA members are rewarded by the knowledge that their work saves lives. The editor and authors urge all who read this book to become active members of SFPE and NFPA. While performing automatic sprinkler system design is rewarding in itself, one also contributes to the profession by serving on NFPA committees and being active in a local chapter or national committee of SFPE. Fire protection codes, standards, and recommended practices are living, vibrant, and relevant statements of one's commitment to making this a better world. More information on SFPE and NFPA can be found on their respective web sites (www.sfpe.org and www.nfpa.org).

Fire protection engineers and engineers in related disciplines such as mechanical, electrical, chemical, HVAC, architectural, petroleum, industrial, nuclear, aeronautical, mining, metallurgical, structural, agricultural, and civil engineering are the primary focus of this book. In addition, it is our hope that technicians, architects, code officials, life safety professionals, installation contractors, insurance underwriters, building inspectors, physical plant managers, facility managers, and authorities having jurisdiction will find this book a useful tool in understanding automatic sprinkler system design and its relationship to other systems and processes in a building. Students of fire science and fire protection engineering and newly hired design professionals will find this book an essential training tool, and practicing professionals will find the book to be a valuable daily professional reference.

NFPA codes, standards, and recommended practices are the world's foremost source of requirements and guidelines on the design of fire protection systems and the building of fire-safe structures. This book is based on NFPA codes, standards, and recommended practices. It is not intended to replace them, but rather to expand upon them. Obtaining the appropriate NFPA documents, such as NFPA 13, *Standard for the Installation of Sprinkler Systems*, and using them in conjunction with this book, is therefore a necessity. All references to NFPA 13 in this book apply to the 2002 edition.

This book tells the story of good automatic sprinkler system design practice in three parts. Part 1, "Automatic Sprinkler Theory and Design Approaches," describes the philosophies, concepts, and approaches that designers are required to take before beginning an automatic sprinkler system design. To understand the fundamentals of control and suppression is to understand the benefits and limitations of automatic sprinkler systems—consequently, Chapter 1 addresses "Fundamentals of Sprinkler Performance." Recognizing our role as ambassadors of the installation of sprinkler systems, Chapter 2, "Impact of Sprinkler Systems on Building Design," is an important guide to the value of sprinkler systems and a potential tool for convincing builders that sprinkler systems are a wise choice. Chapters 3 addresses the basics of "Design Approaches."

Fundamental sprinkler concepts include occupancy (discussed in Chapter 4, "Hazard and Commodity Classification," sprinkler system hardware (covered in Chapter 5, "Selection of Automatic Sprinkler System Materials and Components"), and water supply analysis (Chapter 6, "Evaluation of Water Supply"). The accuracy and relevance of any design are intrinsically linked to these topics. Part 1 closes with the specialized topic of "Impact of Unique Applications on Design Approaches" (Chapter 7).

Based on appropriate selections of occupancy and proper water supply analysis, Part 2, "Design Implementation," is a comprehensive coverage of the analytic machinery that operates during the automatic sprinkler system design process. Part 2 should become an important training tool and reference for designers of automatic sprinkler systems and covers "Overview of Sprinkler System Layout" (Chapter 8), "Automatic Sprinkler System Spacing" (Chapter 9), "Hydraulic Calculations" (Chapter 10), "Backflow Protection for Fire Protection Cross Connections" (Chapter 11), "The Role of Fire Pumps in Automatic Sprinkler Protection" (Chapter 12), and "Protection of Storage" (Chapter 13). Chapter 14, "Specification Writing," and Chapter 15, "Shop Drawings for Automatic Sprinkler Systems," present material that is central to the daily responsibilities of consulting engineers, authorities having jurisdiction, and code officials.

Part 3, "Regulation and Professional Responsibility," is an appropriate way to wrap up a comprehensive coverage of automatic sprinkler system design, with such topics as coordination and professional responsibility. Chapter 16, "Fire Sprinkler Codes and Standards," provides background information on the many documents a designer must consult. Currently expanding in the automatic sprinkler system design profession is the concept of performance-based design, which is given a rigorous treatment in Chapter 17, "The Performance-Based Design Process and Automatic Sprinkler System Design." "Coordination with Other Professionals and Trades" (Chapter 18) deserves coverage in this book because of its prime importance in the building construction process. Part 3 closes with a review of "Professional Issues in Fire Protection Engineering Design" (Chapter 19).

Acknowledgments

Working with the authors of this book, all of whom are SFPE and NFPA members, gave me a revitalized appreciation for the giving of one's time and expertise for the betterment of society, which I have seen exemplified by fire protection professionals throughout my 36-year career in fire protection. The authors represent a cross-section of the pinnacle of SFPE and NFPA leadership, not only in level of expertise, but in the spirit of unselfish giving that attracted me to this profession in the first place. They are all consummate professionals and exceptional citizens, and I am proud to consider them my colleagues and peers.

Three people deserve special thanks for the successful completion of this book. Morgan Hurley, PE, Technical Director of SFPE, was an important driving force who believed in this book from its inception and had supreme confidence in all who carried the flag, even as deadlines got tight and the customary pressures associated with writing and producing a book under a deadline mounted. Pam Powell, Senior Product Manager for NFPA, was our homing beacon and gently helped us set deadlines on a very tight schedule. She invariably provided warm encouragement at the exact time that we needed it and gave us course correction when our bearings required readjustment. In addition, I am grateful to NFPA Senior Fire Protection Engineer Christian Dubay, PE, who carefully reviewed each chapter a number of times—the success and relevance of this book are partly thanks to him.

SFPE, NFPA, the authors, and the editor want you, the reader, to consider this to be your book, in that we want you to take an active role in providing us with helpful suggestions for new or revised material in future editions. Please contact me at 410-442-1600 or via e-mail to robtgagnon@aol.com with the benefit of your expertise and helpful commentary.

Robert M. Gagnon, PE, SET, FSFPE
Editor

Editor Robert M. Gagnon

Robert M. Gagnon, PE, SET, FSFPE

Robert M. Gagnon, PE, SET, FSFPE, is president of Gagnon Engineering. Previous to his founding of Gagnon Engineering, Robert performed sprinkler and special hazard design for "Automatic" Sprinkler Corporation of America, and has 36 years of experience as a technician and an engineer in the design and layout of automatic sprinkler, special hazard, fire alarm and detection systems, and life safety analysis and design. He serves as an adjunct lecturer for courses in sprinkler, special hazard, and fire alarm design at the University of Maryland, and previously taught similar courses at Montgomery College Department of Applied Technologies.

Robert is registered as a fire protection engineer in four states, and is certified as NICET Level IV in automatic sprinkler system layout and NICET Level IV in special hazards system layout. He holds a BA in mathematics from McDaniel College (formerly Western Maryland College), and a BS and MS from the University of Maryland Department of Fire Protection Engineering. He is a member of SFPE, NFPA, the National Fire Sprinkler Association, and the American Fire Sprinkler Association.

His service to the Society of Fire Protection Engineers includes past president and current Awards chair and Registrations chair of the SFPE Chesapeake Chapter, co-chair of the SFPE Engineering Licensing Committee, past chair and current member of the SFPE Careers Task Force, member of the Education Committee and Public Awareness Committee, and mid-Atlantic regional editor of SFPE's *Fire Protection Engineering* magazine. He is the recipient of the SFPE "Hats Off" Award, the Harold E. Nelson Service Award, and was elected as Fellow of SFPE.

His current service to NFPA includes chair of the NFPA 16 Technical Committee, secretary of the NFPA 13, NFPA 15, NFPA 22, NFPA 24, and NFPA 291 Technical Committees, member of the NFPA 214 Technical Committee, and non-voting member of the Technical Correlating Committee on Automatic Sprinkler Systems. He is a current member of the Executive Committee and past vice chair of the Fire Science and Technology Educators Section, and is a member of five additional NFPA sections. He is the author of

"Ultra High Speed Suppression Systems for Explosive Hazards" in NFPA's *Fire Protection Handbook*.

Robert is the author of *Design of Water-Based Fire Protection Systems*, *Design of Special Hazard and Fire Alarm Systems*, co-author of *A Designer's Guide to Fire Alarm Systems*, co-author of *The Engineering Student's Guidebook for Professional Development*, editor of the *Department of Fire Protection Engineering 40th Anniversary History Book*, and has written numerous journal and scholarly articles.

About the Authors

Carl Anderson, PE, is the fire protection engineer for the Tacoma Fire Department, Tacoma, Washington. Carl holds an MS in fire protection engineering from Worcester Polytechnic Institute and is an alternate member of the NFPA Technical Committee on Sprinkler System Discharge Criteria.

Grant Cherkas is employed by the Canadian Nuclear Safety Commission in Ontario, Canada.

Michael A. Crowley, PE, received a BS in fire protection and safety engineering from Illinois Institute of Technology in 1980, and an MBA in 1987 from the University of Houston. He joined Rolf Jensen & Associates, Inc. in 1978 and is currently vice president of the Central region. Mr. Crowley provides professional fire protection engineering and building code consulting services for a variety of project types and sizes, as well as administrative management oversight of the Atlanta, Dallas, Chicago, Houston, Orlando, and Raleigh offices of RJA.

Brian Foster, PE, is an engineering graduate from the University of Central Florida and has been actively working in the fire protection field for 25 years. He has passed the NCEES fire protection engineering exam and maintains engineering registrations in six states. During his career, he has worked for FM Global and HSB-Professional Loss Control, and is currently a principal in Global Fire Engineering. He is the past president of the Florida Chapter of the Society of Fire Protection Engineers and served as chairman of the Florida Engineering Society's Constructed Environment Committee. Mr. Foster is also licensed as a fire safety inspector in the state of Florida.

Scott Futrell, PE, SET, is a fire protection engineer registered in Wisconsin, and a senior engineering technician (NICET IV) with Futrell Fire Consulting & Design, Inc. in Minnesota. His firm provides consulting, project management, and design of fire protection systems for a broad range of building types and occupancies, and he has been involved in the fire protection industry since 1975. Mr. Futrell is active in SFPE as well as the local chapter. He is a member of NFPA, participating in its Education, Industrial Fire Protection, and Architects, Engineers, and Building Officials sections. He is also an instructor for SFPE, Hennepin Technical College, and the Fire Instructors Association of Minnesota.

Cindy Gier, PE, is a registered fire protection engineer and president of CMG Fire Protection Engineering, Inc. and Fire Sprinkler Solutions, Inc., both in

Overland Park, Kansas. She has worked as an electrical design/build engineer and project manager, a city and fire department plans examiner, and a consulting engineer. Cindy is a Certified Building Official, a Master Code Professional and a member of NFPA, SFPE, and the International Code Council. She is an adjunct engineering professor at the University of Kansas and has taught many classes in the areas of performance-based design, code review, and fire protection engineering.

Morgan Hurley, PE, is Technical Director of the Society of Fire Protection Engineers. He is a past member of the technical committee responsible for NFPA 13. He holds a BS and MS from the University of Maryland Department of Fire Protection Engineering.

Al Moore, PE, is a vice president and senior fire protection engineer with Michaud Cooley Erickson Consulting Engineers in Minneapolis, Minnesota, and is a member of NFPA and the SFPE Education Committee.

Jack Poole, PE, received a BS in fire protection engineering from the University of Maryland. He is a licensed fire protection engineer, and is past president of the American Backflow Prevention Association and a member of NFPA, SFPE, the International Code Council, and the AWWA 146 Backflow Standards Committee. Mr. Poole is a principal of Poole Consulting Services, Inc. in Olathe, Kansas, which provides code consulting, fire protection engineering consulting and design services to architectural and engineering firms, consultants, and the industry.

Steven Scandaliato, SET, is a principal partner with Scandaliato Design Group, Inc., and is a NICET IV senior engineering technician. He has 18 years experience in fire protection layout, active and passive system design, code analysis and research, and construction project management. He serves on the Technical Advisory Committee for the AFSA and is a member of NFPA, SFPE, AFSA, and the NFPA 13, 15, and 5000 committees. He has spoken on automatic sprinkler system design for the AFSA, AIA, SFPE, American Society of Plumbing Engineers, and the International Fire Marshals Association and has had articles published in the *NFPA Journal* and *Fire Marshals Quarterly*.

Automatic Sprinkler Theory and Design Approaches

Fundamentals of Sprinkler Performance

Scott Futrell, PE, SET

Automatic sprinkler system design begins with making a determination relative to several important factors. One of the decisions that a fire protection engineer must make is whether the control or suppression philosophy of automatic sprinkler system design would best fit the specific situation or project. This chapter will provide insight into the two philosophies—control and suppression—and the role of automatic sprinklers. This discussion should assist engineers in determining which type of sprinkler is appropriate for their specific project.

Sprinkler systems, from their conception in the late 1800s to the 2002 edition of NFPA 13, *Standard for the Installation of Sprinkler Systems*, have been developed and designed with a performance objective of controlling a fire and limiting it to the room or area of origin until the fire service intervenes to complete the extinguishment of the fire. In general, the design of an automatic sprinkler system in accordance with NFPA 13 does not necessarily include suppression of a fire in advance of fire service involvement. Occasionally, however, the fire is controlled by suppressing it, but prior to the development of the early suppression fast-response (ESFR) sprinkler technology, automatic sprinkler systems were never designed to include suppression as the primary performance objective.

With the development of the ESFR sprinkler in 1988, the performance of suppression can be ensured, provided that the automatic sprinkler system and the occupancy it protects are designed, installed, and maintained in accordance with the requirements for those specific sprinklers. ESFR sprinklers are very specialized sprinklers that have been specifically designed for use primarily in storage occupancies. These sprinklers will be further discussed later in the chapter and in other chapters.

Definitions and Basic Concepts

There are two definitions that need to be evaluated. One is *fire control*: "Limiting the size of a fire by distribution of water so as to decrease the heat release rate and pre-wet adjacent combustibles, while controlling ceiling gas temperatures to avoid structural damage" (NFPA 13, 2002 edition, 3.3.9). In other words, fire control is distributing effective water spray to the area of ignition, while limiting the fire growth or spread through applying enough water to the surrounding fuel load so it will not ignite. Fire control by automatic sprinklers may require the application of water from fire department operations utilizing hose streams.

The other important definition is *fire suppression*: "Sharply reducing the heat release rate (HRR) of a fire and preventing its regrowth by means of direct and sufficient application of water through the fire plume to the burning fuel surface" (NFPA 13, 3.3.10). Fire suppression by automatic sprinklers is a relative term, as fire suppression may also require hose streams from fire department operations especially if it is a shielded fire (where the HRR is drastically reduced, but the sprinklers cannot put out the last embers or coals) or for general overhaul, the process by which fire fighters look for hot spots at a fire scene that may be smoldering or burning. It is a relative term because suppression, by definition, is reducing the HRR, but that does not necessarily mean to zero.

Why, then, would one design an automatic sprinkler system for control if suppression is possible and minimizes fire and smoke damage? The complex answer to that question takes many forms, only a few of which have been listed here. One must realize that suppression generally requires the rapid application of a large amount of water that isn't always available. Suppression may also involve large pipes and fire pumps and must be based on a specific fuel package (commodity, commodity arrangement, commodity height, building height, etc.) that may not be applicable to most general occupancies. Finally, and significantly, suppression mode sprinklers have not been developed for nonstorage occupancies. Some of the design considerations used to answer this question and the theory behind control and suppression are covered in the following sections.

Figure 1.1 shows the relationship between heat release rates and time as they relate to fire suppression and control.

Figure 1.2 is a more detailed look at the concepts in Figure 1.1. Figure 1.2 introduces the concepts of detection and design fire size and explores the relationship between detection and fire growth. Detection of a fire does not

FIGURE 1.1
Simplified Fire Control/Fire Suppression Analogy.

(Source: *Automatic Sprinkler Systems Handbook*, NFPA, 2002, Exhibit 3.2)

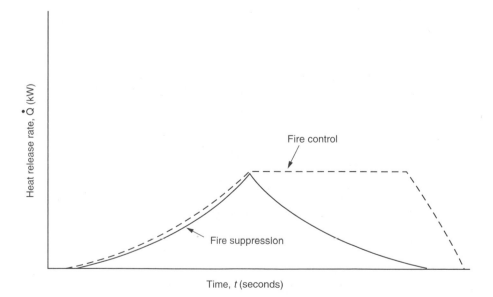

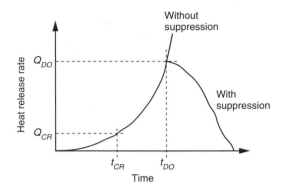

FIGURE 1.2
Critical and Design Objective Heat Release Rates vs. Time.

(Source: *National Fire Alarm Code*®, NFPA, 2002, Figure B.2.3.5)

change the growth rate. Fire growth generally continues until, limited by available oxygen, extinguishment begins. Figure 1.2 shows that over a small change in time, the heat release rate (HRR) can grow rapidly. Q_{DO} represents the HRR of the design fire, at which detection or commencement of suppression is desired. Q_{CR} on the other hand is the point in the fire development where the critical fire size has been reached and detection must have occurred to facilitate the eventual control or suppression of the fire.

Design Considerations

This section will offer definitions and additional information relative to automatic sprinkler system design, as well as on water as an extinguishing agent.

Heat Release Rate (HRR) Curves

The two generic heat release rate curves shown in Figures 1.2 and 1.3 indicate the relation of a fire's rate of heat release history versus time, from the initiation of the fire to its extinction. The HRR is the energy being released in kilowatts (Btu/sec) of a specified fuel. The area under the curve represents the total heat released into a building or area, which is proportional to the damage done by a fire. A system that automatically suppresses a fire minimizes fire and

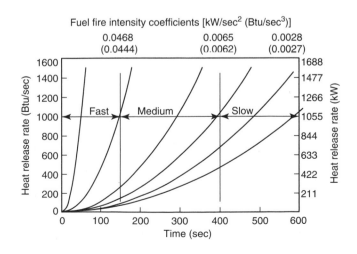

FIGURE 1.3
Power Law Heat Release Rates.

(Source: *National Fire Alarm Code*®, NFPA, 2002, Figure B.2.3.6.2)

smoke damage and should be superior to one that only controls the fire. Another way to look at this is to view different fire growth rates. Figure 1.3, which illustrates the power law heat release rate table, illustrates several fire growth rates as a function of time.

The experienced engineer will need to use the information in Figure 1.3 in conjunction with the definitions from NFPA for occupancy to determine where a product or group of products (furnishings in a room, for example) will fit into the sprinkler system design criteria.

Basic Water Characteristics

Before proceeding to a discussion of the choices of control or suppression, it is important to review some fundamental properties of water as an extinguishing agent. Water can very effectively cool most solid fuels, it can affect the oxygen supply necessary to support combustion, it can cool the fire plume, and it can modify the fuel source. Depending on the form of the water application, it is possible to successfully change the results of the fire from significant damage to control, suppression, or extinguishment.

Heat transfers from the fire to the water that is applied, and when there is an equal amount of heat transferring to the water as there is heat added by the fire, control of the fire begins. If more heat can be transferred to the water than the fire can produce (i.e., HRR), then suppression can be achieved. This heat transfer results in reducing the rate of pyrolysis of the fuel, eventually reducing the HRR of the fire, in effect cooling it. What really causes this cooling? It is the effect of the water in the form of droplets (of many sizes) and steam. The smaller these droplets, the faster they are capable of extracting the heat from the fire and the fire gases. These same small droplets will also evaporate faster, cooling the fire plume but not penetrating to cool the fuel surface. In addition, these small drops may actually be detrimental in a high-challenge fire because they can be "blown" onto adjacent sprinklers, delaying or preventing their operation.

Water droplets in the range of 0.01 to 0.08 in. (0.3 to 2.0 mm) are the most effective in penetrating the fire plume. The larger of these droplets are necessary for penetration of the fire plume developed by higher-challenge fires. Some sprinkler deflectors are developed and tested to distribute the most effective droplets for the application for which the specific sprinkler is designed. One example is that of the large drop automatic sprinkler. These sprinklers were developed specifically for use on high-challenge fires and utilize large drops for fire plume penetration.

Required Delivered Density

To understand how suppression occurs, some new terms need to be introduced and explained. One is the required delivered density (RDD), the minimum rate of water application that, if delivered to the top of the fuel package, is capable of providing early suppression (in gallons per minute per square foot or millimeters per minute) [1]. In other words, the RDD is the quantity of water required at the top surface of the fuel to cool or smother the fire sufficiently such that the fire rapidly goes out. Each fuel has a specific RDD as it burns. By putting together "typical" and "worst-case" scenarios for rooms or areas, the RDD for a given occupancy can be established. Furthermore, as the HRR of a fire increases, the RDD also increases (see Figure 1.4).

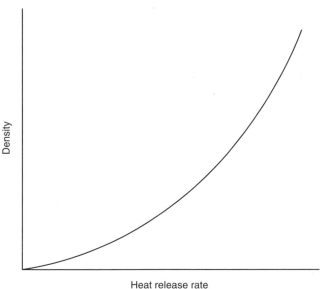

FIGURE 1.4
Generic RDD Curve.

Actual Delivered Density

Another definition is that of the actual delivered density (ADD). ADD is the actual rate of water application that a particular configuration of flowing sprinklers is capable of delivering to the top of a fuel package, depending on the strength of the upward fire plume (in gallons per minute per square foot or millimeters per minute) [2]. When the HRR of a given fire increases, the ADD decreases because as the HRR increases the fire plume becomes significantly stronger, making it much more difficult for water to penetrate the plume and reach the fuel. In addition, some of the water evaporates and the fire plume carries some of the water droplets away from the plume. The ADD is a function of each specific automatic sprinkler, the flow pattern developed, location of the sprinkler in terms of height above the fuel and horizontal distance from the centerline of the fire, upward fire plume velocity, and the burning rate of the fire, as well as the water supply available. Figure 1.5 shows a generic ADD curve.

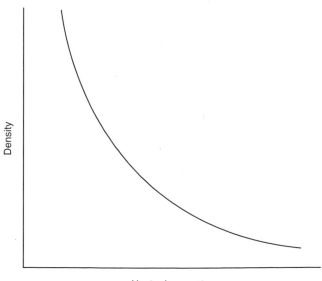

FIGURE 1.5
Generic ADD Curve.

Figure 1.6
Combined RDD and ADD
Curves.

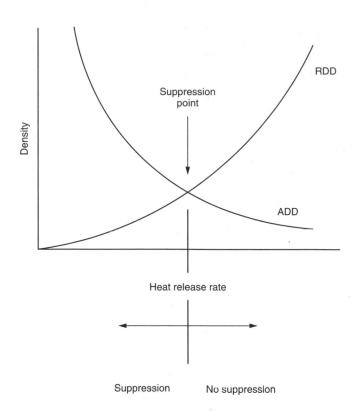

When the ADD and RDD graphs are combined on the same scale (as shown in Figure 1.6), there should be a point where they cross or intersect. At this intersection, called the *suppression point*, there is exactly enough water making it to the surface of the burning fuel to put the fire out. To the left of the suppression point, the ADD is greater than the RDD, with more water reaching the surface of the burning fuel than is necessary for suppression; therefore, the fire will be suppressed. The objective is to design an automatic sprinkler system to react to the fire in a timely fashion, with sufficient water making it to the fire to control or suppress it. On the other hand, if the automatic sprinkler system does not respond until after the suppression point has been reached, the fire will not be suppressed because the RDD exceeds the ADD. In other words, not enough water is getting to the fire to put the fire out. Total suppression requires a sprinkler with a sensitivity capable of discharging early in the fire development with sufficient discharge to extinguish the fire.

Characteristics Affecting Automatic Sprinkler System Performance

Automatic sprinklers and the manner in which the sprinklers are designed play a key role in choosing between control mode and suppression mode design. There are six significant characteristics in the manufacture of automatic sprinklers that can be changed in the manufacturing process to achieve different performance. The six characteristics are the following:

1. Thermal sensitivity
2. Temperature rating

3. Orifice size

4. Installation orientation (deflector)

5. Water distribution characteristics

6. Special service conditions

The engineer needs to understand each of these characteristics and choose the appropriate automatic sprinklers carefully.

Before looking at the six characteristics it is important to have some additional information on the basics of the thermal response of heat detectors or, in this case, automatic sprinklers. *NFPA 72®, National Fire Alarm Code®*, 2002 edition, in Annex B has theoretical background information that has been included as Box 1.1.

BOX 1.1

Automatic Sprinkler Correlations

The heat transfer to a detector or sprinkler can be described by the following equation:

$$Q_{total} = Q_{cond} + Q_{conv} + Q_{rad} \qquad \text{(Equation B.8)}$$

where:

Q_{total} = total heat transfer to a detector in Btu/sec (kW/sec)

Q_{cond} = conductive heat transfer

Q_{conv} = convective heat transfer

Q_{rad} = radiative heat transfer

Because detection typically occurs during the initial stages of a fire, the radiant heat transfer component (Q_{rad}) can be considered negligible. In addition, because the heat-sensing elements of most heat detectors are thermally isolated from the rest of the detection unit, as well as from the ceiling, it can be assumed that the conductive portion of the heat release rate (Q_{cond}) is also negligible, especially when compared to the convective heat transfer rate. Because the majority of the heat transfer to the detection element is via convection, the following equation can be used to calculate the total heat transfer:

$$Q = Q_{conv} = H_c A(T_g - T_d) \qquad \text{(Equation B.9)}$$

where:

Q_{conv} = convective heat transfer in Btu/sec (kW/sec)

H_c = convective heat transfer coefficient for the detector in Btu/ft$^2 \cdot$ sec $\cdot$ °F (kW/m$^2 \cdot$ °C)

A = surface area of the detector's element in ft^2 (m^2)

T_g = temperature of fire gases at the detector in °F (°C)

T_d = temperature rating, or set point, of the detector in °F (°C)

Assuming the detection element can be treated as a lumped mass (m) (lbm or kg) its temperature change can be defined as follows:

$$\frac{dT_d}{dt} = \frac{Q}{mc} \qquad \text{(Equation B.10)}$$

where:

dT_d/dt = change in temperature of detection element (deg/sec)

Q = heat release rate in Btu/sec (kW/sec)

m = detector element's mass (lbm or kg)

c = detector element's specific heat in Btu/lbm · °F (kJ/kg · °C)

Substituting this into the previous equation, the change in temperature of the detection element over time can be expressed as follows:

$$\frac{dT_d}{dt} = \frac{H_c A(T_g - T_d)}{mc} \qquad \text{(Equation B.11)}$$

Note that the variables have been identified previously.

The use of a time constant (τ) was proposed by Heskestad and Smith in order to define the convective heat transfer to a specific detector's heat-sensing element. This time constant is a function of the mass, specific heat, convective heat transfer coefficient, and area of the element and can be expressed as follows:

$$\tau = \frac{mc}{H_c A} \qquad \text{(Equation B.12)}$$

where:

m = detector element's mass (lbm or kg)

c = detector element's specific heat in Btu/lbm · °F (kJ/kg · °C)

H_c = convective heat transfer coefficient for the detector in Btu/ft^2 · sec · °F (kW/m^2 · °C)

A = surface area of the detector's element (ft^2 or m^2)

τ = detector time constant (seconds)

As seen in equation B.12, τ is a measure of the detector's sensitivity. By increasing the mass of the detection element, the time constant, and thus the response time, increases.

Substituting into equation B.11 produces the following:

$$\frac{dT_d}{dt} = \frac{T_g - T_d}{\tau} \qquad \text{(Equation B.13)}$$

Note that the variables have been identified previously.

The convective heat transfer coefficient for sprinklers and heat detection elements are similar to that of spheres, cylinders, and so forth, and is thus approximately proportional to the square root of the velocity of the gases passing the detector. As the mass, thermal capacity, and area of the detection element remain constant, the following relationship can be expressed as the response time index (RTI) for an individual detector:

$$\tau u^{1/2} - \tau_0 u_0^{1/2} = \text{RTI} \qquad \text{(Equation B.14)}$$

where:

τ = detector time constant (seconds)

u = velocity of fire gases (ft/sec or m/sec)

u_0 = instantaneous velocity of fire gases (ft/sec or m/sec)

RTI = response time index

If τ_0 is measured at a given reference velocity (u_0), τ can be determined for any other gas velocity (u) for that detector. A plunge test is the easiest way to measure τ_0. It has also been related to the listed spacing of a detector through a calculation. The RTI value can then be obtained by multiplying τ_0 values by $u_0^{1/2}$.

It has become customary to refer to the time constant using a reference velocity of $u_0 = 5$ ft/sec (1.5 m/sec). For example, where $u_0 = 5$ ft/sec (1.5 m/sec), a τ_0 of 30 seconds corresponds to an RTI of 67 sec$^{1/2}$/ft$^{1/2}$ (or 36 sec$^{1/2}$/m$^{1/2}$). On the other hand, a detector that has an RTI of 67 sec$^{1/2}$/ft$^{1/2}$ (or 36 sec$^{1/2}$/m$^{1/2}$) would have a τ_0 of 23.7 seconds, if measured in an air velocity of 8 ft/sec (2.4 m/sec).

The following equation can therefore be used to calculate the heat transfer to the detection element, and thus determine its temperature from its local fire-induced environment.

$$\frac{dT_d}{dt} = \frac{u^{1/2}(T_g - T_d)}{\text{RTI}} \qquad \text{(Equation B.15)}$$

Note that the variables have been identified previously.

(Source: Adapted from the *National Fire Alarm Code*®, NFPA, 2002, B.3.3.3 through B.3.3.3.10)

The rate of cooling or warming of an object, in this case the detector, is proportional to the temperature difference between it and its surroundings. In accordance with Newton's law of cooling, the following formula is used in conjunction with temperature and heat transfer, where $\Delta T = \Delta T_0$ at time (t) = 0, a relation expressed as Equation 1.1.

$$\Delta T = \Delta T_0 e^{-kt} \qquad (1.1)$$

where:

k = a constant

ΔT = the difference in temperature between the object (automatic sprinkler) and the surroundings and

t = time

The rate at which the temperature of the automatic sprinkler approaches the temperature of the heated fire gases is dependent on the specific properties of the automatic sprinkler and the rate of heat transfer from the fire gases to the automatic sprinkler. These factors are all in the constant k, for the specific automatic sprinkler operating element.

Thermal Sensitivity

The first of the six automatic sprinkler characteristics is thermal sensitivity. Heat detection is the basis for the operation of an automatic sprinkler. The convective heat transfer from the heat released by a fire is the most important factor in the activation of sprinklers [3]. By convection, the heated fire gases transfer heat to the automatic sprinklers' operating element. Convection is the transfer of energy by the movement of heated gases and liquids from the source of heat to a cooler part of the environment [4]. Figure 1.7 shows various types of automatic sprinkler operating elements. The response time index (RTI) of each of these elements may be different.

Figure 1.7
Standard Spray Sprinklers
Showing Various
Arrangements of Releasing
Mechanisms.

(Source: *Fire Protection Handbook*,
NFPA, 2003, Figure 10.10.7)

When developing a new style of sprinkler, the manufacturer can modify the mass and shape of the sprinkler and the activating mechanism to affect the sprinkler's sensitivity. This relationship is more commonly referred to as the *response time* or *thermal sensitivity* and represents how quickly a given thermal element (link or bulb) responds to a change in temperature. Response time is a measure of the rapidity with which the thermal element operates as installed in a specific sprinkler or sprinkler assembly. One measure is the response time index, or RTI. The RTI is the measure of the sensitivity of the thermal element of a specific sprinkler, or a measure of the sensitivity of the sprinkler's thermal element as installed in a specific sprinkler (see NFPA 13, A.3.6.1). There are currently two categories of sprinklers based on response times. One is the standard response sprinkler and the other is the fast response sprinkler.

Evaluating Response Time. Equation 1.2 is the recognized formula for response time index.

$$RTI = \tau u^{1/2} \tag{1.2}$$

where:

τ = a time constant for a specific heat detector

u = the gas velocity

The gas velocity generally referenced is 5.0 ft/sec or 1.5 m/sec, since that is the standard velocity used in sprinkler plunge tests, which measure RTI. In an environment in which the temperature is constantly increasing, the time constant is the amount of time by which the body lags behind its environment after some initial period of time equal to approximately 4 times the time constant [5].

The National Institute for Standards and Technology (NIST) offers two computer programs to calculate the actuation time of heat detectors. In this case, the heat detectors are automatic sprinklers. The two programs are DETACT-QS [Detector Activation, Quasi Steady (QS)] and DETACT-T2 [Detector Activation, Time Square (TS)], both of which are available from NIST through their web site, www.NIST.gov. These computer programs are used to calculate the actuation time of thermal devices below unconfined (smooth, flat) ceilings. They are used to predict the actuation time of automatic sprinklers (and other fixed temperature heat detectors) based on a user-specified fire. The QS program assumes that the fire is a steady-state fire, whereas the TS program assumes that the user-specified fire is growing as the square of time. Both programs assume that the automatic sprinkler is located in a relatively large area and that the only heating is from the fire ceiling flow.

Utilization of the QS program requires inputting the height of the ceiling above the fuel, the radial distance of the automatic sprinkler (or other thermal device) from the axis or centerline of the fire, the actuation temperature of the automatic sprinkler, the response time index (RTI) for the automatic sprinkler, and the rate of heat release of the fuel. The program will report the ceiling gas and the automatic sprinkler temperature, both as a function of time and the time required for actuation. DETACT-QS was written in BASIC by D. D. Evans of NIST. The Society of Fire Protection Engineers has published an evaluation of DETACT-QS, which provides additional information on the model.

The TS program requires inputting the ambient temperature, the RTI for the automatic sprinkler, both the activation temperature and the rate of rise temperature of the automatic sprinkler, the height of the ceiling above the fuel, the device spacing, and the fire growth rate. The results are the time to automatic sprinkler activation and the heat release rate at activation. DETACT-T2 was written in BASIC and FORTRAN by D. W. Stroup of NIST.

The RTI of some sprinklers may not be known by the user. Two automatic sprinkler types for which the RTI may be difficult to obtain are concealed and recessed sprinklers. An accepted practice for modeling concealed sprinklers is to consider them equivalent to pendent sprinklers having a similar thermal response sensitivity installed 12 in. (305 mm) below smooth unobstructed ceilings, and recessed sprinklers can be considered equivalent to pendent sprinklers having a similar thermal response sensitivity installed 8 in. (203 mm) below smooth unobstructed ceilings (see NFPA 13, A.3.6.1).

Fast Response or Quick Response Sprinklers. By definition, fast response or quick response sprinklers are sprinklers that have a thermal element with an RTI of 50 [(meters-seconds)$^{1/2}$] or less. Sprinklers defined as standard response have a thermal element with an RTI of 80 [(m/sec)$^{1/2}$] or more. Sprinklers with an RTI between 50 [(m/sec)$^{1/2}$] and 80 [(m/sec)$^{1/2}$] are considered special sprinklers with the break between the two categories left to clearly distinguish between fast response and standard response sprinklers.

Figure 1.8 is a graphic representation of the specific ranges of thermal sensitivity and conductivity of sprinklers used to create the categories of standard, special, and fast response sprinklers. A "quick response" sprinkler is a listed sprinkler with a "fast response" thermal element. There are sprinklers with fast response elements that are not listed as quick response sprinklers. In general, fast response sprinklers as defined in NFPA 13 are expected to activate faster than standard response sprinklers, with all other things being equal, and are utilized where a lag or delay in operation of the sprinklers is not congruent

Figure 1.8
Proposed International
Sprinkler Sensitivity Chart.

(Source: *Automatic Sprinkler Systems
Handbook*, 7th ed., NFPA, 1996,
Supplement 2, Fast Response
Sprinklers: A Technical Analysis)

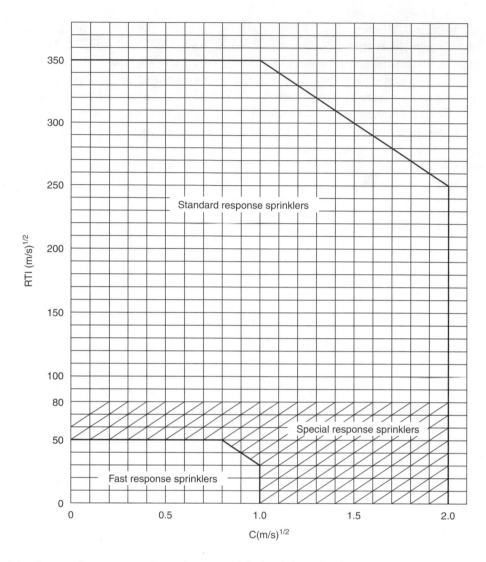

with the performance objectives established for the hazard being protected. However, under certain conditions, it is possible that a standard response automatic sprinkler located directly in the ceiling jet of the fire could respond faster than a quick response concealed automatic sprinkler or a quick response automatic sprinkler that is positioned remote from the plume centerline. Assuming proper sprinkler spacing, both of these situations would be NFPA 13, 2002 edition, compliant.

Temperature Rating

The temperature rating or response temperature of sprinklers is another factor that is essential in the basic design of the automatic sprinkler system. The available temperature rating varies from 135°F to 650°F. Proper selection of a sprinkler from among these predetermined temperatures keep sprinklers from operating accidentally if installed in a high ambient temperature area or control the number of sprinklers operating in the design area in a fire condition. In the special case of the residential sprinklers, the temperature rating along with the thermal sensitivity of the sprinklers is key to commencing fire control at a time in the history of the fire that is capable of preventing fire growth and effectively controlling a fire before it can flashover. Flashover is a relatively rapid (typically less than 1 min) transition from a localized growing fire to a fully developed stage in which all combustibles in the room are involved. When flashover occurs, it is no longer possible to survive in the fire compartment and the fire becomes a major threat beyond the room of origin. Commonly used criteria for the onset of flashover are a hot smoke layer temperature of 1100°F (600°C) and an incident heat flux at floor level of 1.8 Btu/sec ft² (20 kW/m²) [6].

Orifice Size

The orifice size of the sprinkler can be chosen larger or smaller by the engineer to affect the amount of water flow through the sprinkler at a given pressure. Table 1.1 shows the orifice size and the K-factor associated with each orifice for various automatic sprinklers. "K-factor" is a discharge coefficient that is calculated by the manufacturer for a specific orifice, and describes the properties of the orifice, such as smoothness and roundness. Sprinklers less than ½-in. are allowed only for limited use by NFPA 13, 2002 edition (8.3.4) on wet pipe systems in light hazard occupancies and in NFPA 15 water spray systems (NFPA 15, *Standard for Water Spray Fixed Systems for Fire Protection*). A designer may consider using these sprinklers in small spaces such as closets, where a ½-in. orifice sprinkler, even at the minimum operating pressure of

TABLE 1.1

Sprinkler Discharge Characteristics Identification

Nominal K-factor [gpm/(psi)$^{1/2}$]	K-factor Range [gpm/(psi)$^{1/2}$]	K-factor Range [dm³/min/(kPa)$^{1/2}$]	Percent of Nominal K-5.6 Discharge	Thread Type
1.4	1.3–1.5	1.9–2.2	25	½ in. NPT
1.9	1.8–2.0	2.6–2.9	33.3	½ in. NPT
2.8	2.6–2.9	3.8–4.2	50	½ in. NPT
4.2	4.0–4.4	5.9–6.4	75	½ in. NPT
5.6	5.3–5.8	7.6–8.4	100	½ in. NPT
8.0	7.4–8.2	10.7–11.8	140	¾ in. NPT or ½ in. NPT
11.2	11.0–11.5	15.9–16.6	200	½ in. NPT or ¾ in. NPT
14.0	13.5–14.5	19.5–20.9	250	¾ in. NPT
16.8	16.0–17.6	23.1–25.4	300	¾ in. NPT
19.6	18.6–20.6	27.2–30.1	350	1 in. NPT
22.4	21.3–23.5	31.1–34.3	400	1 in. NPT
25.2	23.9–26.5	34.9–38.7	450	1 in. NPT
28.0	26.6–29.4	38.9–43.0	500	1 in. NPT

Source: NFPA 13, 2002, Table 6.2.3.1.

7 psi (0.43 bar), would discharge more water than necessary into the space. Table 1.1 also lists sprinklers with orifice sizes larger than the standard ½-in.-orifice. These sprinklers are much more commonly used than the smaller orifice devices and are used where water pressures may be lower, higher-challenge fires are anticipated, or suppression technology is required.

There have been a number of changes in terminology regarding the orifice of sprinklers. Prior to 1996, sprinklers were given names such as "standard orifice" and "large orifice." NFPA stopped using the names in 1996 because the industry didn't come up with enough names to describe all of the new sprinklers. As larger and larger orifice sprinklers were developed, it became difficult to name them. For example, what describes a sprinkler that is larger than a "very extra large orifice" sprinkler?

In 1996, NFPA started designating the nominal orifice size in inches (e.g., ½ or ¾). Although the change kept us standardized in the United States, it caused a great deal of confusion in places where the metric system is used. The direct metric conversion of ½-in. is 12.7 mm. But the Europeans have been calling the ½-in. sprinkler a 15-mm sprinkler for a long time. The direct metric conversion for a ⅝-in. sprinkler is 15 mm. So when someone specified a 15-mm sprinkler, it was not clear if they wanted a ½-in. or a ⅝-in. sprinkler. In 1999, the problem was solved by moving toward the use of the K-factor instead of the orifice size.

The sprinkler industry now refers to sprinklers as K-5.6, K-8, K-14, and so on. In addition, manufacturers have agreed to standardize their literature to use the same K-factors for the same size sprinklers. These standardized K-factors are required to be used in the hydraulic calculations also, eliminating K-factor confusion (i.e., designing a system with K-5.55 sprinklers and installing K-5.62 sprinklers). Table 1.1 contains the standardized K-factors for the sprinklers currently in production. Additional sprinklers will have K-factors of 22.4 and 28.0, with 25.2 already on the market.

Why are manufacturers developing sprinklers with larger orifices? Why can't the industry stick with the same two orifice sizes that have worked for more than 50 years? The answer is that we are protecting more challenging commodities with sprinklers than we ever have before, such as plastics, big-box retailers, and large quantities of mixed commodities—flammable liquids, pool chemicals, and fertilizers—and the design criteria and sprinkler design densities previously utilized are not adequate for such high-challenge storage scenarios. Flammable liquids, aerosols, and rack storage of plastics at ever-increasing heights demand large quantities of water to absorb the heat from the fire. Full-scale testing has shown, in many cases, that larger-orifice sprinklers provide better performance. They have the potential to deliver more water to the fire, resulting in fewer sprinklers opening and providing exposure protection for adjacent commodities.

Table 1.2 shows the minimum operating pressures that would be necessary from a sprinkler to achieve a flow of 100 gpm (380 L/min). It is easy to see that it would be impractical to design an automatic sprinkler system with such a demand using K-5.6 or even K-8.0 sprinklers. K-14 orifice sprinklers may work, but might need a fire pump to achieve the desired pressure. Further, K-19.6 sprinklers may be more compatible with the available water supply, and might be able to protect the occupancy without a fire pump. Eliminating the fire pump could provide significant cost savings to the owner while also

TABLE 1.2

Selected Orifice Sizes and the Relative Pressure Requirement at 100 gpm

Orifice	Flow gpm (L/min)	Pressure psi (bar)
K-5.6	100 (380)	318 (21.9)
K-8.0	100 (380)	156 (10.7)
K-11.2	100 (380)	79 (5.4)
K-14.0	100 (380)	51 (3.5)
K-16.8	100 (380)	35 (2.4)
K-19.6	100 (380)	26 (1.8)
K-25.2	100 (380)	16 (1.1)

increasing system reliability (because another component, the fire pump, isn't necessary) and decreasing maintenance demands (costs and personnel to maintain a fire and jockey pump). Thus, the proper selection of the orifice size of the sprinkler can be changed to reduce pressure requirements affecting pump and pipe size determination.

For additional information on the development, use, and limitations of larger-orifice sprinklers, NFPA's *Automatic Sprinkler Systems Handbook* (2002 edition) includes an excellent Supplement 2, "The Effect of Large Orifice Sprinklers on High Challenge Fires," that the engineer should read. Figure 1.9 from that supplement is reproduced here to show more detail on the relationship between orifice size, water flow, and operating pressure.

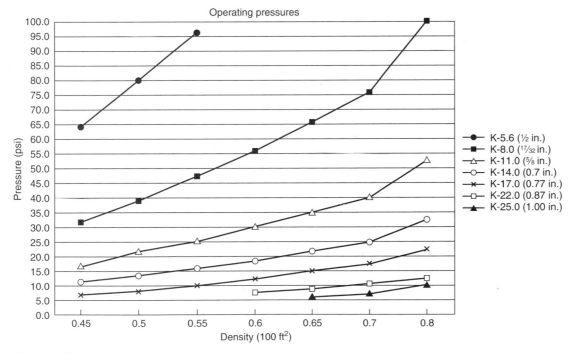

FIGURE 1.9
Pressure Requirements for Various Sized Orifice Sprinklers at Various Densities.
(Source: *Automatic Sprinkler Systems Handbook*, NFPA, 2002, Exhibit S2.1)

Installation Orientation

The next automatic sprinkler attribute that manufacturers can change or engineers can specify is the installation orientation. The three basic sprinkler orientations are the following:

1. Pendent sprinkler (including concealed, pendent, and recessed sprinklers)
2. Upright sprinklers
3. Sidewall sprinklers

Each of these types of sprinklers will be discussed in detail in other chapters of this text. They are defined by the manner in which the sprinkler is installed and each have a component, called the *deflector*, that directs the water distribution pattern to the floor in a manner commensurate to the orientation of the sprinkler. Figure 1.10 shows two of these sprinklers, the pendent on the left and the upright on the right. The deflector is the component on the bottom of the pendent sprinkler and the top of the upright sprinkler.

The original sprinklers developed in the late 1800s are now called "conventional" or "old-style" sprinklers. These sprinklers sent approximately half of their water discharge upward and the other half down to protect items on the floor. It was reasoned that water needed to be discharged up against the ceiling to keep the predominantly wood structural members cool and to prevent the fire from burning across the ceiling (above the sprinklers). Figures 1.11 and 1.12 show the theoretical water distribution patterns from "old-style" sprinklers and present-day sprinklers. With conventional sprinklers, the water spray produced by the deflector developed in two droplet size ranges: medium and large. The large droplets penetrate the fire plume and control or suppress the fire, while the medium-size droplets wet down the adjacent fuel load so that the fire will not spread.

The standard spray sprinkler deflector design for the pendent and upright standard spray sprinkler, developed in the early 1950s, delivers 100 percent of the water discharged toward the floor, in an umbrella pattern. In addition, the

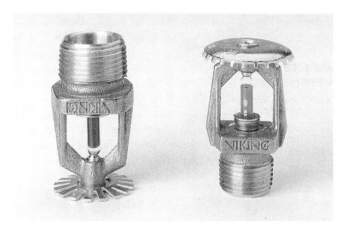

FIGURE 1.10
Viking Model M Glass Bulb Standard Spray Upright and Pendent Sprinklers.

(Source: *Automatic Sprinkler Systems Handbook*, NFPA, 2002, Exhibit 3.26; courtesy of Viking)

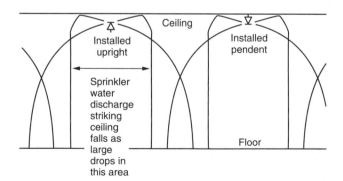

FIGURE 1.11
Principal Distribution Pattern of Water from Old-Style/Conventional Sprinklers (Used Before 1953).

(Source: *Automatic Sprinkler Systems Handbook*, NFPA, 2002, Exhibit 3.21)

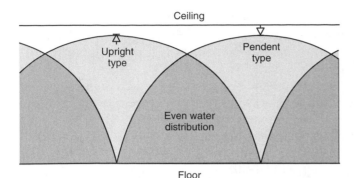

FIGURE 1.12
Principal Distribution Pattern of Water from Spray Sprinklers (Introduced in 1953).

(Source: *Automatic Sprinkler Systems Handbook*, NFPA, 2002, Exhibit 3.22)

current sprinkler deflectors develop droplets in three size ranges, with smaller drops for ceiling cooling, the medium-size drops for prewetting adjacent combustibles, and the larger drops for plume penetration. Small water droplets (almost like a mist) remain near the sprinkler and provide sufficient cooling at the ceiling to protect structural members and prevent a fire from racing across a combustible ceiling above the sprinklers—as long as the fire has not started above the sprinklers. This requirement is valid when sprinklers are installed with deflectors within 12 in. of the (flat smooth) ceiling as required by NFPA 13 for unobstructed construction.

A special type of spray sprinklers—control mode, large drop sprinklers—currently produce the largest size possible droplet of water that will not break into smaller drops while it is falling through the air. Another special type, ESFR sprinklers, on the other hand, operate primarily on the theory that by getting water to the burning fuel faster, while the RDD is still small, they can suppress the fire before it grows to a point where the water cannot penetrate the fire plume.

Figure 1.13 shows four different upright spray sprinklers and the differences in their deflectors. Each of these deflectors and sprinklers develop their own droplet sizes and discharge patterns.

In any fire, there is a battle in momentum occurring between the sprinkler and the fire. The hot gases rising off of the fire have both mass and velocity.

FIGURE 1.13
Upright Sprinklers.

(Source: *Automatic Sprinkler Systems Handbook*, NFPA, 2002, Exhibit 1.5)

The product of these two terms is called *momentum*. Water droplets developing off the sprinkler deflector also have mass and velocity, so they also have momentum. If the momentum of the water droplets is greater than the momentum of the fire gases, the water will reach the fire to achieve control or suppression. But, if the momentum of the fire gases is greater than the water droplets, the water droplets will evaporate or will be carried away from the fire and there may be no control or suppression.

Although the spray sprinkler has become the standard sprinkler used in North America, it did not catch on as well in the rest of the world. In many countries outside North America, "old-style" sprinklers are mandated when combustible ceilings are present. Since the discharge pattern of "old-style" sprinklers is different from standard spray sprinklers (see Figure 1.11), one needs to be cautious in applying sprinkler requirements. Their application rates, distances from the ceiling, and area of coverage are not interchangeable. These sprinklers are still used in North America for fur vaults for piers and wharf protection where upward discharge is necessary, and may be specified for combustible ceilings as deemed necessary.

The deflector can also be changed by the manufacturer and selected by the engineer to affect direction, size of water droplets, and the momentum generated by the spray to produce the characteristics specifically of the sidewall-type sprinklers. Sidewall sprinklers discharge horizontally into a room or area, thus the change in direction of the water spray from standard spray sprinklers. The location of the sprinkler relative to the ceiling or roof and the wall is different for the sidewall sprinklers and the standard spray sprinklers and will be discussed further in other chapters of this book.

Water Distribution Characteristics

The characteristics of water distribution can be a function of the application rate or wall wetting capabilities or lack thereof. The application rate or density, along with other factors, will be critical in determining whether or not control or suppression of specific hazards will be accomplished. The wall wetting characteristic is associated with special sprinklers generally intended for specific applications such as residential sprinklers, which protect rooms where drapes and combustible wall finishes are commonplace.

Special Service Conditions

This aspect of sprinkler design, as the name implies, is for special conditions and for very specific functions. Corrosion-resistant coatings, dry type sprinklers, intermediate-level sprinklers for use in rack storage, window sprinklers, and ornamental or decorative sprinklers all fall into this category.

Corrosion-resistant sprinklers can take on many appearances. Some are coated with Teflon®, in black or white, and can be selected because of their color; others are coated with lead, wax, or wax and lead. The corrosion-resistant coatings are available where sprinklers will be installed in environments that will adversely affect the life expectancy of the automatic sprinkler. Stainless steel sprinklers are another type of corrosion-resistant sprinkler, which could be considered for severe exposures, such as water cooling towers, where exposure to water, atmosphere, and perhaps chemicals could be anticipated, and are available from some manufacturers. Other coatings are also available and should be selected on the basis of the expected environment. Regardless of the corrosion-resistant protection, it must not delay the operation of the sprinkler, interfere with the release of operating parts, or significantly alter the pattern of water distribution. Only the sprinkler manufacturer may apply these coatings.

Dry type sprinklers are another type of special sprinkler. By definition, they are a sprinkler secured in an extension nipple that has a seal at the inlet end to prevent water from entering the nipple until the sprinkler operates (see NFPA 13, 3.6.4.2). These specialty sprinklers are premanufactured or ordered to specific lengths. There are specific installation considerations for these sprinklers regarding the fitting in which they can be installed (a tee only) and a minimum distance from the unheated space to the inlet seat.

Intermediate-level sprinklers are those that are specifically designed for use in storage racks. The intermediate-level sprinkler is defined in NFPA 13 as a sprinkler equipped with integral shields to protect its operating elements from being wetted by the discharge of sprinklers installed at higher elevations (see NFPA 13, 3.6.4.3).

Ornamental or decorative sprinklers are standard sprinklers with specific painting or plating. As with the corrosion-resistant sprinklers, this painting or plating may only be applied by the manufacturer.

Open sprinklers are automatic sprinklers from which the valve cap and heat-responsive elements have been removed and they are used in deluge sprinkler systems.

Each of the six characteristics discussed in this section is not completely independent of the others (the orifice also has some effect on the water droplet size and the temperature rating has some effect on the sensitivity); the discussion has been simplified for ease of comparison of the different types of sprinklers on the market. For additional information on automatic sprinklers used in special service conditions, see Chapter 7, "Impact of Unique Applications on Design Approaches."

Automatic Sprinkler Options

Now it's time to look at some of the different types of sprinklers and the challenges presented by the hazards and occupancies and look more closely at determining control or suppression.

Why are manufacturers developing so many different sprinklers and why are there larger orifices? First, as already mentioned, more challenging commodities are being protected with sprinklers than ever before. The other reason is that expectations have changed. Automatic sprinkler systems are expected to operate better with only a few sprinklers controlling a fire [e.g., 1200 ft² (112 m²) may be an acceptable operating area] compared to a much larger previous fire area where the ½-in. (K-5.6) standard-orifice sprinklers [maybe as much as 5000 ft² (465 m²)] were used.

As was shown in Tables 1.1 and 1.2, larger-orifice sprinklers can both reduce pressure requirements and increase water flow (the density of water application). In addition, with the larger orifice and lower end head pressure, it may be possible to minimize the cost of the automatic sprinkler system by eliminating the need for a fire pump.

These higher-challenge fires produce large quantities of hot gases (mass) at high velocities. Specific sprinklers have been developed to operate under these conditions where the combination of mass at high velocities creates a large momentum. Two of these sprinklers are the large drop sprinkler and the ESFR sprinkler.

Large Drop Sprinklers

In the late 1970s and early 1980s, research was conducted by Factory Mutual Insurance to develop a sprinkler deflector that discharged all of its water droplets downward (similar to the spray sprinkler) and also produced a greater number of large water droplets to handle high-challenge fires. The large drop sprinkler has proven to be effective at protecting many high-challenge commodities. However, it is a control mode sprinkler and not a suppression sprinkler. Because it is not a spray sprinkler, the design of automatic sprinkler systems utilizing large drop sprinklers have their own unique design requirements that must be followed if the sprinkler is to be used correctly.

Selecting a droplet performance objective is an important evaluation for a fire protection engineer. Where an engineer determines that coating of the combustible is the appropriate performance objective, such as might be the case when rolled paper is the protected commodity, the engineer may specify large drop sprinklers. Where evaporation is the performance objective, such as would be expected for the protection of ordinary combustibles, the engineer may specify standard spray sprinklers. The "teeth" of a deflector on a large drop sprinkler are spaced farther apart than those of other sprinklers, permitting the formulation of larger drops, which are more capable of penetrating the rising plume gases and coating the flaming combustible and adjacent commodities.

Early Suppression Fast Response Sprinklers

In the late 1980s, work done at Factory Mutual Insurance led to the development of a sprinkler specifically designed to suppress fires, the early suppression fast response (ESFR) sprinkler. This was the first sprinkler developed to actually suppress fires. The deflector of the ESFR sprinkler produces many large water droplets, and when the sprinkler operates early in the fire, these droplets penetrate the fire plumes of some extremely challenging fires at a high momentum rate, eventually suppressing the fire.

The ESFR sprinkler can be highly effective when compared to any other sprinkler for storage and warehouse occupancies if a designer strictly adheres to all of the special design and installation criteria. The ESFR sprinkler allows, in some cases, for rack storage to be protected without the use of in-rack or intermediate-level sprinklers. Relieving the requirements for in-rack sprinklers has serious positive benefits to the building owner besides the obvious reduction in cost when in-rack sprinklers are not installed. Storage racks can be moved or modified and commodities can be relocated without concern for automatic sprinklers, sprinkler piping, or system design configurations, and fork-lift truck operators don't have to worry about causing water damage by breaking a sprinkler or pipe.

It should be noted that the ESFR sprinkler has some stringent rules for installation, such as ceiling slope, roof construction, installation location, and obstructions, as well as commodity limitations. If all of these prerequisites are met, the building owner will have superior fire protection at a lower cost, but the building design limitations are critical. The building needs to be designed around the ESFR installation criteria or at least with the criteria in mind. There are however, serious obstruction concerns, especially with the K-14 ESFR, and these concerns will be addressed in later chapters of this text.

Of course, there are many other buildings to protect that are not considered high challenge and where suppression is not the goal. The next section will discuss the residential sprinkler occupancy, which differs from the commercial concept of control and suppression, and review the occupancies where control mode and suppression mode sprinklers are possible.

Automatic Sprinklers for Use in Residential Occupancies

One of the more recently developed sprinklers for a specific type of building or occupancy is the residential sprinkler. Residential sprinklers, by their listing, are designed so their spray pattern will by definition enhance survivability in the room of fire origin (see NFPA 13, 3.6.2.10). These sprinklers were specifically developed to aid in the detection and control of residential fires and thus provide improved protection against injury, loss of life, and property damage (see NFPA 13D, 2002 edition, 1.2).

The residential sprinkler is not designed to produce many large water droplets even though the fires in residential occupancies have the potential to release the same amount of heat as the plastic commodities in a warehouse. There is less need for large droplets because the fire in a typical residential occupancy does not have the ceiling height to develop a large fire plume. Instead, heat travels vertically to the ceiling, then moves out horizontally from the fire, across the ceiling. In order to effectively fight a fire in a residential occupancy, water must be delivered quickly to absorb the heat. Full-scale residential fire tests demonstrate that the most efficient way to control residential fires is to quickly deliver water in the form of many small droplets, which increase the surface area of the water in contact with the heat and enhance evaporization as a performance objective. Residential sprinklers were developed with quick response elements and deflectors that produced water droplets with just enough momentum to get to the fire, factoring in the anticipated climate in the room, for example, open windows, and so on. The most important feature related to residential sprinklers is the quick response element. Other important features include low RDD using relatively small quantities of water divided up into many small droplets to absorb the heat and quick response actuation that permits prewetting other combustibles in the area or room.

The general criteria, from the 2002 edition of NFPA 13D, *Standard for the Installation of Sprinkler Systems in One- and Two-Family Dwellings and Manufactured Homes*, for performance of residential sprinklers during a fire test are

1. The maximum gas or air temperature adjacent to the sprinkler—3 in. (76.2 mm) below the ceiling and 8 in. (203 mm) horizontally away from the sprinkler—must not exceed 600°F (316°C).

2. The maximum temperature—5 ft 3 in. (1.6 m) above the floor and half the room length away from each wall—must be less than 200°F (93°C) during the entire test. This temperature must not exceed 130°F (54°C) for more than a 2-minute period.

3. The maximum temperature ¼ in. (6.3 mm) behind the finished surface of the ceiling material directly above the test fire must not exceed 500°F (260°C).

4. No more than two residential sprinklers in the test enclosure can operate [7].

These residential sprinkler criteria lead to residential automatic sprinkler system design requirements that are markedly different from those for industrial or commercial automatic sprinkler systems, with the criteria for a residential occupancy based on egress safety to preserve lives instead of the building protection and associated life safety application in NFPA 13. Design requirements for residential sprinklers are different and vary based on NFPA 13, 13D, or NFPA 13R, *Standard for the Installation of Sprinkler Systems in Residential Occupancies up to and Including Four Stories in Height.*

Commercial Occupancies

There are four main hazard classifications, which are defined and discussed more extensively in Chapter 4, "Hazard and Commodity Classification." For the purposes of this chapter, NFPA 13 and current sprinkler technology permit the use of control mode sprinklers for the design of automatic sprinkler systems in light hazard, ordinary hazard, extra hazard, and storage occupancies. There are a wide variety of sprinklers available from which to choose for control mode design in these commercial occupancies. The criteria that must be considered by the designer or design engineer in making the choice follow:

- Response. Quick response sprinklers are required in light hazard occupancies, whereas either quick response or standard response is allowed in ordinary hazard occupancies and some storage occupancies.
- Hazard. Standard response sprinklers are allowed in ordinary, required in extra hazard, and allowed or required in storage occupancies.
- Water supply.
- Sprinkler spacing.
- Automatic sprinkler orientation.

The choices of sprinkler include the standard response sprinklers (upright and pendent), utilizing standard spacing; or standard response extended coverage sprinklers (upright and pendent), utilizing extended coverage spacing; quick response sprinklers (upright and pendent), utilizing standard or extended coverage spacing; and standard response and quick response sidewall sprinklers, with standard or extended coverage spacing.

Summary

Selecting control or suppression sprinklers is a deliberate decision. However, not every hazard, occupancy, or building can be protected with an automatic sprinkler system designed to suppress a fire. This application is still basically limited to storage occupancies and is not available for the majority of the hazards and occupancies that are currently protected by automatic sprinkler systems.

Controlling a fire to the area of origin through the use of the proper selection of sprinklers, design criteria, water spray, and prewetting of adjacent combustibles is the goal of most automatic sprinkler systems. Suppressing a fire can only be anticipated by the proper use of suppression mode sprinklers and design criteria. However, if suppression is to be achieved, that determination needs to be made very early in the design process of a project and incorporated carefully throughout the building construction or remodeling. Chapter 4, "Hazard and Commodity Classification," will discuss how information is gathered to determine if control or suppression is the goal.

BIBLIOGRAPHY

References Cited

1. Fleming, R. P., "Principles of Automatic Sprinkler System Performance," in Cote, A. E., ed., *Fire Protection Handbook*, 19th edition, NFPA, Quincy, MA, 2003, p. 10-166.
2. Fleming, R. P., "Principles of Automatic Sprinkler System Performance," in Cote, A. E., ed., *Fire Protection Handbook*, 19th edition, NFPA, Quincy, MA, 2003, p. 10-167.
3. Heskestad, G., and Smith, H. F., "Investigation of a New Sprinkler Sensitivity Approval Test: The Plunge Test," FMRC No. 22485, Factory Mutual Research Corp., Norwood, MA, Dec. 1976.
4. Custer, R. L. P., "Dynamics of Compartment Fire Growth," in Cote, A. E., ed., *Fire Protection Handbook*, 19th edition, NFPA, Quincy, MA, 2003, p. 2-74.
5. Fleming, R. P., "Principles of Automatic Sprinkler System Performance," in Cote, A. E., ed., *Fire Protection Handbook*, 19th edition, NFPA, Quincy, MA, 2003, p. 10-160.
6. Janssens, M. L., "Basics of Passive Fire Protection," in Cote, A. E., ed., *Fire Protection Handbook*, 19th edition, NFPA, Quincy, MA, 2003, p. 2-104.
7. *Automatic Sprinkler Systems Handbook*, 9th edition, NFPA, Quincy, MA, 2002, p. 775.

NFPA Codes, Standards, and Recommended Practices

NFPA Publications National Fire Protection Association, 1 Batterymarch Park, Quincy, MA 02169-7471

The following is a list of NFPA codes, standards, and recommended practices cited in this chapter. See the latest version of the *NFPA Catalog* for availability of current editions of these documents.

NFPA 13, *Standard for the Installation of Automatic Sprinkler Systems*
NFPA 13D, *Standard for the Installation of Sprinkler Systems in One- and Two-Family Dwellings and Manufactured Homes*
NFPA 13R, *Standard for the Installation of Sprinkler Systems in Residential Occupancies up to and Including Four Stories in Height*
NFPA 15, *Standard for Water Spray Fixed Systems for Fire Protection*
NFPA 72®, *National Fire Alarm Code®*

2

Impact of Sprinkler Systems on Building Design

Carl Anderson, PE

Installation of an automatic fire sprinkler system in a building can have a profoundly positive impact on the building: Building and fire code requirements change, elements of the site design may change, and tax and insurance incentives can be realized. Installation of a sprinkler system will permit larger, taller buildings utilizing different construction and interior finish materials with reduced fire resistance ratings under the provisions of most model building and fire codes, the fire flow required for manual fire fighting may be reduced, maximum distances to exits may increase, an atrium may be permitted, and corridor fire ratings may in some cases be relaxed.

Perhaps the most important impact of a fire sprinkler system on a building is that the building will be far safer for the occupants and first responders than its unsprinklered counterpart. Reports published by NFPA in 2001 clearly demonstrate a significantly increased level of safety. From 1989 to 1998, NFPA reports a 91 percent decrease in civilian deaths in sprinklered hotel/motel occupancies relative to unsprinklered properties [1]. During the same time, the reduction in direct property loss was 50–70 percent over a range of occupancies. From 1994 to 1998, an 88 percent decrease in deaths in high-rise fires was seen for sprinklered high-rise buildings, with a 44 percent decrease in average dollar loss [2]. NFPA's study notes a reported $41.1 million (U.S.) property lost in high-rise fires in 1998 alone.

Building Code Advantages

General

Building codes establish various combinations of allowable heights and areas based on the occupancy, or use of the building, as well as the type of building construction. When sprinkler systems are installed in a building, they positively affect the design and use. In most circumstances, building codes permit building materials with reduced fire resistance ratings to be used in sprinklered buildings, and the building designs are allowed greater flexibility in maximizing the use of space, rendering the building more profitable and economically attractive for the owner.

The 2003 edition of *NFPA 5000®*, *Building Construction and Safety Code®*, offers a variety of such allowances for sprinklered buildings. Generally, *NFPA 5000* permits building height increases, building area increases, more flexibility in selecting wall and ceiling finishes, increases in exit system travel distances, less restrictive draft stop requirements, inclusion of atria in building design, and larger openings in fire walls. Specific examples of building code provisions favoring the installation of fire sprinkler systems are included in this chapter.

TABLE 2.1

Maximum Allowable Area of Unprotected Openings (percentage of exterior walls)

Horizontal Separation (ft)	Max. Area of Exposing Building Face (ft²)											
	100	150	200	250	300	400	500	600	700	800	900	1000
0	0	0	0	0	0	0	0	0	0	0	0	0
3	0	0	0	0	0	0	0	0	0	0	0	0
4	9	8	8	8	8	7	7	7	7	7	7	7
5	12	11	10	9	9	9	8	8	8	8	8	8
6	18	15	13	12	11	10	10	9	9	9	9	8
7	25	20	17	15	14	12	11	11	10	10	10	9
8	33	25	21	19	17	15	14	13	12	11	11	11
9	43	32	27	23	21	18	16	15	14	13	12	12
10	55	40	33	28	25	21	19	17	16	15	14	13
15	100	96	75	62	54	43	36	32	29	27	25	23
20		100	100	100	97	75	62	54	48	43	39	32
25					100	100	97	83	73	65	59	54
30							100	100	100	92	83	76
35										100	100	100
40												
45												
50												
55												
60												
70												
80												
90												
100												
110												
120												
130												
140												
150												

Note: For SI units, 1 ft = 0.305 m; 1 ft² = 0.093 m².

Source: *NFPA 5000®*, 2003, Table 7.3.5(a).

Exposure Protection

The area of unprotected openings in exterior walls requiring a fire resistance rating permitted by Tables 2.1 and 2.2, is allowed to be doubled when the building is protected throughout with an approved, electrically supervised automatic sprinkler system.

Building Height

Referring to Table 2.3, a sprinklered building is typically permitted to be one story or 20 ft taller than an unsprinklered building of the same occupancy and

1500	2000	2500	3500	5000	10,000	20,000
0	0	0	0	0	0	0
0	0	0	0	0	0	0
7	7	7	7	7	7	7
7	7	7	7	7	7	7
8	8	8	7	7	7	7
9	8	8	8	8	7	7
10	9	9	8	8	7	7
11	10	9	9	8	8	7
12	11	10	9	9	8	7
18	16	14	13	11	9	8
28	23	20	17	14	11	9
40	32	28	23	19	13	11
54	43	37	29	23	16	12
72	57	47	37	29	19	14
92	72	60	46	35	22	16
100	89	74	56	42	24	18
	100	90	67	50	30	20
		100	80	59	35	22
			93	69	40	25
			100	91	51	31
				100	64	37
					78	45
					95	53
					100	62
						72
						83
						94
						100

TABLE 2.2

Maximum Allowable Area of Unprotected Openings (percentage of exterior wall)

Horizontal Separation (ft)	Max. Area of Exposing Building Face (ft²)												
	100	150	200	250	300	400	500	600	700	800	900	1000	
0	0	0	0	0	0	0	0	0	0	0	0	0	
3	0	0	0	0	0	0	0	0	0	0	0	0	
4	4	4	4	4	4	4	4	4	4	4	4	4	
5	6	5	5	5	5	4	4	4	4	4	4	4	
6	9	7	7	6	6	5	5	5	5	4	4	4	
7	12	10	8	8	7	6	6	5	5	5	5	5	
8	17	13	11	9	9	7	7	6	6	6	5	5	
9	21	16	13	12	10	9	8	7	7	7	6	6	
10	27	20	16	14	12	11	9	8	8	7	7	7	
15	69	48	38	31	27	21	18	16	14	13	12	12	
20	100	91	70	57	48	38	31	27	24	22	20	18	
25		100	100	91	77	59	48	41	36	32	29	27	
30				100	100	86	59	46	52	46	42	38	
35						100	96	81	70	62	56	51	
40							100	100	92	81	73	66	
45									100	100	92	84	
50											100	100	
55													
60													
70													
80													
90													
100													
110													
120													
130													
140													
150													
160													
170													
180													
190													
200													
210													

Note: For SI units, 1 ft = 0.305 m; 1 ft² = 0.093 m².

Source: *NFPA 5000®*, 2003, Table 7.3.5(b).

	1500	2000	2500	3500	5000	10,000	20,000
	0	0	0	0	0	0	0
	0	0	0	0	0	0	0
	4	4	4	4	4	4	4
	4	4	4	4	4	4	4
	4	4	4	4	4	4	4
	4	4	4	4	4	4	4
	5	4	4	4	4	4	4
	5	5	5	4	4	4	4
	6	5	5	5	4	4	4
	9	8	7	6	6	5	4
	16	12	10	9	7	6	5
	20	16	14	11	9	7	5
	27	22	18	15	12	8	6
	36	28	24	18	14	10	7
	46	36	30	23	18	11	8
	58	45	37	28	21	13	9
	71	55	45	34	25	15	10
	85	65	54	40	30	17	11
	100	77	63	47	34	20	12
		100	85	62	45	25	15
			100	80	58	32	19
				100	73	39	22
					89	47	26
					100	56	31
						66	36
						77	41
						89	47
						100	53
							60
							67
							75
							83
							91
							100

TABLE 2.3

Height and Area Requirements

Construction Type	TYPE I 442		TYPE I 332		TYPE II 222		TYPE II 111		TYPE II 000	
	S	N	S	N	S	N	S	N	S	N
Maximum Building Height (ft)	UL	UL	420	400	180	160	85	65	75	55
OCCUPANCY										
Assembly > 1000	UL	4	UL	4	12	4	4	3	1	NP
	UL		UL		UL		15500		8500	
Assembly > 300	UL	4	UL	4	12	4	4	3	2	1
	UL		UL		UL		15500		8500	
Assembly ≤ 300	UL	7	UL	7	12	7	4	3	2	1
	UL		UL		UL		15500		8500	
Assembly, outdoor	UL	UL	UL	UL	UL	UL	UL	UL	UL	UL
	UL		UL		UL		UL		UL	
Business	UL	UL	UL	UL	12	11	6	5	5	4
	UL		UL		UL		37500		23000	
Board & care, large	UL	NP	UL	NP	12	NP	3	NP	2	NP
	UL		UL		55000		19000		10000	
Board & care, small	UL	UL	UL	UL	12	11	5	4	5	4
	UL		UL		UL		24000		16000	
Day care	UL	2	UL	2	12	2	6	1	4	1
	UL		UL		60500		26500		13000	
Detention & correctional	UL	7	UL	7	12	7	2	2	2	NP
	UL		UL		UL		15000		10000	
Educational	UL	UL	UL	UL	12	5	4	3	3	2
	UL		UL		UL		26500		14500	
Health care	UL	NP	UL	NP	12	NP	3	NP	1	NP
	UL		UL		UL		15000		11000	
Health care, ambulatory	UL	UL	UL	UL	12	11	6	5	5	1
	UL		UL		UL		37500		23000	
Industrial, ord. hazard	UL	UL	UL	UL	12	11	5	4	3	2
	UL		UL		UL		25000		15500	
Industrial, low hazard	UL	UL	UL	UL	12	11	6	5	4	3
	UL		UL		UL		37500		23000	
Mercantile	UL	UL	UL	UL	12	11	5	4	5	4
	UL		UL		UL		21500		12500	
Residential	UL	UL	UL	UL	12	11	5	4	5	4
	UL		UL		UL		24000		16000	
Residential, 1- & 2-family	UL	UL	UL	UL	12	11	5	4	5	4
	UL		UL		UL		UL		UL	
Storage, ord. hazard	UL	UL	UL	UL	12	11	5	4	4	3
	UL		UL		48000		26000		17500	
Storage, low hazard	UL	UL	UL	UL	12	11	6	5	5	4
	UL		UL		79000		39000		26000	

	TYPE III				TYPE IV		TYPE V			
	211		200		2HH		111		000	
	S	N	S	N	S	N	S	N	S	N
	85	65	75	55	85	65	70	50	60	40
	3	2	NP	NP	3	2	3	2	NP	NP
	14000		NP		15000		11500		NP	
	4	2	1	1	4	2	4	2	1	1
	14000		8500		15000		11500		5500	
	4	3	2	1	4	3	4	3	2	1
	14000		8500		15000		11500		5500	
	4	3	3	2	4	3	3	2	2	1
	UL		UL		UL		UL		UL	
	6	5	5	4	6	5	4	3	3	2
	28500		19000		36000		18000		9000	
	2	NP	1	NP	2	NP	2	NP	1	NP
	16500		10000		18000		10500		4500	
	5	4	5	4	5	4	4	3	3	2
	24000		16000		20500		12000		7000	
	4	1	2	1	2	1	4	1	2	1
	23500		13000		25500		18500		9000	
	2	2	2	NP	2	2	2	2	2	NP
	10500		7500		12000		7500		5000	
	4	3	3	2	4	3	2	1	2	1
	23500		14500		25500		18500		9500	
	1	NP	NP	NP	1	NP	1	NP	NP	NP
	12000		NP		12000		9500		NP	
	6	5	5	1	6	5	4	3	3	1
	28500		19000		36000		18000		9000	
	4	3	3	2	5	4	3	2	2	1
	19000		12000		33500		14000		8500	
	5	4	4	3	6	5	4	3	3	2
	28500		18000		50500		21000		13000	
	5	4	5	4	5	4	4	3	2	1
	18500		12500		20500		14000		9000	
	5	4	5	4	5	4	4	3	3	2
	24000		16000		20500		12000		7000	
	5	4	5	4	5	4	4	3	3	2
	UL		UL		UL		UL		UL	
	4	3	4	3	5	4	4	3	2	1
	26000		17500		25500		14000		9000	
	5	4	5	4	6	5	5	4	3	2
	39000		26000		38500		21000		13500	

TABLE 2.3

(Continued)

HIGH-HAZARD CONTENTS										
Protection Level 1	1	NP	1	NP	1	NP	1	NP	1	NP
	21000		21000		16500		11000		7000	
Protection Level 2	UL	NP	UL	NP	3	NP	2	NP	1	NP
	21000		21000		16500		11000		7000	
Protection Level 3	UL	NP	UL	NP	6	NP	4	NP	2	NP
	UL		UL		60000		26500		14000	
Protection Level 4	UL	NP	UL	NP	8	NP	6	NP	4	NP
	UL		UL		UL		37500		17500	
Protection Level 5	3	NP	3	NP	3	NP	3	NP	3	NP
	UL		UL		UL		37500		23000	

Notes:
1. For SI units, 1 ft = 0.3048 m and 1 ft^2 = 0.093 m^2.
2. Within each occupancy category, the top row refers to the allowable number of stories above grade, and the bottom row refers to allowable area per floor.
3. S = sprinklered maximum building height in feet and maximum number of stories above grade.
4. N = nonsprinklered maximum building height in feet and maximum number of stories above grade.
5. UL = unlimited.
6. NP = not permitted.

Source: *NFPA 5000*®, 2003, Table 7.4.1.

construction type. In some cases, adding the fire sprinkler system allows a building of unlimited height and number of stories. For residential occupancies provided with an approved sprinkler system in accordance with NFPA 13R, *Standard for the Installation of Sprinklers in Residential Occupancies*, the maximum permitted height may be increased by 20 ft and the maximum number of stories may be increased by one story, provided the overall building height does not exceed 60 ft (18 m) and the maximum number of stories does not exceed four stories.

Figure 2.1 displays an example of possible height increase where fire sprinklers are installed. The figure illustrates a business occupancy of unprotected wood frame construction. The occupancy can be three stories with a sprinkler system and two stories without sprinkler protection. Similar or more generous allowances are available for other uses and construction types. For example, the height of a one-story aircraft hangar equipped with a fire sprinkler system is not limited when the building is surrounded by public space, streets, or permanent open yards not less in width than 1½ times the height of the hangar.

Building Area

Buildings protected by an automatic sprinkler system in accordance with *NFPA 5000* are permitted a 200 percent area increase for buildings of two stories or more, and a 300 percent area increase for single-story buildings (see Figure 2.2). This allowance for increased area may yield significant savings, as

1	NP	1	NP	1	NP	1	NP	NP	NP
9500		7000		10500		7500		NP	
2	NP	1	NP	2	NP	1	NP	1	NP
9500		7000		10500		7500		3000	
4	NP	2	NP	4	NP	2	NP	1	NP
17500		13000		25500		10000		5000	
6	NP	4	NP	6	NP	4	NP	3	NP
28500		17500		36000		18000		6500	
3	NP	3	NP	3	NP	3	NP	2	NP
28500		19000		36000		18000		9000	

demonstrated in the following example. Please note that the construction cost values are merely estimated averages.

Construction Alternatives

Table 2.4 illustrates the impact of sprinklers on a 40,000-ft^2 one-story office building classified as a business occupancy, based on requirements from the 2003 edition of *NFPA 5000*. Note that, for this example, sprinklering the building actually saves the owner $715,600 ($4,621,600 − $3,906,000) [3]. This example is somewhat simplified, but it is based on cost factors from the October 2003 *Building Safety Journal* [4] intended for determining permit fees. Actual construction savings might be expected to be similar.

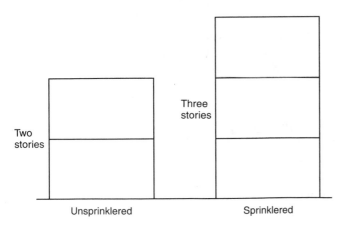

FIGURE 2.1
Unsprinklered and Sprinklered Building Heights in Same Occupancy and Construction Type.

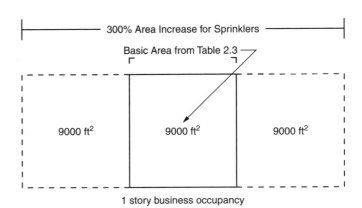

FIGURE 2.2
Area Increase for Single-Story Building.

TABLE 2.4

Construction Alternative Example

Option	Type of Construction	Estimated Construction Cost (per ft²)	Estimated Sprinkler System Cost ($2.50 per ft²)	Total Cost
Unsprinklered	Type II (222)	$115.54	NA	$4,621,600
Sprinklered	Could use any: consider Type III (211)	$95.15	$100,000	$3,906,000

Unlimited Area Building

In some cases, *NFPA 5000* will permit buildings of unlimited area when sprinklered. Several examples follow.

- The area of a one-story building used for business or industrial occupancies with ordinary hazard contents, mercantile occupancies, and assembly uses intended for viewing of indoor sporting events with spectator seating is not limited by *NFPA 5000* when the building is provided with an approved automatic sprinkler system in accordance with NFPA 13, *Standard for the Installation of Sprinkler Systems*, and is surrounded and adjoined by public ways or yards not less than 60 ft wide.

- The total area of a one- or two-story structure used for storage of ordinary hazard contents shall not be limited when the building is protected throughout by an approved automatic sprinkler system installed in accordance with NFPA 13 and the exterior walls face public ways or yards not less than 60 ft wide.

- The area of a two-story building used for business, industrial, or mercantile occupancies shall not be limited when the building is protected by an approved automatic sprinkler system in accordance with NFPA 13 and is surrounded by public ways or yards not less than 60 ft wide.

- The area of one-story buildings used for educational occupancies and of Type II (111), Type II (000), Type III (211), or Type IV construction shall not be limited when all classrooms have not less than two means of egress with one of the means of egress being a direct exit to the outside of the building, the building is surrounded by public ways or yards not less than 60 ft wide, and the building is to be protected by an approved automatic sprinkler system in accordance with NFPA 13.

- The area of a one-story assembly occupancy building used as an auditorium, church, community hall, dance hall, exhibition hall, gymnasium, lecture hall, indoor swimming pool, or tennis court of Type II construction is not to be limited when the building is equipped with an approved automatic sprinkler system in accordance with NFPA 13, the building is surrounded by public ways or yards not less than 60 ft wide, the building does not have a theatrical stage other than a raised platform, and the assembly floor is located at, or within 21 in. of, street or grade level and all exits have ramps to the street or grade level.

Fire Walls

NFPA 5000 limits the number and size of openings in a firewall (sometimes termed *area separation wall*) used to separate two buildings. *NFPA 5000*, paragraph 8.3.3.2, limits the total width of all door and window openings in a firewall to a maximum of 25 percent of the length of the wall in each story and the maximum area of a single opening to 120 ft^2. The 120-ft^2 limit does not apply when both buildings are equipped throughout with an automatic sprinkler system. This provision is of particular importance in a large warehouse separated by firewalls to satisfy the building code area limitations but where a large opening is needed to facilitate material handling equipment.

Vertical Openings

NFPA 5000 permits atria to be situated within sprinklered buildings and permits glass walls and inoperable windows in lieu of atria fire barrier walls with a 1-hour fire resistance rating given specific sprinkler placement. Sprinklers are required to be spaced along both sides of the glass wall and the inoperable window at intervals not to exceed 6 ft, and the sprinklers must be located at a distance from the glass not to exceed 1 ft and be arranged so that the entire surface of the glass is wet upon operation of the sprinklers. In this application, the glass must be tempered, wired, or laminated glass held in place by a gasket system that allows the glass framing system to deflect without breaking the glass before the sprinklers operate. Sprinklers are not required on the atrium side of the glass wall and the inoperable windows where there is not a walkway or other floor area on the atrium side above the main floor level.

In other than large, open areas such as atria and enclosed shopping malls, escalators must have their floor openings enclosed or protected as required for other vertical openings unless the building is protected throughout by an approved automatic sprinkler system. In this case, the opening may be protected in accordance with the method detailed in NFPA 13, as shown in Figure 2.3. This is typically seen at escalators in shopping mall anchor stores, for example.

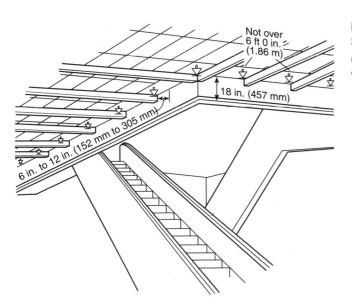

FIGURE 2.3

Sprinklers Around Escalators.

(Source: NFPA 13, 2002, Figure A.8.14.4)

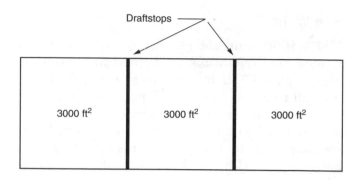

FIGURE 2.4
Draftstops in Nonsprinklered
Attic Space.

Concealed Spaces

Any concealed combustible space in which building materials having a flame spread index less than Class A are exposed, such as attics or floor/ceiling voids with combustible framing, are required to be draftstopped. Attics are required to be draftstopped into areas not exceeding 3000 ft², as shown in Figure 2.4. Floor/ceiling void spaces are required to be draftstopped into areas not exceeding 1000 ft². This draftstopping is not required when the space is protected throughout by an approved automatic sprinkler system.

Interior Finishes

NFPA 5000 permits application of less restrictive interior wall, ceiling, and floor finishes in sprinklered buildings. Class C interior wall and ceiling finish materials are permitted in any location where Class B is otherwise required, and Class B interior wall and ceiling finish materials are permitted in any location where Class A is otherwise required. Class II interior floor finish is permitted in any location where Class I interior floor finish normally is required, and where Class II is normally required, no critical radiant flux rating of the floor material is required.

Assembly Occupancies

The location of an assembly occupancy is limited in *NFPA 5000*. Assembly occupancies are not permitted below the level of exit discharge unless appropriate sprinkler protection is provided. The maximum occupant load of an assembly occupancy at the level of exit discharge is limited in buildings of some construction types. Also, the number of stories and occupant load above the level of exit discharge are limited. Installation of appropriate sprinkler protection generally permits options beyond these limitations. Referring to Table 2.5 the advantages accruing from installation of a sprinkler system are apparent. For example, an assembly occupancy with an occupant load of more than 1000 would only be permitted, if sprinklered, at the level of exit discharge in a building of Type II (000), sometimes called noncombustible, unprotected construction.

Egress

NFPA 5000 permits several egress advantages in sprinklered business and mercantile occupancies. Unrated exit corridors are permitted in mercantile and business occupancies when the occupancy is sprinklered. Dead-end corridors in mercantile and business occupancies are permitted to be a maximum of 50 ft in

TABLE 2.5

Construction Type Limitations

Type of Construction	Below LED	LED	Number of Levels Above LED			
1	2	3	≥4			
I(442)[a][b][c]	Any assembly[d]	Any assembly	Any assembly	Any assembly	Any assembly	Any assembly, if OL > 300[d]
I(332)[a][b][c]						
II(222)[a][b][c]						
II(111)[a][b][c]	Any assembly,[d] limited to 1 level below LED	Any assembly	Any assembly	Any assembly, if OL > 1000[d]	Assembly with OL ≤ 1000[d]	NP
III(211)[b]	Any assembly,[d] limited to 1 level below LED	Any assembly	Any assembly	Any assembly, if OL > 300[d]	Assembly with OL ≤ 1000[d]	NP
IV(2HH)						
V(111)						
II(000)	Assembly with OL ≤ 1000,[d] limited to 1 level below LED	Any assembly, if OL > 1000[d]	Assembly with OL ≤ 300[d]	NP	NP	NP
III(200)	Assembly with OL ≤ 1000,[d] limited to 1 level below LED	Assembly with OL ≤ 1000	Assembly with OL ≤ 300[d]	NP	NP	NP
V(000)						

Notes:
1. LED = level of exit discharge.
2. OL = occupant load.
3. NP = not permitted.
4. For the purpose of this table, a mezzanine is not counted as a level.

[a]Where every part of the structural framework of roofs in Type I or Type II construction is 20 ft (6.1 m) or more above the floor immediately below, omission of all fire protection of the structural members is permitted, including protection of trusses, roof framing, decking, and portions of columns above 20 ft (6.1 m).

[b]Where seating treads and risers serve as floors, such seating treads and risers are permitted to be of 1-hour fire resistance–rated construction. Structural members supporting seating treads and risers are to conform to the requirements of Table 16.1.5.2 [of *NFPA 5000*]. Joints between seating tread and riser units are permitted to be unrated, provided that such joints do not involve separation from areas containing high hazard contents and the facility is constructed and operated in accordance with 16.4.2 [of *NFPA 5000*].

[c]In open-air fixed seating facilities, including stadia, omission of fire protection of structural members exposed to the outside atmosphere is permitted where substantiated by an approved engineering analysis.

[d]Permitted if all the following are protected throughout by an approved, supervised automatic sprinkler system in accordance with Section 55.3 [of *NFPA 5000*]:

(1) The level of the assembly occupancy
(2) Any level below the level of the assembly occupancy
(3) In the case of an assembly occupancy located below the level of exit discharge, any level intervening between that level and the level of exit discharge, including the level of exit discharge

Source: *NFPA 5000®*, 2003, Table 16.1.5.2.

sprinklered business occupancy, whereas only 20-ft dead-ends are permitted in unsprinklered buildings, as illustrated in Figure 2.5. A common path of travel of 100 ft is permitted for exiting from a sprinklered building; otherwise, common path of travel is limited to 75 ft. A 100-ft increase in maximum travel distance to an exit is permitted in a sprinklered business occupancy when protected with an automatic sprinkler system. In industrial occupancies, travel distance to an exit and common path of travel are both increased for a sprinklered building, as seen in Tables 2.6 and 2.7.

FIGURE 2.5
Dead-end Corridors.

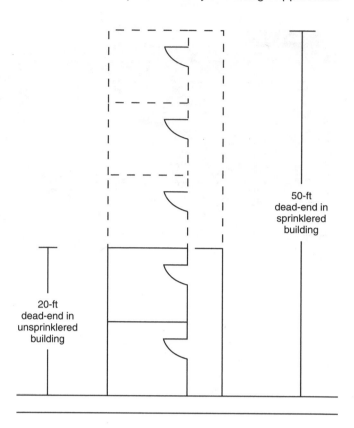

TABLE 2.6

Arrangement of Means of Egress

Arrangement	Industrial Occupancy	Low-Hazard Industrial Occupancy
DEAD-END CORRIDOR LIMITS		
Protected throughout by an approved, supervised automatic sprinkler system in accordance with Section 55.3 and 55.3.2	50 ft (15 m)	50 ft (15 m)
Not protected throughout by an approved, supervised automatic sprinkler system in accordance with Section 55.3 and 55.3.2	50 ft (15 m)	50 ft (15 m)
COMMON PATH OF TRAVEL LIMITS		
Protected throughout by an approved, supervised automatic sprinkler system in accordance with Section 55.3 and 55.3.2	100 ft (30 m)	100 ft (30 m)
Not protected throughout by an approved, supervised automatic sprinkler system in accordance with Section 55.3 and 55.3.2	50 ft (15 m)	50 ft (15 m)

Source: *NFPA 5000®*, 2003, Table 29.2.5.1.

TABLE 2.7

Maximum Travel Distance to Exits

Arrangement	General Industrial Occupancy	Special-Purpose Industrial Occupancy
Protected throughout by an approved, supervised automatic sprinkler system in accordance with Section 55.3 and 55.3.2	250 ft (76 m)[a]	400 ft (122 m)
Not protected throughout by an approved, supervised automatic sprinkler system in accordance with Section 55.3 and 55.3.2	200 ft (60 m)	300 ft (91 m)

[a]In single-story buildings, a travel distance of 400 ft (122 m) is permitted, provided that a performance-based analysis demonstrates that safe egress can be accomplished.

Source: *NFPA 5000®*, 2003, Table 29.2.6.

Delayed-egress locks are commonly utilized as a theft deterrent in "big-box" retail stores, larger bookstores, music stores, and similar occupancies. These delayed-egress locks (an example is shown in Figure 2.6), are only permitted to be installed on doors serving buildings with low and ordinary hazard contents protected throughout by an automatic fire detection system or an automatic sprinkler system.

Where nonrated walls or unprotected openings enclose the exterior of a stairway, and the walls or openings are exposed by other parts of the building at an angle of less than 180°, the building enclosure walls within 10 ft horizontally of the nonrated wall or unprotected opening must be constructed as required for

FIGURE 2.6
Delayed-Egress Lock.

stairway enclosures, including opening protection apparatus such as fire door assemblies. Separation is not required between corridors and outside stairs, provided that the following conditions are met:

- The building, including corridors and stairs, is protected throughout by an automatic sprinkler system in accordance with NFPA 13, or NFPA 13R.
- The corridors have a 1-hour fire rating in compliance with paragraph 11.1.3.1 of *NFPA 5000*, where they serve an occupant load of more than 30.
- The corridors connect on each end to an outside stair complying with 11.2.2.7 of *NFPA 5000*. At any location in the corridor where a change in direction exceeding 45° occurs, a clear opening to the exterior of not less than 35 ft^2, located to restrict the accumulation of smoke and toxic gases, or an outside stair shall be provided.

Smokeproof enclosures using pressurization must have a design pressure difference across the barrier of not less than 0.05 in. water column (12.5 Pa) in sprinklered buildings, or 0.10 in. water column (25 Pa) in nonsprinklered buildings. Equipment and ductwork for pressurization must be located outside of the building and directly connected to the enclosure by ductwork enclosed in noncombustible or limited-combustible construction, with intake and exhaust air directly to the outside or through ductwork enclosed by a 2-hour fire-resistive rating. Such ductwork must be enclosed with a 1-hour fire-resistive rating where the building, including the enclosure, is protected throughout by an approved automatic sprinkler system.

Fire barriers forming horizontal exits may not be penetrated by ducts unless the building is protected throughout by an approved automatic sprinkler system.

An elevator complying with the requirements of Chapter 54 of *NFPA 5000* may be used as a second means of egress from towers as defined in paragraph 3.3.550 of *NFPA 5000* when the tower is not accessible to the general public and certain criteria are met. One of the criteria is the installation of an automatic fire sprinkler system within the tower and any attached structure.

NFPA 5000 includes a variety of specific requirements for areas of refuge in accessible buildings in subsection 11.2.12, including 1-hour rated separation and two-way communication. An area of refuge may also be a story in a building where the building is protected throughout by an approved automatic sprinkler system and at least two accessible rooms or spaces separated from each other by smoke-resisting partitions. In this case, the detailed requirements found in *NFPA 5000* need not be applied.

In health care occupancies, the minimum width of components of the means of egress may be reduced in a sprinklered building. Referring to Table 2.8, note that egress width in sprinklered health care occupancies may be reduced by one-third or more in some instances.

In assembly occupancies, the capacity of egress components may be reduced in accordance with Table 16.4.2.3 of *NFPA 5000*, Capacity Factors for Smoke-Protected Assembly Seating (see Table 2.9). One factor in considering the occupancy to be smoke protected is the installation of a fire sprinkler system where control or suppression of the fire provides increased time for occupants to exit and facilitates the use of narrower egress components.

TABLE 2.8				
Capacity Factors				
Occupancy Area	Stairways		Level Components and Ramps	
	in.*	cm*	in.*	cm*
Board and care	0.4	1.0	0.2	0.5
Health care, sprinklered	0.3	0.8	0.2	0.5
Health care, nonsprinklered	0.6	1.5	0.5	1.3
High-hazard contents exceeding the maximum allowable quantities per control area as set forth in 34.1.3.	0.7	1.8	0.4	1.0
All others	0.3	0.75	0.2	0.5

*Per person.

Source: *NFPA 5000®*, 2003, Table 11.3.

Where two exits or exit access doors are required, *NFPA 5000* requires that they be located at a distance from one another not less than one-half the length of the maximum overall diagonal dimension of the building or area to be served. In buildings protected throughout by an approved automatic sprinkler system, the minimum separation distance between two exits or exit access doors must be not less than one-third the length of the maximum overall diagonal dimension of the building or area to be served, as shown in Figure 2.7.

Mechanical equipment rooms, boiler rooms, furnace rooms, and similar spaces shall be designed to limit the common path of travel to a maximum of 50 ft, whereas 100 ft is permitted in buildings protected throughout by an approved automatic sprinkler system.

TABLE 2.9				
Capacity Factors for Smoke-Protected Assembly Seating				
No. of Seats	Clear Width per Seat Served			
	Stairs		Passageways, Ramps, and Doorways	
	in.	mm	in.	mm
2,000	0.300 *AB*	7.6 *AB*	0.220 *C*	5.6 *C*
5,000	0.200 *AB*	5.1 *AB*	0.150 *C*	3.8 *C*
10,000	0.130 *AB*	3.3 *AB*	0.100 *C*	2.5 *C*
15,000	0.096 *AB*	2.4 *AB*	0.070 *C*	1.8 *C*
20,000	0.076 *AB*	1.9 *AB*	0.056 *C*	1.4 *C*
≥25,000	0.060 *AB*	1.5 *AB*	0.044 *C*	1.1 *C*

Source: *NFPA 5000®*, 2003, Table 16.4.2.3.

Figure 2.7
Reduced Separation
Between Exits in Sprinklered
Building.

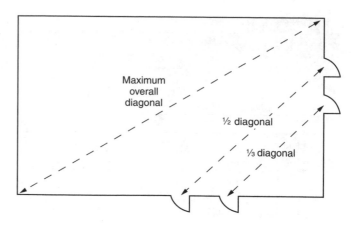

Separation of Occupancies

Buildings of mixed occupancies or with specific potentially hazardous areas such as boiler or storage rooms require the construction of fire-resistance-rated separations. The required fire resistance rating for occupancy separation may be reduced from 3 hour to 2 hour or 2 hour to 1 hour in a sprinklered building. The 1-hour rated separation required for a boiler, furnace, or storage room is eliminated in a sprinklered building.

Fire Code Advantages for Automatic Sprinkler Systems

General

In addition to the incentives for installing an automatic sprinkler system found in *NFPA 5000*, several additional advantages are found in NFPA 1, *Uniform Fire Code™*. Following are some of the advantages found in the 2003 edition of NFPA 1, *Uniform Fire Code*. Some provisions found in NFPA 1 are extracted from *NFPA 5000* and so could also be considered building code advantages.

Fire Alarm

Paragraph 13.3.1.5 of NFPA 1 states, "In areas protected by automatic sprinklers, automatic heat-detection devices required by other sections of this Code shall be permitted to be omitted." *NFPA 72®, National Fire Alarm Code®*, 2002 edition, does not specifically state that sprinkler protection eliminates the need for heat detection. However, in cases where a total- (complete) or partial-coverage fire alarm system is specified, sprinkler protection may be allowed to be substituted for heat detectors in areas that would otherwise have been provided with heat detectors if approved by the authority having jurisdiction (AHJ). The appropriate AHJ(s) should be consulted before making this substitution.

Residential Uses

Generally, new apartment buildings with more than three stories or with more than 11 dwelling units are required to have a fire alarm system; however, a fire alarm system is not required in buildings that are protected throughout by an approved automatic sprinkler system and do not exceed four stories in height or contain more than 16 dwelling units. A corridor smoke detection system is required in new hotels and dormitories other than those protected throughout by an approved, supervised automatic sprinkler system.

Storage and Garage Uses

A fire alarm system is required for storage occupancies exceeding 100,000 ft^2 for storage of ordinary or high-hazard contents. These same buildings, if protected throughout by approved automatic sprinkler system, are not required to have a fire alarm system.

Parking structures exceeding an aggregate floor area of 100,000 ft^2 are required to have a fire alarm system unless they meet the requirements for an open parking structure or are protected by an automatic fire sprinkler system.

Security and Inventory Control

Special locking arrangements are often desired in new retail stores to deter theft by customers and employees or in other occupancies to encourage guests, customers, or tenants to use the main exits rather than less secure side and rear exits intended for emergencies only. Frequently, delayed-egress locks are permitted on required exit doors in mercantile occupancies containing low and ordinary contents. Obviously, delaying egress in the event of a fire is undesirable. Past building codes have called for the installation of a sprinkler system as well as a smoke detection system to protect the occupants and automatically release the egress-control devices. The 2003 edition of NFPA 1 allows for the installation of these devices where the building is protected by either a sprinkler system or an automatic fire detection system.

Fire Flow and Access

Two of the primary concerns addressed in fire codes are water available for fire-fighting operations and access to the site and structures. Fire codes generally relax requirements in both areas where buildings are protected by automatic fire sprinkler systems.

Fire Flow. Table 2.10 outlines the fire flow required by NFPA 1 for buildings of different sizes and construction types. Generally, the larger the building and the less fire-resistive the construction, the more water is required for manual fire-fighting operations. Fire flow is defined as the flow rate of a water supply, measured at 20 psi residual pressure, that is available for fire fighting. Looking at Table 2.11, note that NFPA 1 requirements for the number and spacing of fire hydrants is driven by required fire flow. Larger required fire flow results in a requirement for more hydrants and reduced spacing.

NFPA 1 offers a reduction in required fire flow when a building is sprinklered. A reduction of up to 75 percent to a minimum of 1000 gpm can be allowed by the AHJ for structures other than one- and two-family dwellings; however, not all AHJs will allow the full reduction, so it is imperative that the engineer review this item in detail with all applicable AHJs early in the project, as this will impact the civil engineer's water supply design for the site. This is

TABLE 2.10

Minimum Required Fire Flow and Flow Duration for Buildings

Fire Area ft² ($\times$ 0.0929 for m²)			
I(443),I(332), II(222)[a]	II(111), III(211)[a]	IV(2HH), V(111)[a]	II(000), III(200), III(000)[a]
0–22,700	0–12,700	0–8,200	0–5,900
22,701–30,200	12,701–17,000	8,201–10,900	5,901–7,900
30,201–38,700	17,001–21,800	10,901–12,900	7,901–9,800
38,701–48,300	21,801–24,200	12,901–17,400	9,801–12,600
48,301–59,000	24,201–33,200	17,401–21,300	12,601–15,400
59,001–70,900	33,201–39,700	21,301–25,500	15,401–18,400
70,901–83,700	39,701–47,100	25,501–30,100	18,401–21,800
83,701–97,700	47,101–54,900	30,101–35,200	21,801–25,900
97,701–112,700	54,901–63,400	35,201–40,600	25,901–29,300
112,701–128,700	63,401–72,400	40,601–46,400	29,301–33,500
128,701–145,900	72,401–82,100	46,401–52,500	33,501–37,900
145,901–164,200	82,101–92,400	52,501–59,100	37,901–42,700
164,201–183,400	92,401–103,100	59,101–66,000	42,701–47,700
183,401–203,700	103,101–114,600	66,001–73,300	47,701–53,000
203,701–225,200	114,601–126,700	73,301–81,100	53,001–58,600
225,201–247,700	126,701–139,400	81,101–89,200	58,601–65,400
247,701–271,200	139,401–152,600	89,201–97,700	65,401–70,600
271,201–295,900	152,601–166,500	97,701–106,500	70,601–77,000
295,901–Greater	166,501–Greater	106,501–115,800	77,001–83,700
295,901–Greater	166,501–Greater	115,801–125,500	83,701–90,600
295,901–Greater	166,501–Greater	125,501–135,500	90,601–97,900
295,901–Greater	166,501–Greater	135,501–145,800	97,901–106,800
295,901–Greater	166,501–Greater	145,801–156,700	106,801–113,200
295,901–Greater	166,501–Greater	156,701–167,900	113,201–121,300
295,901–Greater	166,501–Greater	167,901–179,400	121,301–129,600
295,901–Greater	166,501–Greater	179,401–191,400	129,601–138,300
295,901–Greater	166,501–Greater	191,401–Greater	128,301–Greater

[a]Type of construction are based on NFPA 220.
[b]Measured at 20 psi (139.9 kPa).

Source: NFPA 1, 2003, Table H.5.1.

one of many items requiring careful coordination with other professionals involved with any project. Occupancy fire safety must be coordinated by the FPE (fire protection engineer) professional with other requirements from other codes and standards.

Table 2.12 presents an example of the potential advantages of reduced fire flow for sprinklers for a 60,000-ft² building of unprotected steel construction. For this structure without sprinklers, minimum fire flow from NFPA 1 is 5250 gpm for 4 hours (see Table 2.10). Looking then at Table 2.11, NFPA 1 requires a minimum of six hydrants at an average spacing of 300 ft are required. For the same structure with a sprinkler system taking the 75 percent reduction,

V(000)[a]	Fire Flow gpm[b] ($\times$ 3.785 for L/min)	Flow Duration (hours)
0–3,600	1,500	
3,601–4,800	1,750	
4,801–6,200	2,000	
6,201–7,700	2,250	2
7,701–9,400	2,500	
9,401–11,300	2,750	
11,301–13,400	3,000	
13,401–15,600	3,250	
15,601–18,000	3,500	3
18,001–20,600	3,750	
20,601–23,300	4,000	
23,301–26,300	4,250	
26,301–29,300	4,500	
29,301–32,600	4,750	
32,601–36,000	5,000	
36,001–39,600	5,250	
39,601–43,400	5,500	
43,401–47,400	5,750	
47,401–51,500	6,000	4
51,501–55,700	6,250	
55,701–60,200	6,500	
60,201–64,800	6,750	
64,801–69,600	7,000	
69,601–74,600	7,250	
74,601–79,800	7,500	
79,801–85,100	7,750	
85,101–Greater	8,000	

minimum fire flow is 1312.5 gpm for 2 hours. This minimum fire flow yields a minimum of one hydrant.

If this facility relies on on-site water supply tanks, additional advantages in tank and fire pump size may be realized. Without the sprinkler reduction, 1,260,000 gallons of water storage is required. So a tank exceeding this volume and a fire pump capable of delivering 5250 gpm will be needed. Refer to Chapters 13, 18, and Annexes H and I of NFPA 1 and Chapters 6, 12, and 13 of this book regarding water supply and fire pumps for more detail regarding design considerations and safety factors. With a 75 percent reduction for sprinklers, 157,500 gallons of water storage capacity and a pump capable of 1312.5 gpm are required.

TABLE 2.11

Number and Distribution of Fire Hydrants

Fire Flow Requirements (gpm)	Number and Distribution of Fire Hydrants		
	Minimum Number of Hydrants	Average Spacing Between Hydrants [1,2,3](ft)	Maximum Distance from any Point on Street or Road Frontage to a Hydrant[4] (ft)
1750 or less	1	500	250
2000–2250	2	450	225
2500	3	450	225
3000	3	400	225
3500–4000	4	350	210
4500–5000	5	300	180
5500	6	300	180
6000	6	250	150
6500–7000	7	250	150
7500 or more	8 or more[5]	200	120

Note: 1 gpm = 3.8 L/min; 1 ft = 0.3 m.

[1]Reduce by 100 ft (30.5 m) for dead-end streets or roads.

[2]Where streets are provided with median dividers which can be crossed by fire fighters pulling hose lines, or arterial streets are provided with four or more traffic lanes and have a traffic count of more than 30,000 vehicles per day, hydrant spacing shall average 500 ft (152.4 m) on each side of the street and be arranged on an alternating basis up to a fire flow requirement of 7000 gpm (26,500 L/min) and 400 ft (122 m) or higher fire flow requirements.

[3]Where new water mains are extended along streets where hydrants are not needed for protection of structures or similar fire problems, fire hydrants shall be provided at spacing not to exceed 1000 ft (305 m) to provide for transportation hazards.

[4]Reduce by 50 ft (15.2 m) for dead-end streets or roads.

[5]One hydrant for each 1000 gpm (3785 L/min) or fraction thereof.

Source: NFPA 1, 2003, Table I.3.

TABLE 2.12

Impact of Reduced Fire Flow on 60,000-ft² Building of Unprotected Steel Construction [Type II (000), *NFPA 5000*].

	Minimum Fire Flow[a]	Number of Required Hydrants[b]	Hydrant Spacing[b]	Required Water Storage
Unsprinklered	5250 gpm for 4 hr	6	300 ft	NA
Sprinklered	1312.5 gpm for 2 hr[c]	1	500 ft	NA
Unsprinklered with On-Site Water Supply Tanks				1,260,000 gal[d]
Sprinklered with On-Site Water Supply Tanks				157,500 gal[e]

[a]Per Table 2.10

[b]Per Table 2.11

[c]5250 gpm − (5250 gpm × 0.75) = 1312.5 gpm

[d]5250 gpm × (60 min/hr) × 4 hr = 1260,000 gal

[e]1312.5 gpm × (60 min/hr) × 2 hr = 157,500 gal

For one- and two-family dwellings, NFPA 1 sets the minimum fire flow at 1000 gpm for homes less than 3600 ft². This fire flow can be reduced by 50 percent to 500 gpm if the homes are sprinklered. For one- and two-family dwellings in excess of 3600 ft², fire flow requirements are again based on Table 2.10 with a 50 percent reduction permitted for sprinklered homes.

Fire Department Access. In NFPA 1, fire apparatus access roads are required such that any portion of an exterior wall of the first story above grade of the building is located not more than 150 ft from fire apparatus access roads as measured by an approved route around the exterior of the building. However, when buildings are protected by an automatic fire sprinkler system, the distance may be increased to 450 ft if permitted by the AHJ. Coupling the savings associated with smaller water mains and fewer hydrants with less restrictive apparatus access requirements, a significant cost savings should be realized in site and utility design when the building is sprinklered.

Benefits to Residential Developments

At times, benefits may be allowed to residential developments by a jurisdiction when all homes are provided with automatic fire sprinkler systems. These changes could result in an ability to develop a piece of property more efficiently and perhaps more profitably. Site development incentives are sometimes written into law when a jurisdiction passes a specific ordinance. Other times, a planning commission or a city council may evaluate specific incentives on a case-by-case basis. Some potential incentives are reduced street widths, increased allowable distance to a public way, elimination of large cul-de-sacs, fewer fire hydrants, reduced fire flow, smaller water mains, increased building density without sacrificing lot size, and increased response times from the nearest fire station.

The following example demonstrates how installation of sprinklers even in single-family dwellings can potentially save money, create affordable housing, and increase profit for a builder.

The same piece of property can be developed in two ways. Without sprinklers, the builder can fit only five houses on the property and maintain the ½-acre lots. With sprinklers, the road is narrower, the cul-de-sac is eliminated, and the builder can fit six houses on the same property, as shown in Figure 2.8.

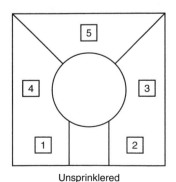
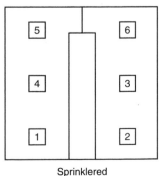

Unsprinklered

Sprinklered

FIGURE 2.8
Property Development, Sprinklered vs. Nonsprinklered.

TABLE 2.13		
Site Development Implications		
	Unsprinklered	Sprinklered
Land	$250,000	$250,000
Houses	$250,000	$336,000
Road	$60,000	$30,000
Other	$50,000	$50,000
Total Cost	$610,000	$666,000
Builder's Profit	$300,000	$350,000
Total Project	$910,000	$1,016,000
Unit Price (i.e., cost to buyer)	$182,000	$169,333

Table 2.13 compares the site development implications for unsprinklered and sprinklered homes. The table is based on the following parameters:

- A piece of property 370 ft × 360 ft (3.1 acres)
- To be developed for single-family homes
- Market and zoning require ½-acre lots
- Unsprinklered
  - Wide street, cul-de-sac, maximum of five homes
- Sprinklered
  - Narrower street, no cul-de-sac, six homes allowed

Based on Table 2.13,

- Sprinklers potentially increase builder's profit by $50,000 (17 percent).
- Sprinklers potentially decrease the unit cost for each house by $12,667 (7 percent), thus helping to reduce the cost of housing, if the savings are passed along to the homeowner.

One Community's Example of Incentives for Residential Sprinklers

The city of Dupont, Washington, provides a specific example of the types of incentives some communities may consider when plans for residential developments include automatic fire sprinklers. By local ordinance, Dupont has required that new homes built in the Northwest Landing Development be equipped with automatic fire sprinkler systems [5]. With this requirement, the city also adopted revisions to hydrant spacing and fire flow requirements. Hydrants are required within 350 ft of the center point of lot frontage rather than 250 ft, and required fire flow is 500 gpm. On dead-end streets not exceeding 600 ft, hydrants are not required if one is provided at the intersection. This can mean a savings of up to 400 ft of large water main installation, as illustrated in Figure 2.9. Other communities have adopted similar requirements and allowances or offer the allowances on a case-by-case basis when sprinklers are installed.

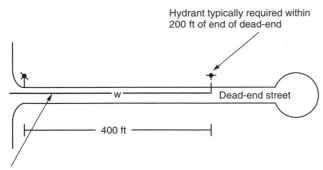

FIGURE 2.9
Hydrant Requirements.

Hydrant typically required within 200 ft of end of dead-end

Dead-end street

w

400 ft

Some AHJs may not require this 400 ft of water main and second hydrant when homes are sprinklered.

Tax and Insurance Incentives

Tax and insurance incentives also make sprinkler system installation more attractive. The cost of a sprinkler system is a deductible item on the federal income tax at the corporate rate. Additionally, the interest on a building improvement loan to install a sprinkler system is completely deductible from federal taxes. Some states and local governments give a straight percentage property tax rebate for buildings with sprinkler systems. Property insurance for building owners and tenants is usually lower with sprinklers, and contents insurance for occupants is usually lower. The actual savings depend on the type of building and type of business and can vary by insurer. The fire sprinkler industry is working to secure additional enhancements to the tax advantages associated with fire sprinkler installation.

Summary

Installation of a fire sprinkler system can positively impact the design and operation of a building in a great variety of ways. Taller buildings of greater area can be built. Floor areas may be more open with relaxed egress requirements, and special features such as atria may be incorporated. Although the addition of a sprinkler system to a building may initially be viewed as an added expense, allowances for more economical construction materials and potential savings in site requirements such as fire flow and the associated water main installation usually result in an overall savings in construction costs when the many advantages permitted by model building and fire codes are incorporated early on into the design of the building.

BIBLIOGRAPHY

References Cited

1. Rohr, K. D., "U.S. Experience with Fire Sprinklers," *NFPA Journal*, Quincy, MA, September 2001.
2. Hall, J. R., "Hi-Rise Building Fires," *NFPA Journal*, Quincy, MA, September 2001.
3. Construction cost information is taken from Marshal and Swift building valuation data service as reported in *Building Standards*, published by the International Code Council.

4. *Building Safety Journal*, October 2003, Vol. I, no. 8, copyright 2003 ICC, 900 Montclair Rd, Birmingham, AL 35213-1206.

5. City of Dupont Municipal Code, Ordinance 04-767, passed June 22, 2004.

NFPA Codes, Standards and Recommended Practices

NFPA Publications National Fire Protection Association, 1 Batterymarch Park, Quincy, MA 02169-7471

The following is a list of NFPA codes, standards, and recommended practices cited in this chapter. See the latest version of the *NFPA Catalog* for availability of current editions of these documents.

NFPA 1, *Uniform Fire Code*™

NFPA 13, *Standard for the Installation of Sprinkler Systems*

NFPA 13R, *Standard for the Installation of Sprinkler Systems in Residential Occupancies up to and Including Four Stories in Height*

NFPA 72®, *National Fire Alarm Code*®

NFPA 5000®, *Building Construction and Safety Code*®

3

Design Approaches

Scott Futrell, PE, SET

The approach to sprinkler system design and the determination of specific criteria cannot be finalized until a number of design factors have been evaluated, such as the following:

- Control versus suppression
- Occupancy or fuel load anticipated for each room, area, and building
- Environment in which the sprinkler system is installed
- Sprinkler system type (wet pipe, antifreeze, dry pipe, preaction, or deluge system)
- One system versus multiple systems
- Special hazards requiring specialty sprinklers or systems
- Water supply and its effect on the type of sprinklers (extended coverage or standard sprinkler, or orifice size), sprinkler spacing, or need for a fire booster pump
- Standpipes, combination sprinkler/standpipes, or inside hose stations
- Building construction
- Special conditions that will affect the design

All of these factors require evaluation and are not in any order since the order of importance may vary by building or room. This chapter will consider some of these criteria.

Occupancy Considerations

Chapter 4, "Hazard and Commodity Classification," includes a detailed discussion of the impact of occupancy on sprinkler system design; this chapter will apply those classifications. It is important to remember that occupancy as used in NFPA 13, *Standard for the Installation of Sprinkler Systems*, is not the same as occupancy as used in the model building codes. Occupancy hazards, as defined in NFPA's *Automatic Sprinkler Systems Handbook*, ". . . provide a convenient means of categorizing the fuel loads and fire severity associated with certain building operations. The classifications also present a relationship between the burning characteristics of these fuels and the ability of a sprinkler system in controlling the associated fires" [1]. On the other hand, by definition from *NFPA 5000®*, *Building Construction and Safety Code®*, occupancy is "the purpose for which a building or other structure, or part thereof, is used or intended to be used. [ASCE 7:1.2]" (*NFPA 5000®*, 2003 edition, 3.3.371). By definition from NFPA *101*, occupancy is "the purpose for which a building or portion thereof is used or intended to be used" (NFPA *101®*, 2003 edition, 3.3). Conversely, NFPA 13 states that "occupancy classifications shall not be intended to be a general classification of occupancy hazards" (NFPA 13, 2002 edition, 5.1.2). For example, the building codes might classify an office building in general as a "B" occupancy, but within that "B" occupancy NFPA 13 will identify light hazard offices, ordinary hazard group one mechanical rooms, and possibly ordinary hazard group two storage areas. The occupancy or fuel load is explained further in Chapter 4.

Overview of Design Approaches

When information about occupancy, environment, water supply, and building construction has been gathered and reviewed, it is time to move on to the design methods. This step in the design process will determine the minimum water demand requirements and compare those demands to the available or potential water supplies.

There are three basic design approaches: the Occupancy Hazard Fire Control Approach, the Special Occupancy Requirements Approach, and the Storage Design Approach. The remainder of this chapter will address those three design approaches.

Chapter 11 of NFPA 13, 2002 edition, details the most common design approach, the Occupancy Hazard Fire Control Approach. This approach is based on the application of specific sprinkler design densities—the amount of water in gallons per minute, applied over each square foot to achieve control of a fire—and design areas—the minimum area to which the water must be applied to achieve control—for a given hazard or class of occupancy.

However, not all of the hazards and occupancies can be appropriately designed using the Occupancy Hazard Fire Control Approach. Beginning with the 1999 edition, NFPA 13 includes Special Occupancy Requirements, now found in Chapter 13. The Special Occupancy Requirements chapter consolidates the sprinkler system design requirements previously found in other NFPA codes and standards. In 2002, NFPA 13 further consolidated the design criteria for the protection of storage from NFPA 231C and other NFPA storage documents into the Storage Design Approach. Chapter 12 of NFPA 13 is devoted to the Storage Design Approach.

Occupancy Hazard Fire Control Approach

As defined in NFPA 13, "Occupancies or portions of occupancies shall be classified according to the quantity and combustibility of contents, the expected rates of heat release, the total potential for energy release, the height of stockpiles, and the presence of flammable and combustible liquids" (NFPA 13, 2002,

11.2.1.3). The classifications and definitions of classifications for light hazard, ordinary hazard group one and two, and extra hazard group one and two are discussed in Chapter 4, "Hazard and Commodity Classification," and may need to be referred to frequently.

Because there are different hazards in a building, it is important to remember that each occupancy or area requiring different design criteria must be separated by a partition capable of delaying the heat from one area fusing sprinklers in the adjacent area. Note that NFPA only requires a "partition capable of delaying heat..." (NFPA 13, 2002, 11.1.2) and does not require a rated wall. Where there isn't a partition, the design criteria for the greater hazard is required to be carried a minimum of 15 ft (4.5 m) into the area of the lesser hazard.

Once again, occupancy classifications related to sprinkler systems are based on a fuel load not necessarily the same as in the occupancy in the model building codes. As used here, occupancy classifications refer to fuel load and fire severity relative to sprinkler system design and installation.

Determination of Sprinkler System Water Demand and Pipe Sizing

The two methods for determining water demand and pipe sizing for a sprinkler system using the Occupancy Hazard Fire Control Approach are the older pipe schedule method and the newer and more widely used hydraulic method.

Pipe Schedule Method

A technique of determining the sprinkler system pipe sizing and water demand that has been employed since the first pipe sizing schedules were published in 1905 is a "cookbook" approach known as the pipe schedule method. The pipe schedule method was used exclusively until the onset of hydraulic calculations and its phaseout started in the 1991 edition of NFPA 13. This method, although still acceptable in limited applications, uses tables with the maximum number of sprinklers on a branch line or main for each size of supply piping. By looking at the appropriate table for each occupancy, the user is able to quickly select pipe sizes based on the number of sprinklers downstream of the given pipe. For example, in a light hazard occupancy, 1-in. pipe is permitted to supply two sprinklers (see Table 3.1) and 1¼ in. is permitted to feed three sprinklers. The pipe schedule method is limited to existing systems and new systems or extensions of existing systems where the requirements in Chapter 11 of NFPA 13 are fulfilled. Use is limited to new occupancies of less than 5000 ft² (465 m²) of light and ordinary hazard occupancies or additions or modifications to existing pipe schedule systems of light, ordinary, or extra hazard occupancies. The use of the pipe schedule system is further limited by the pressure and flow requirements in Chapter 11 of NFPA 13. Chapter 11 does not limit the size of a building for using the pipe schedule method; however, the minimum pressure requirements at the highest elevation of a sprinkler increase from 15 psi (1.0 bar) for light hazard and 20 psi (1.4 bar) for ordinary hazard, to 50 psi (3.4 bar) when the building exceeds 5000 ft² (465 m²). Section 14.5 of NFPA 13 includes Tables 3.1 and 3.2 for use with the pipe schedule method. These tables indicate the required pipe sizing.

TABLE 3.1

Light Hazard Pipe Schedules

Steel		Copper	
1 in.	2 sprinklers	1 in.	2 sprinklers
1¼ in.	3 sprinklers	1¼ in.	3 sprinklers
1½ in.	5 sprinklers	1½ in.	5 sprinklers
2 in.	10 sprinklers	2 in.	12 sprinklers
2½ in.	30 sprinklers	2½ in.	40 sprinklers
3 in.	60 sprinklers	3 in.	65 sprinklers
3½ in.	100 sprinklers	3½ in.	115 sprinklers
4 in.	See Section 8.2 of NFPA 13	4 in.	See Section 8.2 of NFPA 13

For SI units, 1 in. = 25.4 mm.

Source: NFPA 13, 2002, Table 14.5.2.2.1.

It is important to note that the use of the pipe schedule system requires accurate water supply information at the project site as well as a minimum flow and pressure calculation, a limit to the number of sprinklers on a floor, a limit to the number of sprinklers on a riser, and restrictions for slatted floors, large floor openings, mezzanines, large platforms, and stair towers. Also, the pipe schedule method limits the spacing of sprinklers and further limits the designer to using *only* K-5.6 (½-in. standard orifice) sprinklers.

The pipe schedule system is not generally used because of the additional costs associated with the larger pipe sizing, the unknown performance (density of water application), the reduced flexibility for many fuel loads, and potential pressure problems.

TABLE 3.2

Ordinary Hazard Pipe Schedule

Steel		Copper	
1 in.	2 sprinklers	1 in.	2 sprinklers
1¼ in.	3 sprinklers	1¼ in.	3 sprinklers
1½ in.	5 sprinklers	1½ in.	5 sprinklers
2 in.	10 sprinklers	2 in.	12 sprinklers
2½ in.	20 sprinklers	2½ in.	25 sprinklers
3 in.	40 sprinklers	3 in.	45 sprinklers
3½ in.	65 sprinklers	3½ in.	75 sprinklers
4 in.	100 sprinklers	4 in.	115 sprinklers
5 in.	160 sprinklers	5 in.	180 sprinklers
6 in.	275 sprinklers	6 in.	300 sprinklers
8 in.	See Section 8.2 of NFPA 13	8 in.	See Section 8.2 of NFPA 13

For SI units, 1 in. = 25.4 mm.

Source: NFPA 13, 2002, Table 14.5.3.4.

Hydraulic Methods

In hydraulically designed sprinkler systems, the water supply is matched to the water demand necessary to control or suppress the anticipated fire. This method of sprinkler system design evolved through the 1940s and 1950s and first appeared in NFPA 13 in the 1966 edition.

Today, there are two hydraulic design methods for use with the more common sprinkler systems associated with light, ordinary, or extra hazard classifications (i.e., not special-hazard occupancies or storage occupancies). These strategies are the Density/Area Design Method and the Room Design Method. Neither of these strategies applies to specialty sprinkler systems and some special-application sprinklers. Sprinkler systems protecting some storage occupancies utilize the density/area method but don't utilize the same density/area curves.

Density/Area Design Method. The most commonly utilized hydraulic design incorporates the density/area design method with the criteria selected from a table or figure such as the one illustrated in Figure 3.1.

The minimum water supply requirements for a sprinkler system designed by either the density/area method or the room design method are derived from two tables in NFPA 13. The first table is 11.2.3.1.1, reproduced here as Table 3.3, which indicates the hose stream and water supply duration requirements, and the other table is the density/area curves (NFPA 13, Figure 11.2.3.1.5), reproduced as Figure 3.1. The hose stream demand table indicates that 100 gpm (380 L/min) are required for light hazard designs, 250 gpm (946 L/min) for ordinary hazard designs, and 500 gpm (1893 L/min) for extra hazard. It also yields the requirements for duration and inside hose streams. In combination with the density/area table, the minimum water supply for a sprinkler system can be determined.

Figure 3.1 illustrates the primary NFPA 13 Density/Area Curves and the *minimum* water supply required for the sprinkler systems only, or the minimum water demand for the sprinkler systems only. Because the extra hazard curves were not in the early editions, these curves or variations first appeared in NFPA 13 in 1972 and were refined in 1974. The use of the density/area curves does not require the designer to meet every point on the curves (only one point is required) and does not apply readily to all situations. The following section

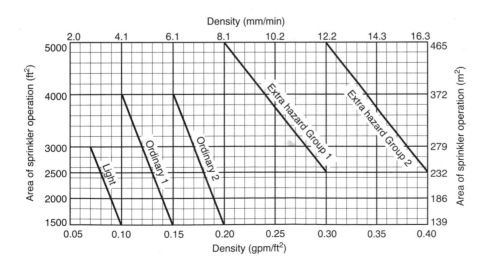

FIGURE 3.1
Density/Area Curves.
(Source: NFPA 13, 2002, Figure 11.2.3.1.5)

TABLE 3.3

Hose Stream Demand and Water Supply Duration Requirements for Hydraulically Calculated Systems

Occupancy	Inside Hose (gpm)	Total Combined Inside and Outside Hose (gpm)	Duration (minutes)
Light hazard	0, 50, or 100	100	30
Ordinary hazard	0, 50, or 100	250	60–90
Extra hazard	0, 50, or 100	500	90–120

For SI units, 1 gpm = 3.785 L/min.

Source: NFPA 13, 2002, Table 11.2.3.1.1.

on special conditions addresses some of these situations. The water supply requirements are then calculated using Chapters 11, 12, or 13 of NFPA 13. However, there are several adjustments that may be made or are required, and there are other circumstances that may require modifications to the density/area curves. These modifications are reviewed in subsequent sections.

Special Conditions or Areas. Examples of the special conditions or areas that affect the water supply include partially sprinklered buildings (NFPA 13, 11.2.3.4.1), building service chutes (maximum of three sprinklers calculated, NFPA 13, 11.2.3.4.1), a single line of sprinklers (all of the sprinklers to a maximum of seven sprinklers calculated, NFPA 13, 11.2.3.4.2), buildings with specific unsprinklered combustible concealed spaces [increase design area to 3000 ft^2 (279 m^2), NFPA 13, 11.2.3.1.8(3)], corridors (with protected openings calculate five sprinklers or without protected openings calculate seven sprinklers, NFPA 13, 11.2.3.3.6), residential sprinklers in NFPA 13 (four hydraulically most demanding sprinklers, NFPA 13, 11.2.3.5.1), special application sprinklers, and so on. The designer or engineer needs to know that carefully perusing Chapter 11 in NFPA 13 is necessary to properly design these special conditions or areas where the basic density/area curves do not apply.

Multihazard Building. Most buildings are multihazard (in other words, there is more than one occupancy or hazard in the building, as in the school example in Chapter 4) and as illustrated in Figure 3.2. For multihazard buildings, several different design criteria must be considered, as well as several of the adjustments. It is very common to find small rooms or areas inside a larger area of a different hazard. The sprinkler system in each room is individually designed because the rooms or areas are separated by walls or partitions from each other. However, it is not unusual to find a small, higher-hazard area located within a lower-hazard area. A mechanical equipment room or a storage room (both ordinary hazard group 1) in an office building (light hazard) is an example. A special condition applies where the two areas are not separated by a floor-to-ceiling/roof partition that is capable of preventing heat from a fire on one side from fusing sprinklers on the other side (see NFPA 13, 11.1.2).

The density for the higher hazard must be calculated for the entire actual area of the higher hazard, plus a minimum of 15 ft (4.5 m) in every direction. Two very important aspects of this are the requirement for a partition only, not a

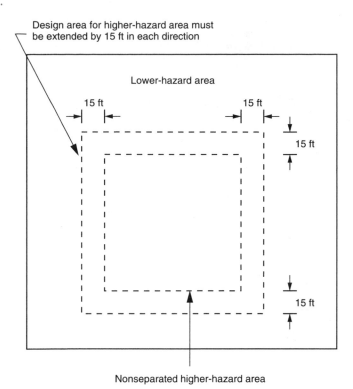

Design area for higher-hazard area must be extended by 15 ft in each direction

FIGURE 3.2
Multihazard Building.

Lower-hazard area

15 ft

15 ft

15 ft

15 ft

15 ft

Nonseparated higher-hazard area

rated wall (see 11.1.2), to alleviate this problem or that the entire required design area must be calculated from the density/area curves of the higher-hazard occupancy when these partitions are not present. If there is a floor-to-ceiling/roof partition that is capable of slowing the flow of the heated fire gases from the fire so that sprinklers in the adjacent area do not operate, then the high-hazard density needs to apply only to the area of the hazard, bounded by the partition. If this design area does not meet the minimum required design area from the density/area curves, then additional sprinklers in the lower hazard must be included until the minimum design area of the higher hazard has been achieved.

EXAMPLE 3.1 Small Retail Store

Consider a small retail store (ordinary hazard group 2—mercantile) with a high-piled storage area where cartoned, solid-pile, group A plastic parts [over 5 ft (1.52 m) high is high-pile storage of plastics] are stored to 10 ft (3.1 m) high covering 600 ft^2 (55.8 m^2). The building is 14 ft (4.3 m) high. In accordance with 12.2.3.1.6 column C of NFPA 13 (reproduced here as Table 3.4), the design criteria for the plastics should be 0.3 gpm/ft^2 [(12.2 (L/min)/m^2] over 2500 ft^2 (232 m^2); see also paragraph 12.2.3.1.7. From Figure 11.2.3.1.5 of NFPA 13 (presented as Figure 3.1) the mercantile is ordinary-hazard group 2, and the design criteria is 0.20 gpm/ft^2 [(8.1 (L/min)/m^2] over 1500 ft^2, (139 m^2) with 1500 ft^2 (139 m^2) as the required minimum design area. Since the plastic storage area is not surrounded and separated from the remainder of the lower-hazard area by a partition, the design area is required to be 2500 ft^2 (232 m^2) because of the higher hazard created by the Group A plastics.

The sprinkler system will need to be designed and hydraulically calculated for the 0.3 gpm/ft^2 (12.2 (L/min)/m^2) density over the storage area plus 15 ft (4.5 m) beyond in all directions. If part of the design area [2500 ft^2 (232 m^2)] is beyond the 600-ft^2 (56-m^2) storage area plus the 15 ft (4.5 m) beyond, then the

TABLE 3.4

Design Densities for Palletized, Solid Piled, Bin Box, or Shelf Storage of Plastic and Rubber Commodities

Storage Height		Roof/Ceiling Height		A		B		
ft	m	ft	m	gpm/ft²	mm/min	gpm/ft²	mm/min	
≤5	1.52	up to 25	up to 7.62	Curve 3	Curve 3	Curve 3	Curve 3	
≤12	3.66	up to 15	up to 4.57	0.2	8.2	Curve 5	Curve 5	
		>15 to 20	>4.57 to 6.1	0.3	12.2	0.6	24.5	
		>20 to 32	>6.1 to 9.75	0.4	16.3	0.8	32.6	
15	4.5	up to 20	up to 6.1	0.25	10.2	0.5	20.4	
		>20 to 25	>6.1 to 7.62	0.4	16.3	0.8	32.6	
		>25 to 35	>7.62 to 10.67	0.45	18.3	0.9	36.7	
20	6.1	up to 25	up to 7.62	0.3	12.2	0.6	24.5	
		>25 to 30	>7.62 to 9.14	0.45	18.3	0.9	36.7	
		>30 to 35	>9.14 to 10.67	0.6	24.5	1.2	48.9	
25	7.62	up to 30	up to 9.14	0.4	16.3	0.75	30.6	
		>30 to 35	>9.14 to 10.67	0.6	24.5	1.2	48.9	

Notes:
1. Minimum clearance between sprinkler deflector and top of storage shall be maintained as required.
2. Column designations correspond to the configuration of plastics storage as follows:
A: (1) Nonexpanded, unstable
   (2) Nonexpanded, stable, solid unit load
B: Expanded, exposed, stable
C: (1) Expanded, exposed, unstable
   (2) Nonexpanded, stable, cartoned
D: Expanded, cartoned, unstable
E: (1) Expanded, cartoned, stable
   (2) Nonexpanded, stable, exposed
3. Curve 3 = Density required by Figure 12.1.10 [of NFPA 13] for Curve 3
   Curve 4 = Density required by Figure 12.1.10 [of NFPA 13] for Curve 4
   Curve 5 = Density required by Figure 12.1.10 [of NFPA 13] for Curve 5
4. Hose streams and durations shall be as follows: ≤5 ft 250 gpm and 90 minutes; >5 ft to ≤20 ft 500 gpm and 120 minutes, >20 ft to ≤25 ft 500 gpm and 150 minutes.

Source: NFPA 13, 2002, Table 12.2.3.1.6.

remainder of the 2500-ft² (232-m²) area needs to be calculated at the lower—in this case, mercantile—density of 0.20 gpm/ft² [(8.1 (L/min)/m²]. In this specific example, the 600-ft² (56-m²) storage area plus 15 ft (4.5 m) beyond in every direction will result in an actual area of approximately 2400 ft² (223 m²), and therefore the calculations will need to include at least 100 ft² (9.3 m²) of the lower density. In addition to this calculation, a second calculation for the remainder of the space that is based on ordinary hazard group 2 occupancy may be required to determine the pipe size of the less-demanding area.

Adjustments that are required or may be used for hydraulically designed systems are discussed below.

Area Reduction for Quick-Response Sprinklers. A quick-response sprinkler is a type of spray sprinkler that meets the criteria of 3.6.1(a)(1) in NFPA 13, that is, has an RTI of 50 (m/sec)$^{1/2}$ and is listed as a quick-response sprinkler for its intended use (NFPA 13, 3.6.2.9). In light and ordinary hazard

Density					
C		D		E	
gpm/ft²	mm/min	gpm/ft²	mm/min	gpm/ft²	mm/min
Curve 3	Curve 3	Curve 3	Curve 3	Curve 3	Curve 3
0.3	12.2	Curve 4	Curve 4	Curve 5	Curve 5
0.5	20.4	Curve 5	Curve 5	Curve 5	Curve 5
0.6	24.5	0.45	18.3	0.7	28.5
0.4	16.3	0.3	12.2	0.45	18.3
0.6	24.5	0.45	18.3	0.7	28.5
0.7	28.5	0.55	22.4	0.85	34.6
0.45	18.3	0.35	14.3	0.55	22.4
0.7	28.5	0.55	22.4	0.85	34.6
0.85	34.6	0.7	28.5	1.1	44.8
0.55	22.4	0.45	18.3	0.7	28.5
0.85	34.6	0.7	28.5	1.1	44.8

occupancies only, the design area may be reduced when listed, quick-response sprinklers are used and several conditions (listed next) are met. This was a change in the NFPA 13, 1996 edition. Figure 11.2.3.2.3.1 from NFPA 13, reproduced here as Figure 3.3, is used to calculate the amount of the reduction in the design area when utilizing quick-response sprinklers.

The conditions that must be satisfied to utilize this area reduction are as follows:

- Light and ordinary hazard occupancies only.
- Buildings or areas must have ceilings less than or equal to 20 ft (6.1 m).

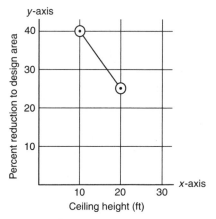

Note: $y = \frac{-3x}{2} + 55$

For ceiling height ≥10 ft and ≤20 ft, $y = \frac{-3x}{2} + 55$

For ceiling height <10 ft, $y = 40$

For ceiling height >20, $y = 0$

For SI units, 1 ft = 0.31 m.

FIGURE 3.3
Design Area Reduction for Quick-Response Sprinklers.
(Source: NFPA 13, 2002, Figure 11.2.3.2.3.1)

- It must be a wet pipe sprinkler system.
- There must not be any unprotected ceiling pockets as allowed by NFPA 13, 8.6.7 and 8.8.7 exceeding 32 ft² (2.9 m²).
- All the sprinklers in the compartment are quick-response.
- Where the ceiling is sloped, the maximum ceiling height is used to determine the percentage of the reduction.
- The number of sprinklers in the design area shall never be less than five.

For example, an office building of light hazard occupancy with a 10-ft ceiling would usually require a density of 0.10 gpm/ft² (4.1 (L/min)/m²) over 1500 ft² (139 m²) from Figure 3.1. Using quick-response sprinklers, the area could be reduced by 40 percent from 1500 ft² (139 m²) to 900 ft² (84 m²). Figure 3.3 can be used to determine the area reduction for any ceiling height up to 20 ft (6.1 m); however, the number of sprinklers calculated must not be less than five, regardless of the coverage area. The modification to the design area for this specific example will reduce the minimum sprinkler system water supply demand from 150 gpm (567 L/min) to 90 gpm (340 L/min) [0.10 gpm per ft² (4.1 (L/min)/m²) multiplied by the area of 1500 ft² (139 m²) equals 150 gpm (567 L/min), and 0.10 gpm per ft² (4.1 (L/min)/m²) multiplied by the area of 900 ft² (84 m²) equals 90 gpm (340 L/min)]. These are the minimum water flow demands for the sprinklers only. The actual demands will be slightly to somewhat higher when the actual hydraulic calculations are completed based on the piping arrangement, sprinkler spacing, friction loss, and end head pressure.

Paragraph 11.2.3.2.2.4 in NFPA 13 requires that when extended coverage sprinklers are used, the minimum design area must be the greater of either that from Figure 11.2.3.1.5 or the area protected by five sprinklers. The reduction allowed for quick-response sprinklers must include the use of extended coverage sprinklers, and since one condition is that the number of sprinklers in the design area must never be less than five, if the extended coverage sprinklers are, for example, each protecting 400 ft² (36 m²), then the minimum design area will be 2000 ft² (186 m²) [5 × 400 ft² (36 m²) = 2000 ft² (186 m²)] without a reduction.

Design Area Reduction for Extra Hazard Occupancies. In extra hazard sprinkler systems, the design area may be reduced by 25 percent from the density/area curves (see Figure 3.1) if high temperature sprinklers are used. However, the area cannot be reduced to less than 2000 ft² (186 m²).

Design Area Increases. The other common design area modifications require increases in the design area, and in some cases multiple adjustments may be required. An increase is required for wet or dry sprinkler systems in buildings with sloped ceilings, in nonstorage occupancies having a slope greater than 2 in. in 1 ft (2/12). The system design area must be increased by 30 percent for the slope when using quick-response sprinklers, spray sprinklers including extended coverage, and large drop sprinklers. It is allowable to increase the design area by 30 percent for the sloped ceiling and then reduce the area in accordance with 11.2.3.2.3.1 (see Figure 3.3) for ceiling heights less than 20 ft (6.1 m) and the use of quick-response sprinklers.

Another required modification applies to dry pipe sprinkler systems and double-interlocked preaction sprinkler systems. These systems are required to have a 30 percent increase in the design area based on the design density/area curves. This increase is based on the delay in getting water to a fire when these two types of sprinkler systems are used. Because it takes time from the operation of the sprinkler until water actually flows from the sprinkler, the fire grows exponentially and will cover more area before the sprinklers can control it. The required increase would modify the minimum 1500 ft² (139 m²) design area to 1950 ft² (181 m²). Noninterlocked and single-interlock preaction sprinkler systems would be exempt from this increase as long as the detection system associated with these types of systems will operate to allow water to enter the piping before the automatic sprinklers themselves operate.

For light and ordinary hazard systems, if a sloped ceiling greater than 2 in 12 is encountered and it is a dry or double-interlock system, both increases apply. When multiple modifiers are applied, the adjustments must be compounded. The minimum design area increases from 1500 ft² (139 m²) to 2535 ft² (236 m²) [1500 ft² × 1.3 × 1.3 = 2535 ft²] with the two required 30 percent increases.

Regardless of the design approach and area increases or reductions selected, in buildings with unsprinklered, combustible concealed spaces, as described in 8.14.1.2 and 8.14.6, the minimum design area must be at least 3000 ft² (279 m²). However, in accordance with 11.2.3.1.8(4), the following unsprinklered combustible concealed spaces shall not require a minimum of sprinkler operation of 3000 ft² (279 m²):

(a) Combustible concealed spaces filled entirely with noncombustible insulation.

(b) Light or ordinary hazard occupancies where noncombustible or limited combustible ceilings are directly attached to the bottom of solid wood joists so as to create enclosed joist spaces 160 ft³ (4.5 m³) or less in volume, including space below insulation that is laid directly on top or within the ceiling joists in an otherwise sprinklered attic.

(c) Concealed spaces where the exposed surfaces have a flame spread rating of 25 or less and the materials have been demonstrated to not propagate fire in the form in which they are installed in the space.

(d) Concealed spaces over isolated small rooms not exceeding 55 ft² (5.1 m²).

(e) Vertical pipe chases under 10 ft² (0.93 m²), provided that in multifloor buildings the chases are firestopped at each floor using materials equivalent to the floor construction. Such pipe chases shall contain no sources of ignition, piping shall be noncombustible, and pipe penetrations at each floor shall be properly sealed.

Room Design Method

The room design method is an alternative to using design density/area curves and is based upon the selection of the room or rooms and communicating spaces that create the greatest demand. This method is typically used with well-subdivided buildings, such as hospitals, office buildings without open office layouts, and in some residential applications. In the room design method, the density is still chosen from the density/area curves (see Figure 3.1), but the design area is required to include the largest room or the largest room and all communicating spaces. The advantage of the room design method is that when the largest room is smaller than the minimum design area of 1500 ft² (139 m²) from the density/area curves, fewer sprinklers are involved in the design area, the pipe sizes are smaller, and the resulting system is theoretically less expensive.

There are design parameters for the room design method. These parameters require that the construction of the walls in *all* rooms must have a fire rating equal to or greater than the minimum required water supply duration (see Table 3.3), and the openings from the room(s) must be protected in accordance with the requirements of 11.2.3.3.5. The minimum water supply duration would be 30 minutes for light hazard occupancies or 60 or 90 minutes for ordinary hazard occupancies, and 90 or 120 minutes for extra hazard occupancies. The lower duration can be utilized in each category where remote station or central station water flow/fire alarm service in accordance with *NFPA 72®*, *National Fire Alarm Code®*, is included.

The rooms must all have automatic or self-closing doors with a fire resistance at least equal to the water supply duration just indicated, with one exception for light hazard occupancies. In a light hazard occupancy, it is permissible to calculate all of the sprinklers in the largest (hydraulically most demanding) room plus two sprinklers in *all* of the communicating spaces—where the openings do not have the required protection (automatic or self-closing doors). Any communicating openings into such a room are to be protected in the same manner as any opening in a fire-resistive partition, with the exception that any nonrated or rated self-closing or automatic doors satisfy the protection requirement for light hazard. It is possible that the largest room might have four sprinklers, but the most demanding area could be a room with only two sprinklers that communicates with other rooms through several unprotected openings. The smaller room, communicating spaces, and adjacent rooms would then all need to be included in the hydraulic calculations. It must be remembered that even if the designer is using the room design method to determine the design area, the minimum design area is 3000 ft^2 (279 m^2) if the building has unsprinklered, combustible concealed spaces.

The room design method might be cost-effective initially, but where change is anticipated, or commonplace, it may be much less attractive in the long term. Where the room design method has been used for the initial or original design, changes in occupancy, remodeling to create larger areas, and changes in water supplies could invalidate this design concept and result in the need for major, costly renovations to the sprinkler system.

Residential Sprinklers

The three design approaches discussed to this point—pipe schedule, density/area, and room design—in general do not apply to residential sprinkler systems. NFPA 13 does permit or require the use of residential sprinklers under some circumstances and the hydraulic design criteria for those occupancies needs to be addressed separately.

NFPA 13 permits the installation of residential sprinklers only in residential portions of occupancies. They are only allowed in dwelling units and their adjoining corridors, and they must be used in wet pipe sprinkler systems unless the residential sprinklers are listed for use in dry pipe systems. As of mid 2004, there are no listed residential sprinklers for use with dry pipe sprinkler systems. NFPA 13 also requires that the design area must include the four hydraulically most demanding sprinklers, regardless of room or corridor configuration (outside of the dwelling units, the NFPA 13 corridor design requirements apply) and the flow shall either be the minimum flow based on the listing of the specific sprinklers chosen for the design or shall be calculated based on 0.1 gpm/ft^2 [4.1 (L/min)/m^2].

Briefly, there are two other standards, NFPA 13D and NFPA 13R, that provide rules for the installation of residential sprinkler systems that can be used instead of NFPA 13 for limited types of occupancies. In NFPA 13D, *Standard for the Installation of Sprinkler Systems in One- and Two-Family Dwellings and Manufactured Homes*, the entire building is the dwelling unit and requires residential sprinklers throughout the dwelling, even in basements, with a few exceptions:

- Bathrooms of 55 ft^2 (5.1 m^2) or less
- Clothes closets, linen closets, and pantries [where the area does not exceed 24 ft^2 (2.2 m^2), the least dimension does not exceed 3 ft (0.9 m), and the walls and ceilings are surfaced with noncombustible or limited-combustible material (see NFPA 220, *Standard on Types of Building Construction*)]
- Garages, open attached porches, carports, and similar structures
- Attics, crawl spaces, and other concealed spaces not used or intended for living purposes
- Covered, unheated projections of the building at entrances and exits as long as these are not the only entrance or exit from the dwelling unit

NFPA 13D requires that all of the sprinklers within a compartment, up to a maximum of two (unless the sprinkler's specific listing requires otherwise, for example, in sloped ceilings), shall be calculated to determine the greatest hydraulic demand. The design flow and pressure for residential sprinklers is based on the listing for the specific sprinklers but shall not be less than 0.5 gpm/ft^2 (2.0 (L/min)/m^2). It is imperative that the most recent manufacturer's product data sheets be used for each sprinkler's specific design criteria.

NFPA 13R, *Standard for Installation of Sprinkler Systems in Residential Occupancies up to and Including Four Stories in Height*, requires residential sprinklers in dwelling units and permits their use in adjoining corridors and lobbies. There is an exception in NFPA 13R that permits quick-response spray sprinklers in dwelling units if the whole dwelling unit can be protected with four sprinklers and is considered a compartment. NFPA 13R requires all of the sprinklers in the most demanding room to be calculated up to four and corridors are considered rooms. As with NFPA 13D systems, the design flow and pressure for residential sprinklers in NFPA 13R systems is based on the listing for the specific sprinklers but shall not be less than 0.5 gpm/ft^2 (2.0 (L/min)/m^2). Again, it is imperative that the most recent manufacturer's product data sheets be used for the specific design criteria of each sprinkler.

Special Considerations

There are a few other special design criteria considerations that are included in the Occupancy Hazard Fire Control Approach of Chapter 11 of NFPA 13. These include the design requirements for protection afforded by exposure systems [7 psi minimum (0.43 bar)] and water curtains [3 gpm per lineal foot

(37 L/min per lineal meter) with 15 gpm (56.8 L/min) minimum per sprinkler], and the calculation of water delivery in dry-pipe sprinkler systems.

Exposure Systems

Deluge, wet-, or dry pipe sprinkler systems can be used to provide protection against an exposure fire. NFPA 80A, *Recommended Practice for Protection of Buildings from Exterior Fire Exposures*, provides assistance in determining the degree of external exposures. However, consideration should also be given to internal exposures, such as high hazard processes within a structure. Exposure protection must be designed to wet the face of the protected wall or structure. The degree of exposure should be considered in terms of the radiant heat generated by the exposure. Where the exposure hazard increases, the minimum design flows from the sprinklers should exceed the flow achieved at 7 psi (0.48 bar) or 14.8 gpm (56 L/min) for K-5.6 sprinklers. However, more specific guidance is not given in the standard.

Water Curtains

Water curtains are created by the spacing of sprinklers [usually 6 ft (1.8 m) on center] adjacent to an opening such as that found around an escalator (see Figure 3.4). NFPA 13 requires a minimum flow rate of 3 gpm per lineal foot (37 L/min per lineal meter) along the length of the adjacent branch lines. For example, if the sprinkler system design area adjacent to the water curtain had a 1500-ft^2 (139-m^2) demand area, the number of closely spaced sprinklers to be calculated in the curtain would be derived from the formula given in 14.4.4.1.1: (1500 ft^2)$^{1/2}$ × 1.2 = 46.5 ft. So the number of sprinklers calculated with a water curtain having sprinklers spaced at 6 ft (1.8 m) on center would be 46.5 ft/6 ft = 7.75 sprinklers. This would be rounded up to eight sprinklers. However, if the water curtain were part of a deluge system, all of the sprinklers would be included in the hydraulic design with all sprinklers flowing. And if

FIGURE 3.4
Sprinklers Around
Escalators.

(Source: NFPA 13, 2002,
Figure A.8.14.4.)

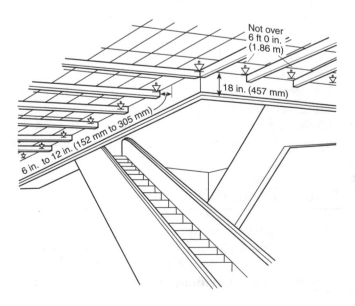

the water curtain could be activated in addition to the adjacent sprinkler system, the hydraulic demands for the adjacent sprinklers and the water curtain sprinklers must be added and hydraulically balanced back to the supply.

A water curtain might be found around an escalator opening in a multistory building. The performance objective of the water curtain is to maintain the fire rating of the floor structure separating one level of the building from another. By providing a water curtain around an opening in a fire-rated ceiling, the designer is permitted to assume the two levels to be separate fire areas, and the designer would be permitted to hydraulically calculate design areas for only one floor operating at a time. Designers must exercise caution in applying the performance objectives and reaping the design benefits of a water curtain to the protection of openings in cases where commodities are capable of becoming projectiles and in cases where the opening is larger than a water curtain is capable of protecting, such as a wide atrium opening. In the case of a wide opening in a floor, water curtains can still be specified, but multilevel actuation of sprinklers should be considered.

Another aspect of closely spaced sprinklers around an unenclosed opening between two or more floors is the relation among the sprinklers, heat from a fire, and the required draft stop—see subsection 8.14.4 of NFPA 13 for additional information. The draft stop is necessary to keep the heat on the lowest level, if that is where the fire started, so the sprinklers on that level will operate. Without the draft stop there is a chance that the sprinklers on an upper level will operate first, denying the sprinkler system on the lower level of its needed water supply.

Water Delivery Times of Dry Sprinkler Systems

The final special condition in Chapter 11 of NFPA 13 to be considered here is the calculation of the water delivery times of dry sprinkler systems. These calculations are necessary because of the limits to the size of dry systems indicated in 7.2.3.1. Calculations now acceptable for water delivery time and guidance are given in 11.2.3.9. Use of a "Listed" calculation program is required.

Additional information on the design criteria for these three special conditions—exposure protection, water curtains, and dry system water delivery times—is found at the end of NFPA 13, Chapter 11.

Special Occupancy Requirements Approach

The next strategy that needs to be considered involves very specific hazards or operations, special conditions, or special occupancy requirements that have not been addressed in either the occupancy hazard or the storage approach. This information is in Chapter 13 of NFPA 13 and is in addition to the requirements of the other approaches and chapters we have reviewed previously. The material contained in Chapter 13 of NFPA 13 has been extracted from other NFPA standards and included in the sprinkler standard to put a majority of the sprinkler system design criteria in NFPA 13 instead having the design criteria in many different standards.

The special occupancy requirements in NFPA 13 are from the following documents:

- NFPA 30B, *Code for the Manufacture and Storage of Aerosol Products*
- NFPA 33, *Standard for Spray Application Using Flammable or Combustible Materials*
- NFPA 40, *Standard for the Storage and Handling of Cellulose Nitrate Film*
- NFPA 42, *Code for the Storage of Pyroxylin Plastic*
- NFPA 51, *Standard for the Design and Installation of Oxygen–Fuel Gas Systems for Welding, Cutting, and Allied Processes*
- NFPA 51A, *Standard for Acetylene Cylinder Charging Plants*
- NFPA 55, *Standard for the Storage, Use, and Handling of Compressed Gases and Cryogenic Fluids in Portable and Stationary Containers, Cylinders, and Tanks*
- NFPA 59, *Utility LP-Gas Plant Code*
- NFPA 59A, *Standard for the Production, Storage, and Handling of Liquefied Natural Gas (LNG)*
- NFPA 75, *Standard for the Protection of Information Technology Equipment*
- NFPA 80A, *Recommended Practice for Protection of Buildings from Exterior Fire Exposures*
- NFPA 82, *Standard on Incinerators and Waste and Linen Handling Systems and Equipment*
- NFPA 86, *Standard for Ovens and Furnaces* (incorporates NFPA 86C and 86D)
- NFPA 96, *Standard for Ventilation Control and Fire Protection of Commercial Cooking Operations*
- NFPA 99, *Standard for Health Care Facilities*
- NFPA 130, *Standard for Fixed Guideway Transit and Passenger Rail Systems*
- NFPA 150, *Standard on Fire Safety in Racetrack Stables*
- NFPA 214, *Standard on Water-Cooling Towers*
- NFPA 220, *Standard on Types of Building Construction*
- NFPA 307, *Standard for the Construction and Fire Protection of Marine Terminals, Piers, and Wharves*
- NFPA 318, *Standard for the Protection of Semiconductor Fabrication Facilities*
- NFPA 409, *Standard on Aircraft Hangars*
- NFPA 415, *Standard on Airport Terminal Buildings, Fueling Ramp Drainage, and Loading Walkways*
- NFPA 423, *Standard for Construction and Protection of Aircraft Engine Test Facilities*
- NFPA 430, *Code for the Storage of Liquid and Solid Oxidizers*
- NFPA 432, *Code for the Storage of Organic Peroxide Formulations*

- NFPA 804, *Standard for Fire Protection for Advanced Light Water Reactor Electric Generating Plants*
- NFPA 805, *Performance-Based Standard for Fire Protection for Light Water Reactor Electric Generating Plants*
- NFPA 850, *Recommended Practice for Fire Protection for Electric Generating Plants and High Voltage Direct Current Converter Stations*
- NFPA 851, *Recommended Practice for Fire Protection for Hydroelectric Generating Plants*
- NFPA 909, *Code for the Protection of Cultural Resource Properties—Museums, Libraries, and Places of Worship*

The sprinkler design criteria in Chapter 13 of NFPA 13 include those for aerosols, spray applications using flammable and combustible materials, special occupancies such as cooling towers, chemistry labs, kitchen hoods and many, many more. Hazardous materials such as oxidizers are also in Chapter 13 of NFPA 13.

Storage Design Approach

Chapter 12 of the 2002 edition of NFPA 13 deals with the Storage Design Approach, a discussion of design criteria for storage occupancies. Specific design criteria will be covered much more extensively in other chapters of this book. This chapter of design approaches will review differences between the Occupancy Hazard Approach and the Storage Design Approach, look at a few of the special designs and specialty sprinklers that could be used, and examine how determining the hazard is a prerequisite to choosing a sprinkler system design.

First, compare the general requirements from NFPA 13, Chapters 11 and 12. The general requirements in Chapter 11 apply with additional requirements in Chapter 12 for roof vents, draft curtains, building height, hose connections, and specific density and design areas for storage occupancies. These requirements are specific to the hazard associated with high-piled storage and may be further restricted or required by the model building codes. As with all sprinkler system designs, wet pipe sprinkler systems are preferred, but of course they can't be used where ambient temperatures are below 40°F (4°C). In those cases, the types of systems and/or the design criteria will be altered. The basic requirements for high-piled storage are included in this design approaches chapter, but the design criteria and hazard classification requirements are discussed much more extensively in another chapter of this book.

Roof Vents and Draft Curtains

Roof vents and draft curtains are building features that may have an impact on the design and even the operation of a sprinkler system. The sprinkler system criteria in NFPA 13 are based on the assumption that roof vents (smoke and heat vents) and draft curtains are not present. The *Automatic Sprinkler Systems Handbook* directs the reader to Technical Reports of the International

Fire Sprinkler/Smoke and Heat Vent/Draft Curtain Fire Test Project published by the National Fire Protection Research Foundation (NFPRF).

The tests that were conducted by the NFPRF used open eave-line windows and louvers to simulate the smoke and heat vents. Data from those tests (the test information is available through NFPA) resulted in 87.5 percent to 91 percent more sprinklers operating than comparative tests with the eave-line windows and louvers closed. The conclusion states that during fire department operations for clean up after a fire, ventilating systems should be capable of manual exhaust operations (see NFPA 13, C.6 [12.1.1]).

Similarly, draft curtains are not recommended in conjunction with standard spray sprinklers, but are required to separate standard spray sprinklers from early suppression fast response (ESFR) sprinklers. NFPA 13, in 8.4.6.4.1, indicates that where the ESFR sprinklers are installed adjacent to standard response sprinklers, a draft curtain is required. These draft curtains must be at least 2 ft (0.6 m) in depth and be of noncombustible construction. In addition, the draft curtain must be centered over an aisle that is at least 4 ft (1.2 m) wide.

In storage occupancies, as found earlier with the density/area design method, when adjacent occupancies or hazards are not separated by a partition or barrier, the design criteria for the higher hazard must be carried at least 15 ft (4.5 m) into the lower hazard. This also applies to where high-temperature sprinklers are installed at the roof or ceiling for storage protection. The high-temperature sprinklers must continue 15 ft (4.5 m) past the perimeter of the storage requiring those high-temperature sprinklers. Also, when dry pipe sprinkler systems are designed, the design area must be increased by 30 percent. However, an additional stipulation to the storage approach requires that the area of operation after the 30 percent increase cannot be greater than 6000 ft^2 (558 m^2).

Unlike the previous design criteria for systems in accordance with the density/area curves in NFPA 13, Chapter 11, sprinkler systems designed for storage with roofs or ceilings having a slope greater than 2 in 12 are beyond the scope of the storage chapter. In other words, if the roof or ceiling slope is greater than 2 in 12, an alternative design is required and NFPA 13 does *not* provide those guidelines. One possible alternative approach is to install a suspended ceiling below the sloped ceiling that permits installation of sprinklers in accordance with NFPA 13.

The design criteria in the storage design approach are based on correctly identifying and classifying the hazard. However, under the storage approach the hazard not only includes the commodity but also its packaging material and arrangement, its storage arrangement, and the storage height. Some commodities, such as plastic, rubber, roll paper, plastic pallets, wood pallets, aerosols, flammable and combustible liquids, and so on, create greater control issues than ordinary combustibles and are addressed separately. Unlike the density/area curves, the sprinkler system design criteria for storage occupancies are based on the commodity classification (I, II, III, IV, or Plastics), storage height, storage arrangement (solid pile, rack storage, palletized, bin box, or shelf storage, for example), building height, and building construction. The engineer must accurately identify each of these components and apply the proper control or suppression strategy based on the comprehensive protection guidelines in NFPA 13, Chapter 12.

EXAMPLE 3.2 Heated Warehouse of Metal Construction

The design criteria for a 30-ft-high (9.1-m), heated warehouse of metal construction, storing Group A expanded, stable plastics, in cartons, on pallets, nonencapsulated, to a maximum height of 25 ft (7.62 m) in solid piles, utilizing control mode sprinklers would require protection in accordance with NFPA 13, 12.2.3.1.6, column E, which specifies a minimum design of 0.70 gpm ft^2 (28.5 mm/min) over the most demanding 2500 ft^2 (232 m^2). In addition, 500 gpm (1900 L/min) for inside and outside hose streams would be required and the minimum size of the automatic sprinklers would be $K = 11.2$ (see 12.1.13.3 in NFPA 13).

Another storage configuration that is addressed in Chapter 12 is rack storage. Rack storage is "any combination of vertical, horizontal, and diagonal members that supports stored materials" (NFPA 13, 3:10.8). Rack storage of commodities may require in-rack or intermediate-level sprinklers in addition to the roof or ceiling sprinklers. These are sprinklers installed within the rack structure itself to further protect the building and its contents and may be inadvertent targets for fork trucks and other loading equipment. As the storage path of the design approach for a new or existing facility is taken, all these additional items need to be considered.

Pallets

Although there are many hazards with unique fire sprinkler system requirements, one that should be at least briefly addressed is pallet storage. NFPA 13 refers to the pallets that products rest upon as idle pallets when they are not in use. Storage of idle pallets introduces a significant fire hazard and a serious fire condition. Idle wood pallets, stacked, are one of the greatest challenges to sprinklers because of their ease of ignition and ability to rapidly spread fire with a high rate of heat release. The construction of the pallets allows a fire to grow easily and quickly and shields the growing fire from sprinkler discharge. The NFPA preferred arrangement for storage of these pallets is outside or in a detached building. If that cannot be accomplished, then protection schemes should be provided using standard sprinklers or ESFR sprinklers. "Surround and drown" may best sum up the strategy, but more specific guidance is given in 12.1.9.1.2 for idle wood pallet storage.

The use and storage of plastic pallets is another concern. Although storing them outside or in a detached building is also recommended, 12.1.9.2 gives guidance for the design of sprinkler systems when they are stored inside a building. There are many types of plastic pallets and many different, possible storage configurations. Protection criteria for some pallets and some storage arrangements must come from test data instead of NFPA 13. Plastic pallets are addressed further in other chapters of this book, but Chapter 5, subsection 5.6.2, of NFPA 13 states that sprinkler system design criteria are based on wooden or metal pallets. Unless the plastic pallets have a demonstrated fire hazard (by full-scale fire testing) less than or equal to that of wooden pallets, and are "listed" as such, additional sprinkler system design requirements are imposed on the Class I through IV commodities that rest upon these pallets. In general, the hazard classification is increased either one or two classes (from Class I to Class II or III, from Class II to Class III or IV, or from Class III to Class IV) when these pallets are used.

Miscellaneous Storage

Miscellaneous storage is defined in NFPA 13 as "storage that does not exceed 12 ft (3.66 m) in height and is incidental to another occupancy use group. Such storage shall not constitute more than 10 percent of the building area or 4000 ft² (372 m²) of the sprinklered area, whichever is greater. Such storage shall not exceed 1000 ft² (93 m²) in one pile or area, and each such pile or area shall be separated from other storage areas by at least 25 ft (7.62 m)" (NFPA 13, 3.3.15). Miscellaneous storage applies to buildings where only a relatively small portion of another occupancy is used for storage and where that storage is not excessively high. Examples might be a manufacturing facility that stores a very small amount of raw materials and/or finished goods or a mercantile occupancy with a limited amount of backroom storage for its immediate-use inventory. The design requirements for miscellaneous storage are also included in Chapter 12 of NFPA 13. Specifically, miscellaneous storage design criteria from Table 12.1.10.1.1 is for Classes I through IV, Group A plastics, rubber tires, and rolled paper to 12 ft (3.7 m) high and idle pallets to 6 ft (1.4 m) high.

Specialty Sprinklers

There are many different specialty sprinklers. This section will briefly look at two of them—the large drop and the early suppression fast-response (ESFR) sprinklers.

Large Drop Sprinklers. The first of two specialty storage sprinklers are the large drop sprinklers. Large drop sprinklers are a special type of control mode sprinkler that produces large water droplets. See Section 8.11 in NFPA 13 for additional information on these sprinklers and their use. Large drop sprinklers can be effective in controlling high-challenge fires.

In recent years, alternatives such as the extra-large orifice (K-11.2), the very-extra-large orifice (K-14.0), K-17 sprinklers, and ESFR sprinklers have minimized the use of the large drop sprinkler. Design criteria for the number of sprinklers (15 or 25) and minimum end head pressure [50 psi (3.4 bar) to 95 psi (6.5 bar)] for large drop sprinklers are found in Chapter 12 of NFPA 13. Large drop sprinklers do not use the design density tables and cannot be used in a pipe schedule system.

Early Suppression Fast Response Sprinkler. Early suppression fast response (ESFR) sprinklers are another special type of sprinkler. These sprinklers are suppression mode sprinklers and are discussed much more extensively in another section of this manual. There have been two generations of ESFR sprinklers since their inception in the late 1980s. The original K-14 was subject to many building construction and obstruction restrictions as well as high-pressure requirements. The more recent K-25 is considerably less restrictive in application, obstruction criteria, where and when it can be used, and has reduced pressure requirements.

ESFR sprinklers can only be used in wet-pipe sprinkler systems. Additional information on ESFR sprinklers can be found in Chapters 8 and 12 of NFPA 13. The key to the success of the ESFR sprinkler is the rapid and early application of large quantities of high-momentum water droplets. Anything that might delay the operation of the ESFR sprinkler or interrupt the water spray pattern must be avoided. With the larger orifice ESFR sprinklers (K-25), it is possible to minimize or eliminate the need for fire pumps in certain applications due to the much lower pressure requirements resulting from a larger orifice opening. The use of these sprinklers generally is based on spacing of 80 to 100 ft^2 (7.4 to 10.7 m^2) per sprinkler and minimum operating pressures of 25 to 125 psi (1.72 to 8.6 bar).

These two specialty types of sprinklers, and particularly the ESFR sprinklers, have to be chosen early in the design process and can have implications for the building design and the water supply. Those implications will limit construction types and obstructions created by construction and will require the use of water supplies with high pressures and great volumes available. Either may end up requiring fire booster pumps and/or water storage tanks.

Summary

The design approach for sprinkler systems needs to consider many aspects of building construction, the environment that the sprinkler system will be installed in, occupancy of all areas in the building, the water supply for the building, and whether fire control or fire suppression is desired. The building construction, the environment, the occupancy, and the water supply all can have implications for the type of sprinkler system, the type of sprinklers, and the spacing of the sprinklers, and vice versa. The engineer's approach to designing every sprinkler system has to include the conscientious research and decisions detailed here, which, along with the information from Chapter 4, "Hazard and Commodity Classification," will assist the engineer in evaluating the key factor in the design approach—that of determining the minimum amount of water needed to effectively control or suppress a fire.

BIBLIOGRAPHY

Reference Cited

1. *Automatic Sprinkler Systems Handbook*, 9th edition, NFPA, Quincy, MA, 2002, p. 79.

NFPA Codes, Standards and Recommended Practices

NFPA Publications National Fire Protection Association, 1 Batterymarch Park, Quincy, MA 02169-7471

The following is a list of NFPA codes, standards, and recommended practices cited in this chapter. See the latest version of the *NFPA Catalog* for availability of current editions of these documents.

NFPA 13, *Standard for the Installation of Sprinkler Systems*

NFPA 13D, *Standard for the Installation of Sprinkler Systems in One- and Two-Family Dwellings and Manufactured Homes*

NFPA 13R, *Standard for the Installation of Sprinkler Systems in Residential Occupancies up to and Including Four Stories in Height*

NFPA 30, *Flammable and Combustible Liquids Code*

NFPA 30B, *Code for the Manufacture and Storage of Aerosol Products*

NFPA 33, *Standard for Spray Application Using Flammable or Combustible Materials*

NFPA 40, *Standard for the Storage and Handling of Cellulose Nitrate Film*

NFPA 42, *Code for the Storage of Pyroxylin Plastic*

NFPA 51, *Standard for the Design and Installation of Oxygen–Fuel Gas Systems for Welding, Cutting, and Allied Processes*

NFPA 51A, *Standard for Acetylene Cylinder Charging Plants*

NFPA 55, *Standard for the Storage, Use, and Handling of Compressed Gases and Cryogenic Fluids in Portable and Stationary Containers, Cylinders, and Tanks*

NFPA 59, *Utility LP-Gas Plant Code*

NFPA 59A, *Standard for the Production, Storage, and Handling of Liquefied Natural Gas (LNG)*

NFPA 72®, *National Fire Alarm Code*®

NFPA 75, *Standard for the Protection of Information Technology Equipment*

NFPA 80A, *Recommended Practice for Protection of Buildings from Exterior Fire Exposures*

NFPA 82, *Standard on Incinerators and Waste and Linen Handling Systems and Equipment*

NFPA 86, *Standard for Ovens and Furnaces* (incorporates NFPA 86C and 86D)

NFPA 96, *Standard for Ventilation Control and Fire Protection of Commercial Cooking Operations*

NFPA 99, *Standard for Health Care Facilities*

NFPA 101®, *Life Safety Code*®

NFPA 130, *Standard for Fixed Guideway Transit and Passenger Rail Systems*

NFPA 150, *Standard on Fire Safety in Racetrack Stables*

NFPA 214, *Standard on Water-Cooling Towers*

NFPA 220, *Standard on Types of Building Construction*

NFPA 307, *Standard for the Construction and Fire Protection of Marine Terminals, Piers, and Wharves*

NFPA 318, *Standard for the Protection of Semiconductor Fabrication Facilities*

NFPA 409, *Standard on Aircraft Hangars*

NFPA 415, *Standard on Airport Terminal Buildings, Fueling Ramp Drainage, and Loading Walkways*

NFPA 423, *Standard for Construction and Protection of Aircraft Engine Test Facilities*

NFPA 430, *Code for the Storage of Liquid and Solid Oxidizers*

NFPA 432, *Code for the Storage of Organic Peroxide Formulations*

NFPA 804, *Standard for Fire Protection for Advanced Light Water Reactor Electric Generating Plants*

NFPA 805, *Performance-Based Standard for Fire Protection for Light Water Reactor Electric Generating Plants*

NFPA 850, *Recommended Practice for Fire Protection for Electric Generating Plants and High Voltage Direct Current Converter Stations*

NFPA 851, *Recommended Practice for Fire Protection for Hydroelectric Generating Plants*

NFPA 909, *Code for the Protection of Cultural Resource Properties—Museums, Libraries, and Places of Worship*

NFPA 5000®, *Building Construction and Safety Code*®

4

Hazard and Commodity Classification

Scott Futrell, PE, SET

No concept is more important to sprinkler design than the proper classification of the hazards. The hazards must be properly determined to match the sprinkler system and its components with the water supply and achieve the goal of either fire control or fire suppression. Chapter 1, "Fundamentals of Sprinkler Performance," and Chapter 3, "Design Approaches," include information on hazard classification, fuel load, and occupancy and should be used in conjunction with this chapter. This chapter looks into the process of classifying the hazard.

The classification of the hazard posed by the contents of the building may be the most critical decision the engineer makes—if the hazards are not identified properly, the sprinklers, the sprinkler spacing, the pipe sizes, and water supplies (pumps and tanks), can all be incorrect. The following is a partial list of the factors that rely on the hazard or commodity classification being correct:

- The amount of water that will be needed to fight the fire, with sprinklers (density), hose (inside and outside hose streams), hydrants (outside hose streams), and needed fire flows (NFF). This is a function of the heat that is expected to be released by the fire—the heat release rate (HRR).
- The number of sprinklers that will open. This is a function of the fire size and growth prior to activation of the sprinklers, sprinkler density, activation temperature, and sprinkler spacing, and is a consideration in control mode or suppression mode sprinklers and systems.
- The maximum allowable coverage area for each sprinkler.
- The maximum allowable sprinkler system area.
- The need or use of certain types of sprinklers or specialty sprinklers.
- The need for in-rack sprinklers to protect storage as the height of the commodities increases.

Occupancy Classification

General Considerations

The design of the sprinkler system, as indicated in Chapter 3, "Design Approaches," is predicated on several assessments. One of these important assessments is the proper fire hazard classification or fire load of each of the individual rooms or areas in the building. It is very important to understand that the occupancy in NFPA 13, *Standard for the Installation of Sprinkler Systems*, is *not* the same as occupancy in the building codes. The *Automatic Sprinkler Systems Handbook*, 2002 edition, states, "The proper occupancy hazard classification for a given building operation should be determined by carefully reviewing the descriptions of each occupancy hazard and by evaluating the quantity, combustibility, and heat release rate of the associated contents" [1]. The *Handbook* continues, "The occupancy hazard classifications are presented as qualitative description rather than quantifiable measurements. Ideally, quantification of key factors such as the fire safety goals of the standard, the likely hazards contained within a specific space, the effect of building geometry and ventilation, and the interaction of sprinkler discharge with the fire would form the basis for NFPA 13" [1]. In most cases, when using NFPA 13 the entire building is classified not as a single hazard as it may be in the model building codes, but as many individual areas of differing fire hazards (occupancies).

There are several factors affecting the classifications of the hazards. These include the combustibility of the product, the amount of combustible product present, the volume of the product, the storage height, and the heat release rate (HRR). Other factors affecting the classification include the decision to classify the hazard or classify the occupancy and, in the case of storage occupancies, to classify the commodity, with the full knowledge and consent of the building owner or occupant as to how they are going to use the space. The building owner or occupant needs to be made aware of any limitations regarding the occupancy or commodity classification decision and must be able to live within these limitations. In addition, NFPA 13, 2002 edition, Section 4.3, requires the owner of the building to provide the sprinkler system installer with critical information prior to the layout of the sprinkler system. It is imperative that the engineer or specifier of the sprinkler system design should have this information before the sprinkler system installers begin their design work. Section 4.3, "Owner's Certificate" (see Figure 4.1 for an Owner's Information Certificate), states:

1. Intended use of the building including the materials within the building and the maximum height of any storage.
2. A preliminary plan of the building or structure along with the design concepts necessary to perform the layout and detail for the fire sprinkler system.
3. Any special knowledge of the water supply including known environmental conditions that might be responsible for microbiologically influenced corrosion (MIC).

For speculative buildings, without a specific occupant in mind when the building is constructed, the limitations need to be made apparent not only to the owner but to all prospective tenants prior to their occupation so that changes can be made if the tenant introduces greater fuel loads with different sprinkler system design requirements. Sometimes, when speculative buildings anticipate light or ordinary hazard occupancies (light and ordinary hazard occupancies will be defined later), but a storage occupancy actually moves in, there are building construction features as well as sprinkler system design and components that are not compliant. For example, a building constructed for office or mercantile use may not be appropriately designed for storing pool chemicals or arranging plastic parts in stacks, piles, or racks greater than 5 ft (1.52 m) high. The plastic

OWNER'S INFORMATION CERTIFICATE

Name/address of property to be protected with sprinkler protection _____

_____

Owner _____

Existing or planned construction is:
- ❑ Fire resistive or noncombustible
- ❑ Wood frame or ordinary (masonry walls with wood beams)
- ❑ Unknown

Is the system installation intended for one of the following special occupancies:

Aircraft hangar	❑ Yes	❑ No	Airport terminal	❑ Yes	❑ No
Fixed guideway transit system	❑ Yes	❑ No	Aircraft engine test facility	❑ Yes	❑ No
Race track stable	❑ Yes	❑ No	Power plant	❑ Yes	❑ No
Marine terminal, pier, or wharf	❑ Yes	❑ No	Water-cooling tower	❑ Yes	❑ No

If the answer to any of the above is "yes," the appropriate NFPA standard should be referenced for sprinkler density/area criteria.

Indicate whether any of the following special materials are intended to be present:

Flammable or combustible liquids	❑ Yes	❑ No	Compressed or liquefied gas cylinders	❑ Yes	❑ No
Aerosol products	❑ Yes	❑ No	Liquid or solid oxidizers	❑ Yes	❑ No
Nitrate film	❑ Yes	❑ No	Organic peroxide formulations	❑ Yes	❑ No
Pyroxylin plastic	❑ Yes	❑ No	Idle pallets	❑ Yes	❑ No

If the answer to any of the above is "yes," describe type, location, arrangement, and intended maximum quantities. _____

_____

Indicate whether the protection is intended for one of the following specialized occupancies or areas:

Spray area or mixing room	❑ Yes	❑ No	Commercial cooking operation	❑ Yes	❑ No
Solvent extraction	❑ Yes	❑ No	Class A hyperbaric chamber	❑ Yes	❑ No
Laboratory using chemicals	❑ Yes	❑ No	Cleanroom	❑ Yes	❑ No
Oxygen–fuel gas system for welding or cutting	❑ Yes	❑ No	Incinerator or waste-handling system	❑ Yes	❑ No
			Linen-handling system	❑ Yes	❑ No
Acetylene cylinder charging	❑ Yes	❑ No	Industrial furnace	❑ Yes	❑ No
Production or use of compressed or liquefied gases	❑ Yes	❑ No	Water-cooling tower	❑ Yes	❑ No

If the answer to any of the above is "yes," describe type, location, arrangement, and intended maximum quantities. _____

_____

Will there be any storage of products over 12 ft (3.6 m) in height?   ❑ Yes   ❑ No

If the answer is "yes," describe product, intended storage arrangement, and height. _____

_____

Will there be any storage of plastic, rubber, or similar products over 5 ft (1.5 m) high except as described above?   ❑ Yes   ❑ No

If the answer is "yes," describe product, intended storage arrangement, and height. _____

_____

I certify that I have knowledge of the intended use of the property and that the above information is correct.

Signature of owner's representative or agent _____ Date _____

Name of owner's representative or agent completing certificate (print) _____

Relationship and firm of agent (print) _____

© 2002 National Fire Protection Association

FIGURE 4.1
Owner's Information Certificate.
(Source: Adapted from NFPA 13, 2002, Figure A.14.1(b))

parts become high-piled storage and there may be issues with smoke and heat venting, as well as with the underdesign of the sprinkler system.

The engineer must balance the possible future uses of the building, the flexibility that the owner or occupant may need, and the cost of the system. Sprinkler system design and code enforcement would both be easy if all buildings were sprinklered using the extra hazard rules and if all storage occupancies were protected as if they were storing Group A plastics (a high-hazard commodity that will be defined and discussed shortly), but that is not possible in the real world. Instead, we must tailor the sprinkler system for the most likely use intended in the space. A well-designed sprinkler system will be one where the engineer has worked with the building owners to understand their needs and will produce a design flexible enough to take the owner's needs into account for the foreseeable future. To a building owner, the value of a sprinkler system could, in part, be attributable to a fire protection engineer's design for the probable uses of the building during its life span. The owner and the fire protection engineer must carefully evaluate short-term as well as long-term potential uses of the building. It would certainly not be cost-efficient to reevaluate and modify the system every time a new tenant signs a lease. Unfortunately, this is a common occurrence caused by not including the proper thought process in the original design but instead being influenced by the minimum possible sprinkler system design criteria.

Light Hazard Occupancy

Light hazard occupancies are defined in NFPA 13 as occupancies or portions of other occupancies where the quantity and/or combustibility of contents is low, and fires with relatively low rates of heat release are expected (see NFPA 13, Section 5.2). Churches, educational facilities, hospitals, offices, residential facilities, and restaurant seating areas are the occupancies that NFPA 13 considers to be light hazard. These examples are in Annex paragraph A.5.2 of NFPA 13, 2002 edition, and not in the body of the standard because every office is not light hazard, just like every school is not necessarily light hazard. NFPA 13 does not attempt to legislate hazard classifications; instead, the design professional needs to make a survey of the use and contents of the building prior to designing the sprinkler system. If the office building is typical of low-combustibility buildings, then it can be considered light hazard. If there is going to be a greater fuel load or more storage than usual, then it might be more appropriate to use the ordinary hazard criteria.

The use of light hazard design criteria is among the most debated topics in the fire protection community. Some insurance companies do not permit light hazard requirements to be followed because light hazard does not provide the minimum level of protection that the insurer requires. Consequently, in addition to the codes and standards, the engineer is advised to check with the insurance company for a project before finalizing the sprinkler system design.

A designer should recognize that NFPA 13 is a general document that establishes the minimum level of protection and cannot possibly address every conceivable use for a building, leaving significant latitude for the design professional. Insurance companies who primarily insure highly protected risks are likely to have requirements that exceed NFPA 13.

Recently—with the advent of composite or plastic desks, chairs, office equipment, and computers, the shielding of computers by desks and other plastic components by the office furniture, and the presence of multiple sources of ignition—there is some thought that office occupancies may no longer contain light hazard fuel loads. Full-scale office fire research by National Institute of Standards and Technology (NIST) has concluded that the heat release from typical office fuel packages can be significant. Although there have been several studies done to determine the fuel load of office furnishings, specifically workstations, this chapter will provide information available from one of those studies. In his work for the NIST, Daniel Madrzykowski conducted a study of workstations and examined their heat release rates. The peak HRR ranged from 2.8 megawatts (MW) to 6.9 MW. Fire growth for different configurations of office furnishings followed "slow"–"medium" curves all the way to "fast"–"ultra-fast" rates [2]. Additional information is available directly from NIST.

Many years ago, before the light hazard rules existed, only factory and industrial occupancies were sprinklered. The sprinkler committee was asked to relax some of the water delivery and sprinkler spacing rules in order to create a simpler and cheaper system to install in buildings like hospitals and schools that could not afford what is now known as ordinary hazard sprinkler systems. After reviewing test data and the spray pattern development of sprinklers, the committee agreed in 1930 to develop a different set of rules, actually a separate standard, for these light hazard occupancies that were not being sprinklered at that time. The original term for these sprinkler systems was "Class B" systems. The Class B rules were incorporated into NFPA 13 in 1931 and have evolved into what we know today as light hazard.

Although some practitioners might be concerned about the relaxation of the rules of ordinary hazard to create the light hazard category, the fact remains that with over 70 years of real fire experience, the light hazard rules are appropriate and there are sufficient safety factors in place to maintain the reliability of fire sprinkler systems that are properly designed, installed, and maintained in light hazard occupancies. One of those safety factors is the higher water flow, or densities, that are produced by the first few sprinklers that operate in a fire.

Ordinary Hazard Occupancy

The ordinary hazard classification is broken into two groups—ordinary hazard group 1 and ordinary hazard group 2.

Ordinary Hazard Group 1. Ordinary hazard group 1 occupancies are occupancies or portions of other occupancies where the following conditions exist (see NFPA 13, 5.3.1):

- The combustibility is low.
- Quantity of combustibles is moderate.
- Stockpiles of combustibles do not exceed 8 ft (2.4 m) in height.
- Fires with moderate rates of heat release are expected.

From NFPA 13 (2002) Annex A, the following are examples of typical ordinary hazard group 1 occupancies: automobile parking and showrooms, bakeries, beverage manufacturing, canneries, dairy product manufacturing, glass product manufacturing, laundries (NOT dry cleaners), and restaurant service areas. Generally speaking, manufacturing is considered ordinary hazard and if the

FIGURE 4.2
Power Law Heat Release
Rates.

(Source: *National Fire Alarm Code®*,
NFPA, 2002, Figure B.2.3.2.6.2)

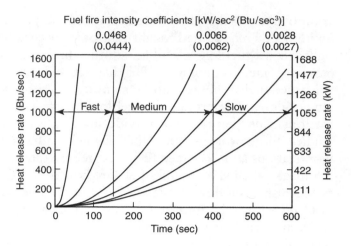

manufactured goods are noncombustible it is probably ordinary hazard group 1. If the manufactured goods are combustible, it is probably ordinary hazard group 2.

The parameters for each of the two ordinary hazard occupancies are not specific or detailed enough for the necessary classification. Because of the lack of definition, most engineers and designers base their selection on the limited number of examples in Annex A in NFPA 13 which are recommendations, not requirements. This method is not always the appropriate or correct way to establish the classification.

Ordinary Hazard Group 2. Ordinary hazard group 2 occupancies are occupancies or portions of other occupancies where the quantity and combustibility of contents is moderate to high, stockpiles do not exceed 12 ft (3.6 m) in height, and fires with moderate to high rates of heat release are expected (see NFPA 13, 5.3.2).

Figure 4.2, which is from *NFPA 72®*, *National Fire Alarm Code®*, displays the Power Law Heat Release Rates that may be useful to the engineer in determining design criteria.

Slow, medium, and fast heat release rates do not directly correlate to the density/area curves; however, in general, the slow release rate curve would be associated most closely with light hazard sprinkler system design criteria. The medium heat release curve would be associated with some of the Class I through IV commodity and ordinary hazard sprinkler system design criteria, and the fast heat release curve would be associated with Group A plastics, flammable and combustible liquids, and the extra hazard sprinkler system design criteria.

The following are examples of typical ordinary hazard group 2 occupancies from Annex A of NFPA 13: cereal mills, chemical plants—ordinary, confectionary products, distilleries, dry cleaners, feed mills, horse stables, libraries—large stack rooms, machine rooms, metal working, mercantile, paper and pulp mills, paper processing plants, piers and wharves, post offices, printing and publishing, repair garages, resin application areas, stages, textile manufacturing, tire manufacturing, tobacco products manufacturing, wood machining, and wood product assembly.

Note that one of the examples is tire manufacturing. A fire in rubber tires is one of the most difficult fires to fight. It is a very intense shielded fire where the tires themselves shield the fire from direct water spray. However, the machines used to make the tires are a much less significant hazard because they hold only a limited quantity of rubber. The rest of the machine is generally noncombustible and therefore the manufacture of tires is considered ordinary hazard group 2. As long as manufacturing is the only process in the area, (i.e., there is no storage of any kind), then the classification is ordinary hazard group 2. In this example, if the finished tires are stored in the room after the manufacturing is complete, then the room needs to be treated as rubber tire storage and it must, at a minimum, conform to the rules of Miscellaneous Rubber Tire Storage in NFPA 13 or follow the requirements of NFPA 230, *Standard for the Fire Protection of Storage*.

At one time there were three groups in the ordinary hazard classification. However, in 1991, the sprinkler committee decided that there was too little difference between the densities for ordinary hazard group 2 and ordinary hazard group 3 and settled on the two classifications we have today. This change has made the designer's life a little easier and made the standard more user-friendly.

Extra Hazard Occupancy

In 1940, the occupancy classes expanded, from the two now known as light and ordinary to three with the development of the extra hazard classification. Like the ordinary hazard classification, the extra hazard classification is also divided into two groups—extra hazard group 1 and extra hazard group 2.

Extra Hazard Group 1. Extra hazard group 1 occupancies are occupancies or portions of other occupancies where the quantity and combustibility of contents is very high and dust, lint, or other materials are present, introducing the probability of rapidly developing fire with high rates of heat release but with little or no combustible or flammable liquids (see NFPA 13, 5.4.1).

From NFPA 13, Annex A, the following are examples of typical extra hazard group 1 occupancies: aircraft hangars (except as governed by NFPA 409, *Standard on Aircraft Hangars*); combustible hydraulic fluid use area; die casting; metal extruding; plywood and particle board manufacturing; printing [using inks with a flash point below 100°F (38°C)]; rubber reclaiming/vulcanizing; saw mills; textile picking, opening, blending, garnetting, or carding, combining of cotton, synthetics, wool shoddy, or burlap; and upholstering with plastic foams.

Extra Hazard Group 2. Extra hazard group 2 occupancies are occupancies or portions of other occupancies with moderate to substantial amounts of flammable or combustible liquids or occupancies where shielding of combustibles is extensive (see NFPA 13, 5.4.2).

From NFPA 13, Annex A, the following are examples of typical extra hazard group 2 occupancies: asphalt saturating, flammable liquids spraying, flow coating, manufactured home or modular building assembly, open oil quenching, plastics processing, solvent cleaning, and varnish and paint dipping.

Mixed Occupancies

A good example of the various occupancies in a building might be a school building. Generally, the school includes many light hazard areas—classrooms, offices, toilets, and locker rooms. Then there may be the ordinary hazard group 1 areas in the building that would include storage areas, mechanical rooms, the kitchen food preparation area, and janitor's closets. There might also be ordinary hazard group 2 areas such as a wood shop, metal shop, biology and chemistry rooms, and stages. Sometimes, multipurpose rooms or the gymnasium are used for community fairs, exhibitions, and similar activities, and these would then be considered ordinary hazard 2. Finally, it is also possible to have an extra hazard group 1 area such as a paint storage area and possibly extra hazard group 2 if there is a paint spray booth (flammable liquids spraying). Two hazards that sometimes are overlooked in a school are the shielding by movable, often Group A plastic, bleachers and the high hazard, exposed, expanded plastics, associated with the mats that are attached to walls in gymnasiums and on the floors and walls in wrestling or similar rooms. Many buildings have these different occupancy classifications, and this example should point out the necessity for the sprinkler system design drawings to indicate all of the possible information regarding the occupancy of each and every room.

Commodity Classification

The occupancy classifications we have looked at to this point in this chapter form the basis for the occupancy hazard fire control approach of Chapter 3, "Design Approaches," and apply to nonstorage occupancies. The next classifications that need to be defined and understood relate to the storage design approach of Chapter 12 of NFPA 13, 2002 edition.

The storage design approach uses the definitions for these hazards found in NFPA 13, Section 5.6, "Commodity Classification," where the general classifications of many products are indicated. As with the occupancy approach, this list includes some of the possible commodity combinations, but there are many more. The engineer is required to attempt to classify products by using the information provided in the body of NFPA 13 or in its Annex. However, those products that cannot be readily classified might require actual testing to provide realistic protection schemes based on the actual HRR of each specific commodity.

There are seven different and distinct classifications into which a commodity could fall. Class I, Class II, Class III, and Class IV represent general combustible commodities. Plastics or rubber products are placed in Group A Plastics, Group B Plastics, or Group C Plastics. Then there are classifications for rubber tires, rolled paper, baled cotton, and mixed commodities. Although there are several different classifications, there are only five different sets of sprinkler requirements to protect those classifications. These five classifications are described in the section ("Storage Classification") that follows. Some of the rubber and plastic commodities can be adequately protected by the general combustible rules. Storage occupancies have different requirements for sprinkler system design depending on the type of commodity being stored and frequently, the height and arrangement of the storage and the spacing between the top of the storage and the sprinklers. The classifications are based primarily on the types and amounts of materials that make up the product and the primary packaging materials used with that specific product.

Storage Classification

The remainder of this chapter deals with the often difficult decision of how to classify storage commodities. Some special commodities such as flammable and combustible liquids, aerosols, and so on have their own NFPA standards. If one of these special commodities is encountered, the engineer will need to utilize the requirement in the NFPA standard that applies to the special commodity.

Table 4.1 from the Annex of NFPA 13 lists commodities and their classifications.

TABLE 4.1	

Alphabetized Listing of Commodity Classes

Commodity	Commodity Class
Aerosols	
Cartoned or uncartoned	
- Level 1	Class III
Alcoholic Beverages	
Cartoned or uncartoned	
- Up to 20 percent alcohol in metal, glass, or ceramic containers	Class I
- Up to 20 percent alcohol in wood containers	Class II
Ammunition	
Small arms, shotgun	
- Packaged, cartoned	Class IV
Appliances, Major (e.g., stoves, refrigerators)	
- Not packaged, no appreciable plastic exterior trim	Class I
- Corrugated, cartoned (no appreciable plastic trim)	Class II
Baked Goods	
Cookies, cakes, pies	
- Frozen, packaged in cartons[1]	Class II
- Packaged, in cartons	Class III
Batteries	
Dry cells (nonlithium or similar exotic metals)	
- Packaged in cartons	Class I
- Blister-packed in cartons	Class II
Automobile	
- Filled[2]	Class I
Truck or larger	
- Empty or filled[2]	Group A plastics
Beans	
Dried	
- Packaged, cartoned	Class III
Bottles, Jars	
Empty, cartoned	
- Glass	Class I
- Plastic PET (polyethylene terephthalate)	Class IV
Filled noncombustible powders	
- Plastic PET	Class II
- Glass, cartoned	Class I
- Plastic, cartoned [less than 1 gal (3.8 L)]	Class IV
- Plastic, uncartoned (other than PET), any size	Group A plastics
- Plastic, cartoned or exposed [greater than 1 gal (3.8 L)]	Group A plastics
- Plastic, solid plastic crates	Group A plastics
- Plastic, open plastic crates	Group A plastics
Filled noncombustible liquids	
- Glass, cartoned	Class I
- Plastic, cartoned [less than 5 gal (18.9 L)]	Class I
- Plastic, open or solid plastic crates[3]	Group A plastics
- Plastic, PET	Class I
Boxes, Crates	
- Empty, wood, solid walls	Class II
- Empty, wood, slatted[4]	Outside of scope

continues

TABLE 4.1

(Continued)

Commodity	Commodity Class
Bread	
Wrapped, cartoned	Class III
Butter	
Whipped spread	Class III
Candles	
Packaged, cartoned	
- Treat as expanded plastic	Group A plastics
Candy	
Packaged, cartoned	Class III
Canned Foods	
In ordinary cartons	Class I
Cans	
Metal	
- Empty	Class I
Carpet Tiles	
Cartoned	Group A plastics
Cartons	
Corrugated	
- Unassembled (neat piles)	Class III
- Partially assembled	Class IV
Wax coated, single walled	Group A plastics
Cement	
Bagged	Class I
Cereals	
Packaged, cartoned	Class III
Charcoal	
Bagged	
- Standard	Class III
Cheese	
- Packaged, cartoned	Class III
- Wheels, cartoned	Class III
Chewing Gum	
Packaged, cartoned	Class III
Chocolate	
Packaged, cartoned	Class III
Cloth	
Cartoned and not cartoned	
- Natural fiber, viscose	Class III
- Synthetic[5]	Class IV
Cocoa Products	
Packaged, cartoned	Class III
Coffee	
- Canned, cartoned	Class I
- Packaged, cartoned	Class III
Coffee Beans	
Bagged	Class III
Cotton	
Packaged, cartoned	Class III

TABLE 4.1	
(Continued)	

Commodity	Commodity Class
Diapers	
- Cotton, linen	Class III
- Disposable with plastics and nonwoven fabric (in cartons)	Class IV
- Disposable with plastics and nonwoven fabric (uncartoned), plastic wrapped	Group A plastics
Dried Foods	
Packaged, cartoned	Class III
Fertilizers	
Bagged	
- Phosphates	Class I
- Nitrates	Class II
Fiberglass Insulation	
- Paper-backed rolls, bagged or unbagged	Class IV
File Cabinets	
Metal	
- Cardboard box or shroud	Class I
Fish or Fish Products	
Frozen	
- Nonwaxed, nonplastic packaging	Class I
- Waxed-paper containers, cartoned	Class II
- Boxed or barreled	Class II
- Plastic trays, cartoned	Class III
Canned	
- Cartoned	Class I
Frozen Foods	
Nonwaxed, nonplastic packaging	Class I
- Waxed-paper containers, cartoned	Class II
- Plastic trays	Class III
Fruit	
Fresh	
- Nonplastic trays or containers	Class I
- With wood spacers	Class I
Furniture	
Wood	
- No plastic coverings or foam plastic cushioning	Class III
- With plastic coverings	Class IV
- With foam plastic cushioning	Group A plastics
Grains—Packaged in Cartons	
- Barley	Class III
- Rice	Class III
- Oats	Class III
Ice Cream	Class I
Leather Goods	Class III
Leather Hides	
Baled	Class II
Light Fixtures	
Nonplastic	
- Cartoned	Class II

continues

TABLE 4.1

(Continued)

Commodity	Commodity Class
Lighters	
Butane	
- Blister-packed, cartoned	Group A plastics
- Loose in large containers (Level 3 aerosol)	Outside of scope
Liquor	
100 proof or less, 1 gal (3.8 L) or less, cartoned	
- Glass (palletized)[6]	Class IV
- Plastic bottles	Class IV
Marble	
Artificial sinks, countertops	
- Cartoned, crated	Class II
Margarine	
- Up to 50 percent oil (in paper or plastic containers)	Class III
- Between 50 percent and 80 percent oil (in any packaging)	Group A plastics
Matches	
Packaged, cartoned	
- Paper	Class IV
- Wood	Group A plastics
Mattresses	
- Standard (box spring)	Class III
- Foam (in finished form)	Group A plastics
Meat, Meat Products	
- Bulk	Class I
- Canned, cartoned	Class I
- Frozen, nonwaxed, nonplastic containers	Class I
- Frozen, waxed-paper containers	Class II
- Frozen, expanded plastic trays	Class II
Metal Desks	
- With plastic tops and trim	Class I
Milk	
- Nonwaxed-paper containers	Class I
- Waxed-paper containers	Class I
- Plastic containers	Class I
- Containers in plastic crates	Group A plastics
Motors	
- Electric	Class I
Nail Polish	
- 1-oz to 2-oz (29.6-ml to 59.1-ml) glass, cartoned	Class IV
- 1-oz to 2-oz (29.6-ml to 59.1-ml) plastic bottles, cartoned	Group A plastics
Nuts	
- Canned, cartoned	Class I
- Packaged, cartoned	Class III
- Bagged	Class III
Paints	
Friction-top cans, cartoned	
- Water-based (latex)	Class I
- Oil-based	Class IV
Paper Products	
- Books, magazines, stationery, plastic-coated paper food containers, newspapers, cardboard games, or cartoned tissue products	Class III
- Tissue products, uncartoned and plastic wrapped	Group A plastics

TABLE 4.1

(Continued)

Commodity	Commodity Class
Paper, Rolled	
In racks or on side	Class III
- Medium- or heavyweight	
In racks	Class IV
- Lightweight	
Paper, Waxed	
Packaged in cartons	Class IV
Pharmaceuticals	
Pills, powders	
- Glass bottles, cartoned	Class II
- Plastic bottles, cartoned	Class IV
Nonflammable liquids	
- Glass bottles, cartoned	Class II
Photographic Film	
- Motion picture or bulk rolls of film in polycarbonate, polyethylene, or metal cans; polyethylene bagged in cardboard boxes	Class II
- 35-mm in metal film cartridges in polyethylene cans in cardboard boxes	Class III
- Paper, in sheets, bagged in polyethylene, in cardboard boxes	Class III
- Rolls in polycarbonate plastic cassettes, bulk wrapped in cardboard boxes	Class IV
Plastic Containers (except PET)	
- Noncombustible liquids or semiliquids in plastic containers less than 5 gal (18.9 L) capacity	Class I
- Noncombustible liquids or semiliquids (such as ketchup) in plastic containers with nominal wall thickness of ¼ in. (6.4 mm) or less and larger than 5 gal (18.9 L) capacity	Class II
- Noncombustible liquids or semiliquids (such as ketchup) in plastic containers with nominal wall thickness greater than ¼ in. (6.4 mm) and larger than 5 gal (18.9 L) capacity	Group A plastics
Polyurethane	
- Cartoned or uncartoned expanded	Group A plastics
Poultry Products	
- Canned, cartoned	Class I
- Frozen, nonwaxed, nonplastic containers	Class I
- Frozen (on paper or expanded plastic trays)	Class II
Powders	
Ordinary combustibles—free flowing	
- In paper bags (e.g., flour, sugar)	Class II
PVA (polyvinyl alcohol) Resins	
PVC (polyvinyl chloride)	
- Flexible (e.g., cable jackets, plasticized sheets)	Class III
- Rigid (e.g., pipe, pipe fittings)	Class III
- Bagged resins	Class III
Rags	
Baled	
- Natural fibers	Class III
- Synthetic fibers	Class IV
Rubber	
- Natural, blocks in cartons	Class IV
- Synthetic	Group A plastics

continues

TABLE 4.1

(Continued)

Commodity	Commodity Class
Salt	
- Bagged	Class I
- Packaged, cartoned	Class II
Shingles	
- Asphalt-coated fiberglass	Class III
- Asphalt-impregnated felt	Class IV
Shock Absorbers	
- Metal dust cover	Class II
- Plastic dust cover	Class III
Signatures	
Books, magazines	
- Solid array on pallet	Class II
Skis	
- Wood	Class III
- Foam core	Class IV
Stuffed Toys	
Foam or synthetic	Group A plastics
Syrup	
- Drummed (metal containers)	Class I
- Barreled, wood	Class II
Textiles	
Natural fiber clothing or textile products	Class III
Synthetics (except rayon and nylon)—50/50 blend or less	
- Thread, yarn on wood or paper spools	Class III
- Fabrics	Class III
- Thread, yarn on plastic spools	Class IV
- Baled fiber	Group A plastics
Synthetics (except rayon and nylon)—greater than 50/50 blend	
- Thread, yarn on wood or paper spools	Class IV
- Fabrics	Class IV
- Baled fiber	Group A plastics
- Thread, yarn on plastic spools	Group A plastics
Rayon and nylon	
- Baled fiber	Class IV
- Thread, yarn on wood or paper spools	Class IV
- Fabrics	Class IV
- Thread, yarn on plastic spools	Group A plastics
Tobacco Products	
In paperboard cartons	Class III
Transformers	
Dry and oil filled	Class I
Vinyl-Coated Fabric	
Cartoned	Group A plastics
Vinyl Floor Coverings	
- Tiles in cartons	Class IV
- Rolled	Group A plastics
Wax-Coated Paper	
Cups, plates	
- Boxed or packaged inside cartons (emphasis on packaging)	Class IV
- Loose inside large cartons	Group A plastics

TABLE 4.1

(Continued)

Commodity	Commodity Class
Wax	
Paraffin/petroleum wax, blocks, cartoned	Group A plastics
Wire	
- Bare wire on metal spools on wood skids	Class I
- Bare wire on wood or cardboard spools on wood skids	Class II
- Bare wire on metal, wood, or cardboard spools in cardboard boxes on wood skids	Class II
- Single- or multiple-layer PVC-covered wire on metal spools on wood skids	Class II
- Insulated (PVC) cable on large wood or metal spools on wood skids	Class II
- Bare wire on plastic spools in cardboard boxes on wood skids	Class IV
- Single- or multiple-layer PVC-covered wire on plastic spools in cardboard boxes on wood skids	Class IV
- Single, multiple, or power cables (PVC) on large plastic spools	Class IV
- Bulk storage of empty plastic spools	Group A plastics
Wood Products	
- Solid piles—lumber, plywood, particleboard, pressboard (smooth ends and edges)	Class II
- Spools (empty)	Class III
- Toothpicks, clothespins, hangers in cartons	Class III
- Doors, windows, wood cabinets, and furniture	Class III
- Patterns	Class IV

[1]The product is presumed to be in a plastic-coated package in a corrugated carton. If packaged in a metal foil, it can be considered Class I.

[2]Most batteries have a polypropylene case and, if stored empty, should be treated as a Group A plastic. Truck batteries, even where filled, should be considered a Group A plastic because of their thicker walls.

[3]As the openings in plastic crates become larger, the product behaves more like a Class III commodity. Conversely, as the openings become smaller, the product behaves more like a plastic.

[4]These items should be treated as idle pallets.

[5]Tests clearly indicate that a synthetic or synthetic blend is considered greater than Class III.

[6]When liquor is stored in glass containers in racks, it should be considered a Class III commodity; where it is palletized, it should be considered a Class IV commodity.

Source: NFPA 13, 2002, Table A.5.6.3.

Class I Commodities

The first of the seven classifications is the Class I Commodities. Noncombustible products placed directly on wooden pallets, in single-layer corrugated cartons with or without single-thickness cardboard dividers, or shrink-wrapped or paper-wrapped as a unit with or without pallets are Class I commodities. From NFPA 13, Annex A, typical examples of Class I commodities are metal and glass products without combustible packaging material, noncombustible foods, noncombustible products placed directly on wooden pallets, and noncombustible products shrink-wrapped on the sides only or paper-wrapped as a unit load (with or without pallets). The entire list is given in Table 4.2.

The insurance carrier FM Global is also a fire protection product "approvals" laboratory and fire protection testing facility. Many of the product fire tests and product approvals that are used in fire protection have come from Factory Mutual Insurance Company, now named FM Global. FM

TABLE 4.2
Examples of Class I Commodities

Alcoholic Beverages
  Cartoned or uncartoned
  - Up to 20 percent alcohol in metal, glass, or ceramic containers

Appliances, Major (e.g., stoves, refrigerators)   - Not packaged, no appreciable plastic exterior trim

Batteries
  Dry cells (nonlithium or similar exotic metals)
  - Packaged in cartons
  Automobile
  - Filled*

Bottles, Jars
  Empty, cartoned
  - Glass
  Filled noncombustible liquids
  - Glass, cartoned
  - Plastic, cartoned [less than 5 gal (18.9 L)]
  - Plastic, PET
  Filled noncombustible powders
  - Glass, cartoned

Canned Foods   In ordinary cartons

Cans
  Metal
  - Empty

Cement   Bagged
Coffee   Canned, cartoned

Fertilizers
  Bagged
  - Phosphates

File Cabinets
  Metal
  - Cardboard box or shroud

Fish or Fish Products
  Frozen
  - Nonwaxed, nonplastic packaging
  Canned
  - Cartoned

Frozen Foods   Nonwaxed, nonplastic packaging

Fruit
  Fresh
  - Nonplastic trays or containers
  - With wood spacers

Ice Cream

Meat, Meat Products
  - Bulk
  - Canned, cartoned
  - Frozen, nonwaxed, nonplastic containers

Metal Desks   - With plastic tops and trim

TABLE 4.2

(Continued)

Milk
- Nonwaxed-paper containers
- Waxed-paper containers
- Plastic containers
Motors
- Electric
Nuts
- Canned, cartoned
Paints
Friction-top cans, cartoned
- Water-based (latex)
Plastic Containers
- Noncombustible liquids or semiliquids in plastic containers less than 5 gal (18.9 L) capacity
Poultry Products
- Canned, cartoned
- Frozen, nonwaxed, nonplastic containers
Salt
Bagged
Syrup
Drummed (metal containers)
Transformers
Dry and oil filled
Wire
Bare wire on metal spools on wood skids

*Most batteries have a polypropylene case and, if stored empty, should be treated as a Group A plastic. Truck batteries, even where filled, should be considered a Group A plastic because of their thicker walls.

Source: NFPA 13, 2002, Table A.5.6.3.1.

Global has developed benchmark tests for the commodity classifications we are discussing here. These tests or definitions are used to establish two things: (1) to give designers an idea as to what type of commodity the rules should apply to; (2) to establish an easily reproducible fire for testing purposes. Over the years, many fire tests with the same commodities in different configurations or with different sprinkler protection criteria can be and have been analyzed to show differing methods of fire protection.

For example, the FM Global benchmark for a Class I commodity is glass jars in compartmented cardboard cartons. Other examples of Class I commodities from NFPA 13, Annex A, include alcoholic beverages (up to 20 percent alcohol) in metal, glass, or ceramic containers; appliances (stoves, refrigerators) not packaged or with no appreciable plastic trim; bottles and jars, empty glass, and plastic PET (polyethylene terephthalate); filled noncombustible plastic cartons of less than 5 gal [18.9 L] (filled with noncombustible liquid); frozen foods (nonwaxed, nonplastic packaging); paints [friction-top cans, cartoned, water-based (latex)]; and poultry products (frozen with nonwaxed, nonplastic containers) to list a few. Figure 4.3 illustrates another sample Class I commodity.

Class II Commodities

Class II commodities are defined by NFPA 13 as "noncombustible product that is in slatted wooden crates, solid wood boxes, multiple-layered corrugated cartons, or equivalent combustible packaging material, with or without pallets" (NFPA 13, 5.6.3.2). Essentially, Class II commodities are the same as Class I commodities, but in more substantial packaging.

The FM Global benchmark (sometimes referred to as Factory Mutual Research Standard Class 2) for this commodity is metal-lined, double tri-wall cartons [3]. Other examples of Class II commodities from NFPA 13, Annex A, include alcoholic beverages (up to 20 percent alcohol) in wooden containers, appliances (stoves, refrigerators) in corrugated cartons with no appreciable plastic trim, baked goods (cookies, cakes, and pies) frozen and in plastic-coated packages, frozen foods (waxed-paper containers, cartoned), plastic containers (except PET) filled with noncombustible liquids or semiliquids (e.g., ketchup) in plastic containers with nominal wall thickness of ¼ in. (6.4 mm) or less and larger than 5 gal [18.9 L] capacity, and poultry (frozen on paper or expanded plastic trays) to list a few. See Table 4.1 for the entire list. Figure 4.4 illustrates a typical Class II commodity.

Class III Commodities

Class III commodities are defined by NFPA 13, 2002 edition, as "a product fashioned from wood, paper, natural fibers, or Group C plastics with or without cartons, boxes, or crates and with or without pallets" (NFPA 13, 5.6.3.3). These

commodities are allowed to include a limited amount of Group A or B plastics. A limited amount is defined as 5 percent or less by weight or volume. This definition is not the same for all codes and standards (i.e., FM Global and the International Fire Code have different limits) and the engineer should be careful to evaluate the commodity based on the appropriate governing document.

The FM Global benchmark for this commodity is paper cups in compartmented cartons. Other examples of Class III commodities from NFPA 13, Annex A, include Level 1 aerosols, baked goods (cookies, cakes, pies) packaged in cartons, grains (packaged in cartons), charcoal (bagged), frozen foods (on plastic trays), wood furniture (no plastic covers or foam plastic cushioning), textiles (natural fiber clothing or textile products), lumber; and textiles—thread, yarn on wood or paper spools (synthetics—except rayon and nylon, 50-50 blend or less) to list a few. See Figure 4.1 for the entire list.

Figure 4.5 illustrates a wooden chair, a Class III commodity. The plastic fittings shown in Figure 4.6 are also a Class III commodity.

Class IV Commodities

Class IV commodities are defined by NFPA 13, 2002 edition, as "a product, with or without pallets, that meets one of the following criteria: (1) constructed partially or totally of Group B plastics; (2) consists of free-flowing Group A plastic materials; or (3) contains within itself or its packaging an appreciable amount (5 percent to 15 percent by weight or 5 percent to 25 percent by volume) of Group A plastics" (NFPA 13, 5.6.3.4.).

FIGURE 4.5
A Typical Class III
Commodity.

FIGURE 4.6
Plastic Fittings, a Class III
Commodity.

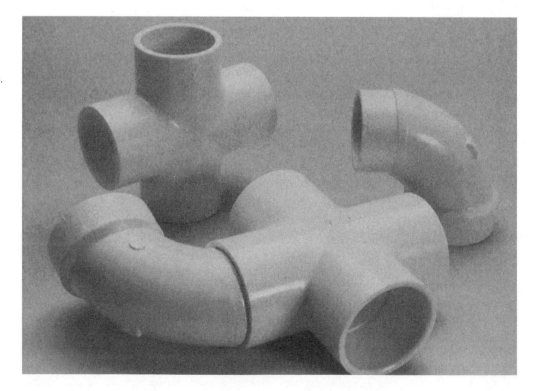

FIGURE 4.7
FM Benchmark for Class IV Commodity.

The FM Global benchmark for this commodity is unexpanded polystyrene (15 percent by weight) and paper cups in compartmented cartons, as illustrated by Figure 4.7. Other examples of Class IV commodities from NFPA 13, Annex A, include ammunition (small arms or shotgun packaged in cartons), empty and cartoned bottles and jars (plastic PET), liquor [100 proof or less in 1 gal (3.8 L) or less glass or plastic bottles], paints (friction-top cans, cartoned, oil-based), and vinyl floor coverings (tiles in cartons) to list a few. See Figure 4.1 for the entire list. Also see Figure 4.8, whose plastic wrapping places the commodity in Class IV.

Plastics, Elastomers, and Rubber Commodities

Finally, there are the plastics, elastomers, and rubber commodities that are divided into three categories depending on their burning characteristics: Group A, Group B, and Group C. Manufacturers can add fire or flame retardants to lower a product's category from A to B or B to C. The FM Global benchmark (sometimes referred to as Factory Mutual Research Standard Plastic) for Group A plastic commodities is unexpanded polystyrene cups in compartmented cartons [3]. (See Figure 4.8.) Other examples of Group A plastic commodities from NFPA 13, Annex A, include acrylic, butyl rubber, FRP (fiberglass-reinforced polyester), natural rubber (if expanded), polyethylene, polypropylene, and polystyrene to list a few. See Figure 4.1 for the entire list. Two examples of expanded plastic commodities appear in Figures 4.9 and 4.10.

FIGURE 4.8
Plastic Wrapping in a Class
IV Commodity.

FIGURE 4.9
Sample Expanded Plastic
Commodity.

Then there are products that burn like Group A plastics. Some of these include truck or larger batteries (empty or full), candles (packaged, cartoned—treat as expanded plastic), diapers (disposable with plastics and nonwoven fabric—uncartoned, plastic-wrapped), margarine (between 50 and 80 percent oil in any packaging), mattresses (foam—in finished form), milk (containers in plastic crates), and nail polish [1- to 2-oz (29.6-ml to 59.1-ml) plastic bottles in cartons] to list a few.

Group B plastics are protected the same as Class IV commodities. Some of the Group B plastics include fluoroplastics, nylon 6, silicone rubber, and natural rubber (not expanded), from NFPA 13, Annex A.

Group C plastics have the lowest heat release rates of the plastics and are protected the same as Class III commodities. A few examples of the Group C plastics include PVC (polyvinyl chloride—flexible), melamine (melamine formaldehyde), and urea (urea formaldehyde) from NFPA 13, Annex A.

Encapsulation

It has already been seen that to some extent the packaging and packaging material can affect the classification. Another term for a method of packaging that has not yet been discussed is *encapsulation*. Encapsulation is defined by NFPA 13, 2002 edition, as "a method of packaging consisting of a plastic sheet completely enclosing the sides and top of a pallet load containing a combustible commodity or combustible package or a group of combustible commodities or combustible packages" (NFPA 13, 3.9.8). Banding or stretch-wrapping around the sides (and not the top) is not considered encapsulation. Likewise, where there are holes or openings in a plastic or weatherproof cover on the top of a carton or container that exceed more than half the area of the cover, the carton or container is not considered encapsulated. Encapsulation may affect the design of the sprinkler system but does not change the commodity classification.

Full-scale tests have shown that for high-pile storage to 15 ft (4.5 m), there is no appreciable difference in the number of sprinklers that open for either nonencapsulated or encapsulated products. Test data is not available for encapsulated products stored higher than 15 ft (4.5 m). However, test data for rack storage involving encapsulated storage 20 ft (6.1 m) high and higher require increased protection over that for nonencapsulated storage according to NFPA 13. The reason for this increased protection is primarily because of the inability of water to penetrate the encapsulation and do what it does best—cool and prewet, the two significant factors in the successful operation of control mode sprinkler systems.

Classification of Difficult or Unknown Commodities

The design engineer for sprinkler systems in storage occupancies needs to take great care to properly identify and classify **all** of the commodities. Because of the large, complex number of plastics, and the ease of changing their burning characteristics with additives, plastics can be very difficult to classify, and there may be many different components and sprinkler system design requirements in a given room or area. The heat release rate of plastics can be 3 to 5 times greater than for ordinary combustibles with the same storage configuration.

The heat of combustion of ordinary combustibles (e.g., wood or paper) generally ranges between 6,000 and 8,000 Btu/lb (14,000–19,000 kJ/kg). The heat of combustion for plastics generally ranges between 12,000 and 20,000 Btu/lb (28,000–47,000 kJ/kg) [3].

Commodity classifications are primarily done by comparing a product to the definition of a known product. Sometimes a product can be tested to determine its specific HRR and subsequently its classification and protection requirements. This testing must involve sprinklers and must be to a scale large enough to provide meaningful results. Other times, the designer may be able to find specific test data from nationally recognized testing laboratories to assist in determining the classification.

Mixed Commodities

When attempting to classify materials for sprinkler system design or other reasons, there are several other requirements that need to be considered. One of those is that the protection requirements are not based on the commodity that predominantly occupies the area or facility, but are based on the commodity with the highest classification. Buildings or areas can be protected for the highest, most demanding commodity; they can be protected by confining the highest-hazard material to a designated area, with the sprinkler system design criteria extending 15 ft (4.5 m) past in all directions; or they can be protected in accordance with the mixed commodities section in NFPA 13, 2002 edition, paragraph 5.6.1.2. Model building and fire codes often indicate different requirements from NFPA 13 and usually are more restrictive.

Paragraph 5.6.1.2, for mixed commodities, permits for the random spacing of a few pallet loads of the higher commodity without changing the commodity classification, with limitations. Those limitations include placing up to 10 pallet loads in any 40,000 ft² (3716 m²) of a higher-hazard commodity as long as they are randomly spaced and there are no adjacent higher-hazard pallet loads in any direction, including diagonally. However, when the fire protection system is designed for Class I or Class II protection, the number of pallet loads of Class IV or Group A plastics allowed is reduced from 10 to 5.

Pallets

Another consideration in classifying commodities is the pallet the commodity or load rests on. The commodity classifications we have reviewed to this point are based on the use of wooden or metal pallets. There are, however, also reinforced and unreinforced plastic pallets.

Plastic pallets are permitted by paragraph 5.6.2, but the fire protection system design is required to be increased by one commodity classification when the unreinforced, polypropylene or high-density polyethylene pallets are used. The fire protection design is required to be increased by two commodity classifications when reinforced polypropylene or high-density polyethylene plastic pallets are used. However, the use of K-17 sprinklers in the sprinkler system design can eliminate this increase in many cases.

There are pallets that aren't made of wood or metal that do not require an increase in the commodity classification. These pallets have demonstrated a fire hazard equal to or less than that of the wood or metal pallets during testing or "listing" and are "Classified Fire Retardant Plastic Pallets." These pallets are required to be identified with a label or mark.

One final example will look at the possible classifications found in a truck repair and parts facility. The facility performs repair service on large trucks and has several bays or stations where the trucks are serviced or repaired. In addition, it may have an office area, a reception area, parts storage on shelves, rack storage of parts, rubber tire storage, parts washing, lubrication areas, lubricants, aerosol storage, and drums or barrels of flammable or combustible liquids. To properly classify this facility and the different areas within the facility, the engineer will need to evaluate each of those areas and gather information on the commodities and fuel loads. The trucks are one hazard, the repair bays another, the parts washing may be an open-use flammable liquids hazard, storage will be several different and distinct hazards (rubber tires, plastics, aerosols, etc.), and the shelf storage for parts and office areas are other hazards. It is possible there will be mixed commodity storage and storage of waste materials that pose not only a fire hazard but also a health hazard and a potential hazard to fire fighters.

Summary

Hazard classification is an extremely important aspect of the design of a sprinkler system. It becomes more and more complex as the occupancy and use of the building becomes varied or increases in fuel loading. The use of available test data, Material Data Safety Sheets (MSDS), Hazardous Material Inventory

Summaries, Owner's Certificate, and publications by other testing and listing organizations may be necessary to determine the classification of many commodities and building operations.

These lists of preclassified hazards and occupancies are a convenient way to determine design criteria for typical fuel loads or fire hazards. However, matching the proper fuel load and hazard to the specific building or area of a building takes careful consideration and is essential in designing the sprinkler system to adequately protect that building. Hasty decisions or choices based on the fuel load that is anticipated because of the occupancy of the building but not verified can lead to the failure of the sprinkler system during a fire either to control or suppress the fire as it was intended to do.

BIBLIOGRAPHY

References Cited

1. *Automatic Sprinkler Systems Handbook*, 9th edition, NFPA, Quincy, MA, 2002, p. 80.
2. Madrzykowski, D., "Office Work Station Heat Release Rate Study: Full Scale vs. Bench Scale," Building and Fire Research Laboratory, National Institute of Standards and Technology, Gaithersburg, MD 20899 from Interflam "96," 7th International Interflam Conference, March 26–28, 1996, Cambridge, England. Sponsored by Interscience Communications Ltd., National Institute of Standards and Technology; Building Research Establishment; Society of Fire Protection Engineers; and Swedish National Testing and Research Institute. Interscience Communications Ltd., London, England, Franks, C. A., and Grayson, S., eds., pp. 47–55.
3. Factory Mutual Insurance Company, FM Global Resource Collection, *FM Global Property Loss Prevention Sheets,* 8-1 May 2001, Norwood, MA, May, 2003.

NFPA Codes, Standards and Recommended Practices

NFPA Publications National Fire Protection Association, 1 Batterymarch Park, Quincy, MA 02169-7471

The following is a list of NFPA codes, standards, and recommended practices cited in this chapter. See the latest version of the *NFPA Catalog* for availability of current editions of these documents.

NFPA 13, *Standard for the Installation of Sprinkler Systems*
NFPA 72®, *National Fire Alarm Code®*
NFPA 230, *Standard for the Fire Protection of Storage*
NFPA 409, *Standard on Aircraft Hangars*

5

Selection of Automatic Sprinkler System Materials and Components

Carl Anderson, PE

An automatic fire sprinkler system is designed to control or suppress a fire, thereby providing a reasonable degree of protection against the loss of life and property. The ability of a sprinkler system to achieve this purpose is multifaceted. This chapter will address one of the many factors influencing the performance and reliability of a sprinkler system—the selection of materials and components—and provides an overview of the various components of a fire sprinkler system, the material options available, and some basic selection considerations. Selecting the appropriate type of system, system components, and materials is critical if the sprinkler system is to meet its performance objectives properly and predictably over the desired life cycle. Some factors to consider when selecting the materials and components are reliability, predictability, compatibility with the environment in which they are installed, and, of course, cost.

NFPA 13, *Standard for the Installation of Sprinkler Systems*, and NFPA 24, *Standard for the Installation of Private Fire Service Mains and Their Appurtenances*, require that most system components be listed, and this helps to keep system performance more predictable. Listed components are tested to a consistent standard, and quality assurance is provided by the listing agency's quality control oversight. Listing and approval agencies are discussed further in the next section.

Other topics addressed in this chapter are several sprinkler system components, including pipe, fittings, hangers, bracing, meters, strainers, and valves. Information on the corrosion resistance ratio (CRR) and microbiologically influenced corrosion (MIC) is included in the section on aboveground piping and appurtenances.

Sprinklers and cross-connection control (backflow prevention) are not discussed in detail in this chapter. Refer to Chapter 1, "Fundamentals of Sprinkler Performance," and Chapter 11, "Cross Connections and Backflow Prevention," for more information on these subjects.

The bottom line is that it is the task of the engineer to design/specify a system that performs its intended function in a consistent and reliable manner without undue expense either up front or over the life of the system.

Listings and Approvals

NFPA 13 and NFPA 24 require that all materials and devices essential to successful system operation be listed, and both standards also include some important exceptions to the listing requirement. These exceptions are discussed later in this chapter. The term *listed* as it applies to sprinkler systems is defined in NFPA 13 as "equipment, materials, or services included in a list published by an organization that is acceptable to the authority having jurisdiction and concerned with evaluation of products or services, that maintains periodic inspection of production of listed equipment or materials or periodic evaluation of services, and whose listing states that either the equipment, material, or service meets appropriate designated standards or has been tested and found suitable for a specified purpose" (NFPA 13, 2002 edition, 3.2.3).

In the United States, materials and components are often listed by Underwriters Laboratories (UL). Underwriters Laboratories Canada performs similar services in that country. UL publishes a number of directories covering a variety of products and assemblies. Some of the UL directories are the *Building Materials Directory, Electrical Appliances and Utilization Equipment Directory, Fire Resistance Directory*, and *Fire Protection Equipment Directory*. The *Fire Protection Equipment Directory* is of primary interest with regard to sprinkler system materials and components.

The term *approved* as it relates to sprinkler system means that a material, component, procedure, or system as a whole is acceptable to the authority having jurisdiction (AHJ). Examples of approvals often required are larger insurers such as FM Global, the local fire or building authority, state fire marshal, or local water purveyor. As noted in Annex A of NFPA 13, the term *AHJ* is used in NFPA documents in a broad manner, since jurisdictions and approval agencies vary, as do their responsibilities. Where public safety is primary, the authority having jurisdiction may be a federal, state, local, or other regional department or individual such as a fire chief; fire marshal; chief of a fire prevention bureau, labor department, or health department; building official; electrical inspector; or others having statutory authority. For insurance purposes, an insurance inspection department, rating bureau, or other insurance company representative may be the authority having jurisdiction. In many circumstances, the property owner or the owner's designated agent assumes the role of the authority having jurisdiction; at government installations, the commanding officer or departmental official may be the authority having jurisdiction.

Simply because a particular item or installation bears the designation of *listed* as defined in NFPA 13 does not guarantee that that item is acceptable to the AHJ. The engineer should investigate what authorities will impact a given project and determine to the extent possible what materials and components of the sprinkler system will need specific approval by any of those authorities. For example, a local fire department may prohibit the use of plastic sprinkler pipe in buildings over a certain size although the specified pipe may hold a listing from Underwriters Laboratories for the intended use; a water purveyor may allow only ductile iron pipe for underground mains even though PVC pipe might perform the intended function; or an insurer may require that sprinklers installed meet its approval and may have requirements beyond that of a UL listing.

Sprinkler System Materials: General

Both NFPA 13 and NFPA 24 require that piping materials meet or exceed certain recognized standards or that they be specifically listed for fire protection service. Pipe to be installed underground generally needs to meet or exceed American Water Works Association (AWWA) standards or American Society for Testing and Materials (ASTM) standards for copper pipe and tube. Pipe to be installed above ground generally needs to meet or exceed certain ASTM or American National Standards Institute (ANSI) standards or be specifically listed for automatic sprinkler service.

Underground Piping and Appurtenances for Sprinkler Systems

Manufacturing Standards

Table 5.1 lists the manufacturing standards for underground piping. Tables 5.2(a) and 5.2(b) give similar information for underground fittings.

TABLE 5.1

Manufacturing Standards for Underground Pipe

Materials and Dimensions	Standard
Ductile Iron	
Cement Mortar Lining for Ductile Iron Pipe and Fittings for Water	AWWA C104
Polyethylene Encasement for Ductile Iron Pipe Systems	AWWA C105
Ductile Iron and Gray Iron Fittings, 3-in. Through 48-in., for Water and Other Liquids	AWWA C110
Rubber-Gasket Joints for Ductile Iron Pressure Pipe and Fittings	AWWA C111
Flanged Ductile Iron Pipe with Ductile Iron or Gray Iron Threaded Flanges	AWWA C115
Thickness Design of Ductile Iron Pipe	AWWA C150
Ductile Iron Pipe, Centrifugally Cast for Water	AWWA C151
Standard for the Installation of Ductile Iron Water Mains and Their Appurtenances	AWWA C600
Steel	
Steel Water Pipe 6 in. and Larger	AWWA C200
Coal-Tar Protective Coatings and Linings for Steel Water Pipelines Enamel and Tape—Hot Applied	AWWA C203
Cement-Mortar Protective Lining and Coating for Steel Water Pipe 4 in. and Larger—Shop Applied	AWWA C205
Field Welding of Steel Water Pipe	AWWA C206
Steel Pipe Flanges for Waterworks Service—Sizes 4 in. Through 144 in.	AWWA C207
Dimensions for Fabricated Steel Water Pipe Fittings	AWWA C208
A Guide for Steel Pipe Design and Installation	AWWA M11
Concrete	
Reinforced Concrete Pressure Pipe, Steel-Cylinder Type, for Water and Other Liquids	AWWA C300
Prestressed Concrete Pressure Pipe, Steel-Cylinder Type, for Water and Other Liquids	AWWA C301
Reinforced Concrete Pressure Pipe, Non-Cylinder Type, for Water and Other Liquids	AWWA C302

TABLE 5.1

(Continued)

Materials and Dimensions	Standard
Concrete	
Reinforced Concrete Pressure Pipe, Steel-Cylinder Type, Pretensioned, for Water and Other Liquids	AWWA C303
Standard for Asbestos-Cement Distribution Pipe, 4 in. Through 16 in., for Water and Other Liquids	AWWA C400
Standard Practice for the Selection of Asbestos-Cement Water Pipe	AWWA C401
Cement-Mortar Lining of Water Pipe Lines 4 in. and Larger—in Place	AWWA C602
Standard for the Installation of Asbestos-Cement Water Pipe	AWWA C603
Plastic	
Polyvinyl Chloride (PVC) Pressure Pipe, 4 in. Through 12 in., for Water and Other Liquids	AWWA C900
Copper	
Specification for Seamless Copper Tube	ASTM B 75
Specification for Seamless Copper Water Tube	ASTM B 88
Requirements for Wrought Seamless Copper and Copper-Alloy Tube	ASTM B 251

Source: NFPA 24, 2002, Table 10.1.1.

TABLE 5.2(a)

Fittings Materials and Dimensions

Materials and Dimensions	Standard
Cast Iron	
Cast Iron Threaded Fittings, Class 125 and 250	ASME B16.4
Cast Iron Pipe Flanges and Flanged Fittings	ASME B16.1
Malleable Iron	
Malleable Iron Threaded Fittings, Class 150 and 300	ASME B16.3
Steel	
Factory-Made Wrought Steel Buttweld Fittings	ASME B16.9
Buttwelding Ends for Pipe, Valves, Flanges, and Fittings	ASME B16.25
Specification for Piping Fittings of Wrought Carbon Steel and Alloy Steel for Moderate and Elevated Temperatures	ASTM A 234
Steel Pipe Flanges and Flanged Fittings	ASME B16.5
Forged Steel Fittings, Socket Welded and Threaded	ASME B16.11
Copper	
Wrought Copper and Bronze Solder Joint Pressure Fittings	ASME B16.22
Cast Bronze Solder Joint Pressure Fittings	ASME B16.18

Source: NFPA 24, 2002, Table 10.2.1(a).

TABLE 5.2(b)

Specially Listed Fittings Materials and Dimensions

Materials and Dimensions	Standard
Chlorinated Polyvinyl Chloride (CPVC) Specification for Schedule 80 CPVC Threaded Fittings	ASTM F 437
Specification for Schedule 40 CPVC Socket-Type Fittings	ASTM F 438
Specification for Schedule 80 CPVC Socket-Type Fittings	ASTM F 439

Source: NFPA 24, 2002, Table 10.2.1(b).

Selection Considerations

Important considerations when selecting underground pipe are the maximum system working pressure, soil conditions, susceptibility to corrosion, and potential for external loading of the pipe, such as vehicle loads or installation under buildings.

Working Pressure. NFPA 24 requires that pipe be designed for a minimum working pressure of 150 psi. Where higher working pressure is expected, pipe selection should account for this increase.

Soil Conditions/Corrosion. External (soil) corrosion is accelerated in conditions where metallic salts, acids, or other substances in the soil combine with moisture and result in an electrochemical reaction. This reaction causes iron ions to separate from the iron or steel pipe. The mass of the metal at the pipe surface is diminished, and the surface becomes pitted or corroded. External corrosion occurs where iron or steel pipe is installed under coal piles, in cinder fill, or wherever acids, alkalis, or other potential corrosives penetrate the soil. Installation of iron or steel pipe in these circumstances should be avoided or suitable external protection must be provided [1].

Stray electrical currents from an exterior source many reach the pipe and then travel along the pipe until a point of lesser resistance to ground is encountered. At this point, the current leaves the pipe and ionization occurs. Electrical corrosion is similar to soil corrosion. To avoid corrosion due to stray electrical current, the source of the current should be eliminated or bonding and grounding should be provided.

Cathodic methods may be used as a means of protecting against corrosion. Cathodic protection involves the application of direct current from a galvanic cathode to the pipe. This type of application is uncommon in fire service underground applications. Therefore, an alternative pipe material might be appropriate.

External Loading. External loading may impact the type of pipe selected for underground installation. Although steel pipe is generally not installed underground due to corrosion concerns, specially listed steel pipe is recognized in NFPA 13 and 24 for underground use and may be useful if the main is subject to high external loads [2]. Installations under railroad tracks or heavy vehicle loading areas are examples. Internally galvanized, externally wrapped and coated steel pipe may be used between the check valve and outside hose coupling for a fire department connection.

Piping Materials

Piping materials typical of underground use are ductile iron, PVC (polyvinyl chloride), HDPE (high-density polyethylene), CPVC (chlorinated polyvinyl chloride), PVCO (molecularly oriented polyvinyl chloride), reinforced concrete, and steel. Asbestos is also permitted, although its use is not common.

Ductile Iron. Ductile iron pipe is probably the most commonly used pipe in underground installations. Ductile iron is available in diameters from 3 in. (76 mm) to 64 in. (163 mm) but is most commonly seen in 4- to 24-in. (100–610 mm) diameter in fire service applications. Typically, ductile iron pipe is supplied with a cement-mortar lining yielding a Hazen–Williams coefficient of 140 at the time of installation, with this factor diminishing to a lower value after years of service due to corrosion and accumulated deposits. The cement-mortar lining provides a barrier between the inside pipe wall and the water and creates a high pH condition at the wall, thereby aiding in the prevention of tuberculation. Coal-tar lining is also available, but it is less common. Ductile iron pipe is highly resistant to corrosion and rarely needs coating or cathodic protection. In cases of extremely corrosive conditions, field-applied polyethylene encasement is possible to increase corrosion resistance.

Fittings for ductile iron pipe are provided in cast and ductile iron. Listed retainer-type fittings are available and may be used where thrust restraint is needed. NFPA 24 requires that fittings meet the requirements of AWWA C110/A21.10-03, *ANSI Standard for Ductile-Iron and Gray-Iron Fittings, 3 in.–48 in. (76 mm–1,219 mm), for Water and Other Liquids* or AWWA C153/A21.53-00, *ANSI Standard for Ductile-Iron Compact Fittings for Water Service.*

A brochure from the Ductile Iron Pipe Research Association entitled *Ductile Iron Pipe and Design of Ductile Iron Pipe* offers an overview of the history and use of ductile iron pipe [3].

PVC (Polyvinyl Chloride). Another common material for fire protection pipe installed underground is PVC (polyvinyl chloride). Class 150 or Class 200 PVC meeting or exceeding AWWA C900, *Polyvinyl Chloride (PVC) Pressure Pipe, 4 in. Through 12 in. (100 mm through 300 mm) for Water and Other Liquids*, is acceptable for fire service use and is commonly available in 4-in. (100-mm), 6-in. (150-mm), 8-in. (200-mm), 10-in. (250-mm), and 12-in. (300-mm) diameters. PVC for fire service use is manufactured with cast iron outside diameter and is compatible with standard cast iron mechanical-joint and slip-joint fittings; PVC fittings are also available.

Mechanical restraint collars are available for C900 pipe. These collars attach behind the bell of one pipe and grip the spigot of the next and are used to provide a level of thrust restraint. PVC offers good impact strength, and its smooth interior surface provides a Hazen–Williams coefficient of 150. Although PVC is highly resistive to corrosion, it should not be installed where there is a potential for aromatic hydrocarbon (toluene, benzene, etc.) spills, as these substances can dissolve PVC. The *Uni-Bell Handbook of PVC Pipe* offers a great deal more detail on the subject of PVC pipe [4].

HDPE (High-Density Polyethylene). HDPE piping, although not as common in fire service installations as PVC, is approved by FM Global in sizes 4 in. (100 mm) through 20 in. (500 mm). HDPE does not tuberculate and offers excellent resistance to damage by chemical exposure and corrosion. HDPE has excellent flow characteristics, with a Hazen–Williams coefficient of 150. Since HDPE pipe is joined by butt-fusion, it is fully restrained without the use of thrust blocks or tie rods.

CPVC (Chlorinated Polyvinyl Chloride). CPVC pipe is listed by UL for underground installation and is available in ¾-in. (20-mm) through 2-in. (50-mm) sizes. CPVC is not commonly used for underground fire service due to its limited size, but may be appropriate for smaller systems where a 2-in. (50-mm) service is sufficient.

PVCO (Molecularly Oriented Polyvinyl Chloride). PVCO pressure pipe in 4-in. (100-mm) through 12-in. (300-mm) diameter meeting AWWA C909 *Molecularly Oriented Polyvinyl Chloride (PVCO) Pressure Pipe, 4 in.–12 in. (100 mm–300 mm), for Water Distribution* is available. PVCO is manufactured by a process that reorients the molecules of conventionally extruded PVC pipe. PVCO can be manufactured with outside diameters to match ductile iron pipe, so no transition gaskets or special fittings are required when connecting to ductile iron. PVCO has a larger inside diameter than ductile iron or conventional PVC, resulting in less friction loss. PVCO is lighter weight than ductile iron and conventional PVC and has up to 4 times higher impact strength than conventional PVC.

Reinforced Concrete. Reinforced concrete underground piping is available in diameters larger than 24 in. (610 mm) should large-diameter pipe be needed. Diameters of this size are not typically required for fire sprinkler supply, but might be needed in a main system for larger industrial facilities.

Steel. Steel pipe, when properly lined and coated, may be used for fire protection service. The high tensile strength of steel pipe may make it an appropriate choice for installations subject to shock or heavy vehicle impact loads (e.g., under railroad tracks or highways). Steel may also be considered where steep slopes or unstable soils are a concern.

Asbestos. Additionally, asbestos cement pipe is acceptable for fire service underground installation. Asbestos pipe is rarely used due to health concerns caused by dusts that are created when cutting asbestos pipe. OSHA work practices must be followed when working with asbestos cement pipe as with any material containing asbestos.

Meters, Valves, and Strainers

Some other sprinkler system components associated with the underground supply are check valves and meters, backflow preventers, control valves, and strainers.

Detector Check Valves and Meters. Where the local water purveyor requires metering, detector check valves are the most typical means. Detector check valves consist of a check valve with a weighted clapper in the main passage and a small meter in a bypass. The small meter accurately measures small flows such as from leaks or improper connections to the fire system. The clapper opens under larger flow conditions such as a large number of open sprinklers. Meters may be installed by the water purveyor or required by the purveyor and installed privately. This requirement varies from AHJ to AHJ and affects how the specification is written. Listed devices are required. UL 312, *Check Valves for Fire-Protection Service*, covers detector check valves for fire service. Figure 5.1 shows an example of a typical detector check valve.

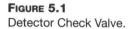

Figure 5.1
Detector Check Valve.

Detector check valves contribute to friction loss in underground pipe and must be considered in hydraulic calculations. Friction loss is minimal for most detector checks, 5 psi or less, at rated flows. Refer to the manufacturer's literature for specific friction loss and rated/actual flow data.

It is less common, but some water purveyors may require full registration meters. These meters more accurately record water usage but have greater friction loss characteristics and are more costly. Specific metering requirements should be confirmed with the appropriate AHJ.

Backflow Preventers. Backflow prevention is required by most if not all water purveyors and is discussed in detail in Chapter 11.

Control Valves. Where control valves are installed in fire service mains, NFPA 13 and NFPA 24 require listed indicating-type valves. Nonindicating valves in roadway boxes are permitted when accepted by AHJ and are typically found at the connection to the public main, on each side of a meter or check valve installed underground. NFPA 13 and 24 require control valves on each side of all check valves except for the check valve in the fire department connection (FDC) line. Control valves are not permitted on the FDC line; however, it is acceptable in NFPA 13 and NFPA 24 to have a control valve between the FDC and the remainder of the system if the FDC is connected directly to the yard main. Check valves are required on the FDC line and where more than one source of supply is provided, as shown in Figure 5.2.

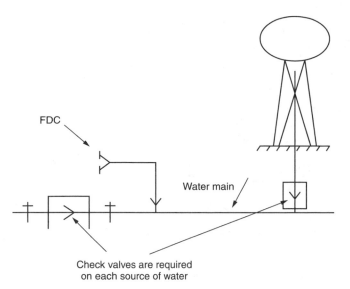

FIGURE 5.2
Check Valve on FDC Line.

FDC

Water main

Check valves are required
on each source of water

Strainers. A raw water source, such as a lake or reservoir, may contain materials that if admitted to a sprinkler system could block water flow through pipes or sprinklers. Therefore, NFPA 13 and 24 require that water supply connections from penstocks, flumes, rivers, lakes, or reservoirs be arranged to avoid mud and sediment. Approved strainers or approved, double, removable screens are required. UL-listed strainers are available.

Above Ground Piping and Appurtenances for Sprinkler Systems

Manufacturing Standards

NFPA 13 requires that all components of a fire sprinkler system essential to its successful operation be listed. Exceptions to the listing requirement are pipe as described in Table 5.3, fittings as described in Table 5.4, mild steel hanger rods, hangers certified by a professional engineer to meet the criteria found in NFPA 13, and components not essential to system performance, such as drain piping, drain valves, and signage. Table 5.3 lists the manufacturing standards for sprinkler system piping installed above ground and not specifically listed for sprinkler system use. Table 5.4 addresses similar standards for fittings not specifically listed for sprinkler system use.

All pipe for fire sprinkler applications, including specially listed pipe, is required to be properly marked continuously along its length by the manufacturer. Pipe marking is to include the manufacturer's name, model designation, or schedule.

Listed Pipe

Pipe not meeting one of the standards noted in NFPA 13 is required to be listed. Alternative types of pipe are chlorinated polyvinyl chloride (CPVC) and steel pipe other than Schedule 40 and Schedule 10. Various manufacturers produce listed steel pipe for fire sprinkler systems. Examples of listed steel pipe

TABLE 5.3

Pipe or Tube Materials and Dimensions

Materials and Dimensions	Standard
Ferrous Piping (Welded and Seamless)	
Specification for black and hot-dipped zinc-coated (galvanized) welded and seamless steel pipe for fire protection use	ASTM A 795
Specification for welded and seamless steel pipe	ANSI/ASTM A 53
Wrought steel pipe	ANSI/ASME B36.10M
Specification for electric-resistance-welded steel pipe	ASTM A 135
Copper Tube (Drawn, Seamless)	
Specification for seamless copper tube	ASTM B 75
Specification for seamless copper water tube	ASTM B 88
Specification for general requirements for wrought seamless copper and copper-alloy tube	ASTM B 251
Fluxes for soldering applications of copper and copper-alloy tube	ASTM B 813
Brazing filler metal (classification BCuP-3 or BCuP-4)	AWS A5.8
Solder metal, 95-5 (tin-antimony-Grade 95TA)	ASTM B 32
Alloy materials	ASTM B 446

Source: NFPA 13, 2002, Table 6.3.1.1.

include Super XL®, Dyna-Light©, and Eddylite©. Refer to the *UL Fire Protection Equipment Directory* under the heading "Steel Sprinkler Pipe" for information on all UL-listed steel sprinkler pipe and its manufacturers.

Steel. The thinner-wall steel pipes are lighter and typically less expensive than their Schedule 40 steel counterparts, so there may be cost savings in both material and labor. Also, the thinner-wall steel pipes have larger inside diameters, giving

TABLE 5.4

Fittings Materials and Dimensions

Materials and Dimensions	Standard
Cast Iron	
Cast iron threaded fittings, Class 125 and 250	ASME B16.4
Cast iron pipe flanges and flanged fittings	ASME B16.1
Malleable Iron	
Malleable iron threaded fittings, Class 150 and 300 steel	ASME B16.3
Factory-made wrought steel buttweld fittings	ASME B16.9
Buttwelding ends for pipe, valves, flanges, and fittings	ASME B16.25
Specification for piping fittings of wrought carbon steel and alloy steel for moderate and elevated temperatures	ASTM A 234
Steel pipe flanges and flanged fittings	ASME B16.5
Forged steel fittings, socket welded and threaded copper	ASME B16.11
Wrought copper and copper alloy solder joint pressure fittings	ASME B16.22
Cast copper alloy solder joint pressure fittings	ASME B16.18

Source: NFPA 13, 2002, Table 6.4.1.

an advantage in hydraulic calculations. Hydraulic calculations are covered in more detail in Chapter 10. The downside to thinner-wall steel pipe is a potential for reduced life cycle due to less resistance to corrosion. This may be of greater concern in corrosive environments or dry-pipe systems than in typical wet-pipe installations. Not all listed steel pipe is less resistant to corrosion than Schedule 40 pipe. A more detailed discussion of the corrosion resistance ratio occurs in a later section.

CPVC (Chlorinated Polyvinyl Chloride). CPVC is another type of specially listed pipe for fire sprinkler systems, and its most typical uses seem to be in residential applications, but it is also listed for use in other light-hazard occupancies. Advantages to the use of CPVC may include resistance to corrosion, lower material cost, and lower labor costs due to easier installation. It is important to refer to the most current listing conditions prior to specifying the use of CPVC. The UL listing includes very specific guidelines and limitations for proper application of CPVC pipe and fittings. CPVC pipe is only permitted for wet-pipe systems and must be protected in accordance with its listing. Some AHJs will not permit CPVC in certain applications even when used in accordance with its listing. For example, some cities may allow CPVC only in residential occupancies, or only in NFPA 13R or 13D systems, or only in buildings up to a specified maximum square footage or height. All appropriate AHJs need to be consulted prior to specifying CPVC pipe.

Copper Tube. Copper tube may also be used in fire sprinkler systems. Although not commonly used, copper may be the answer to an aesthetic need, where exposed pipe is necessary, but the appearance of steel or plastic is undesirable. Copper pipe may also be more suitable than steel pipe for certain types of corrosive environments.

Fittings and Couplings

Grooved-end, rubber-gasketed couplings are commonly used to connect grooved-end pipe and grooved-end fittings. Both rigid and flexible couplings are available. Flexible couplings are used to provide flexibility and to allow sections of pipe to move differentially with the section of the building to which they are attached. NFPA 13 lists locations where flexible couplings are required, such as to provide flexibility in risers and at seismic joints in areas subject to earthquakes. Figure 5.3 shows a typical coupling.

Fittings and couplings are generally cast iron, malleable iron, steel, or other material noted in NFPA 13 or specifically listed materials such as CVPC. Fittings may be threaded for threaded pipe or grooved, or plain-end couplings may be used to join plain- or grooved-end pipe.

Plain-end couplings are less common but are allowed by NFPA 13. These couplings join plain-end pipe and fittings. Proper strength at these fittings depends on meeting manufacturer-specified torque requirements; however, many of these couplings have no integrated visual means of verifying that the appropriate torque has been applied. Some AHJs may not allow the use of this type of coupling in areas subject to earthquakes.

FIGURE 5.3
Coupling.

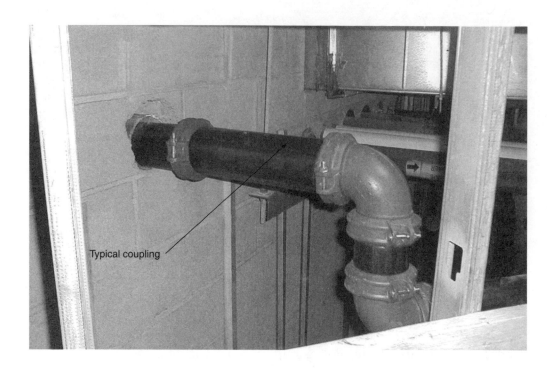

Figures 5.4 and 5.5 are two seismic zone maps from NFPA 13. They are examples of maps from model building codes used in making a determination as to whether an area is subject to earthquakes. Once it is determined that seismic requirements apply, NFPA 13 provides the applicable seismic design and installation requirements. For some projects it will be important to determine

FIGURE 5.4
Seismic Zone Map of the United States.

(Source: NFPA 13, 2002, Figure A.9.3.1(a))

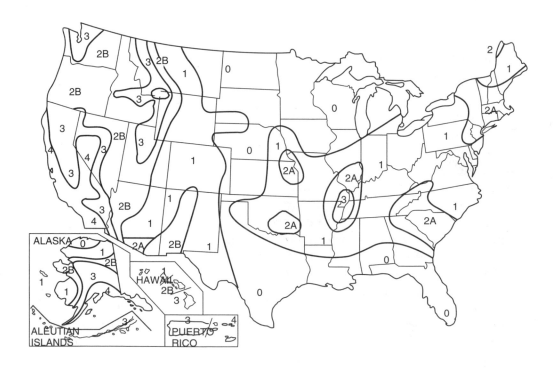

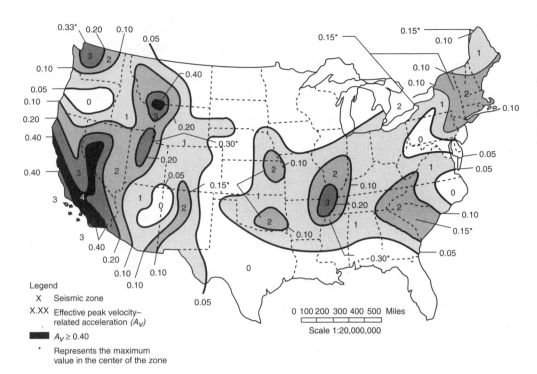

FIGURE 5.5

Map of Seismic Zones and Effective Peak Velocity-Related Acceleration (A_V) for the Contiguous 48 States; Linear Interpolation Between Contours Is Acceptable.

(Source: NFPA 13, 2002, Figure A.9.3.1(b))

what additional earthquake provisions are required by insurance providers. For example, FM Global has additional installation requirements as well as its own seismic zone map.

Pipe Hangers

Hanger components, which attach directly to the pipe or to the structure, must either be listed or be certified by a professional engineer or structural engineer. Certification requirements include the following:

- All ferrous components
- Designed to support 5 times the weight of the water-filled pipe plus 250 lb (113 kg)
- Adequate points of support

In some instances, a professional engineer's services are required in hanger design, such as with hanger configurations that are not commercially purchased and whose load capacities are not listed in a manufacturer's catalog (welded angle iron supports, pipe stands, etc.). A fire protection engineer should enlist the assistance of a structural engineer and not attempt structural design of hanger components if this work is not within his or her area of expertise.

In some facilities, NFPA 13 requirements must be exceeded, such as in nuclear power plants. Some owners will have specific requirements unique to their industry, and some insurance providers will have requirements beyond the minimums detailed in NFPA 13. The engineer will need to consider these issues and research what areas of system design will need to exceed the minimum standard. An example would be where the performance objective of a sprinkler hanger is not only to support sprinkler pipe but also to prevent damage to adjacent cables and mechanical components.

Selection of the type of hanger depends on the type of structure. In corrosive environments, protective coating should be applied to hangers so that they can be expected to have a life cycle similar to the pipe they are supporting. An example might be galvanized, painted, or coated hangers to support piping under a pier.

Bracing Components

Where water-based fire protection systems are required to be protected against damage from earthquakes, the requirements of NFPA 13 apply. Allowable loads and types of fasteners can be determined from NFPA 13 or can be in accordance with manufacturer's requirements for listed bracing components. Arrangements other than as called for in NFPA 13 are acceptable when certified by a professional engineer to support the loads noted in NFPA 13. The system piping is braced to resist both lateral and longitudinal horizontal seismic loads and to prevent vertical motion resulting from seismic loads.

Typical sway bracing consists of Schedule 40 steel pipe in conjunction with listed components used to attach the bracing pipe to the structure and to the sprinkler pipe. These components typically include a simple visual means of verifying that the proper attachment has been made, such as clamp ears that touch when properly tightened or nut ends that break off when proper torque is applied. Figure 5.6 shows a typical sway brace. Figure 5.7 shows a typical sway brace attachment.

FIGURE 5.6
Sway Brace.

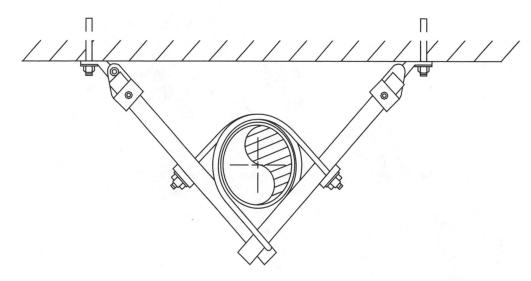

FIGURE 5.7
Sway Brace Attachment.

Generally, sway braces must be designed to resist loads in tension and compression; however, tension-only bracing may be utilized. Tension-only bracing is sometimes referred to as cable bracing, and such systems are required to be listed. Figure 5.8 shows a cable bracing detail.

Seismic separation assemblies with flexible fittings must be installed where sprinkler piping, regardless of size, crosses building seismic separation joints above ground level. A typical seismic separation assembly is shown in Figure 5.9.

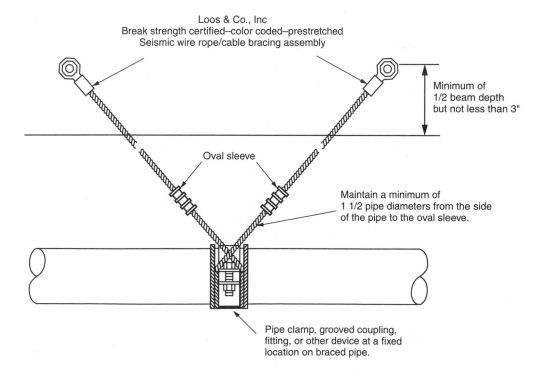

FIGURE 5.8
Seismic Wire Rope/Cable Bracing (Longitudinal Bracing).

(Courtesy Loos and Co., Inc.)

FIGURE 5.9
Seismic Separation
Assembly.

Some insurers have sway bracing requirements in excess of those found in NFPA 13. The engineer should verify any such requirements for inclusion in the project specifications.

Control Valves, Backflow Preventers, and Fire Department Connections (FDCs)

A few other typical system components for above ground piping are control valves, backflow preventers, and fire department connections (FDCs).

Control Valves. Control valves are required to be listed, indicating-type valves. Typical indicating valves are outside screw and yoke (OS&Y), indicating butterfly valves, and post indicating valves (PIV). Cost and space available for installation are considerations in selecting the appropriate valve. For instance, it is not typical in new installations to find OS&Y valves at floor control assemblies in stair towers. Indicating butterfly valves, often with integrated tamper switches, are more common in this application as they are more compact. Figures 5.10, 5.11, and 5.12 show an OS&Y, PIV, and butterfly valve.

Backflow Preventers. Backflow prevention is discussed in detail in Chapter 11.

Fire Department Connection. A fire department connection (FDC) allows the responding fire fighters to boost sprinkler system pressure and deliver more

FIGURE 5.10
OS&Y Valve.

water to operating sprinklers. The first consideration in selecting an FDC is thread type. Many fire departments simply require that the FDC have 2½-in. inlets with National Standard Hose Thread. Some jurisdictions may require a different hose thread or a quick-connect device such as a Storz coupling. Figures 5.13 and 5.14 show two types of FDC arrangements.

FIGURE 5.11
PIV.

FIGURE 5.12
Butterfly Valve.

Corrosion Resistance Ratio

The resistance of steel pipe to corrosion is measured relative to that of threaded Schedule 40 pipe. This value is referred to as the corrosion resistance ratio, or CRR. Pipe with a CRR of <1.0 is less resistant to corrosion

FIGURE 5.13
Chromed FDC for Use Where Aesthetics Are Important.

Figure 5.14
Older Fire Department
Connection.

than threaded Schedule 40 pipe. The UL *Fire Protection Equipment Directory* lists a CRR for all specially listed steel for sprinkler pipe included in the directory. In the 2000 edition of the *Fire Protection Equipment Directory*, CRR values ranging from 0.14 to 19.63 are noted. CRR is expressed mathematically as CRR = $(X/X_{40})^3$, where X is the thickness of the listed pipe measured under the first exposed thread or at the thinnest wall section for unthreaded pipe, and X_{40} is the thickness of Schedule 40 pipe under the first exposed thread.

Care should be exercised in selecting Schedule 10 pipe or listed pipe with CRR <1.00 for systems subject to water with corrosives, corrosive atmospheres, elevated temperatures, high humidity, frequent draining and filling, or in dry-pipe systems.

Corrosion may be of greater concern in dry-pipe systems. Each time the dry-pipe valve is tripped, the system piping is filled with water and then subsequently drained, leaving steel pipe exposed to moisture and oxygen. Even in normal operation, a dry-pipe system may encounter condensate in the piping, which may become a source of corrosion. In such instances, filtered air or nitrogen may be advised as a corrosion prevention measure. In a wet-pipe system, where the pipe is generally left filled with water, the oxygen supply is depleted and less corrosion may be expected.

Galvanized pipe is more resistant to corrosion than black steel pipe and may be required by some insurance companies or other AHJs in dry-pipe systems. Galvanized pipe should also be considered for systems exposed to external corrosive environments such as the underside of piers, open parking structures, or exterior canopies. Where ambient temperatures exceed 130°F (54°C), such as in lumber kilns, galvanized pipe should not be used. Corrosion of zinc is accelerated at temperatures exceeding 130°F (54°C). Schedule 40 black steel should be used if temperatures are going to exceed 130°F (54°C).

An extra measure of protection under severely corrosive environments such as water-cooling towers or chemical plants may be in order. Stainless steel pipe and field-applied coatings for pipe, fittings, and hangers, such as a rubberized paint to coat pipes and components as an extra measure of protection, may be considered. This type of additional protection should also be considered for applications where sprinkler systems are exposed to salt water, such as the underside of piers and wharves.

Microbiologically Influenced Corrosion (MIC)

Microbiologically influenced corrosion is a corrosion process caused by microbes and is characterized by failure of sprinkler systems from the tuberculation and pitting of pipe and the associated loss of pipe integrity [5, 6]. Microbes, in conjunction with tuberculation and corrosion, cause minute leaks in carbon steel as well as copper and galvanized piping, often characterized by pinhole leaks after only a few years of service. Besides pinholes, MIC can also form growths on the interior of a pipe, which reduce its diameter, increase its roughness, and reduce the c-factor of the pipe. MIC has been known to almost completely fill the cross section of a pipe. Figure 5.15 shows the results of MIC in some 2-year-old sprinkler piping at the United States Air Force Academy.

Bacteria related to MIC can be aerobic (thrive in oxygenated water) or anaerobic (die when exposed to oxygen). Two common anaerobic types are sulfate-reducing bacteria (SRB) or sulfide-producing bacteria (SPB) and acid-producing

FIGURE 5.15
MIC in Sprinkler Piping.

bacteria (APB) commonly found in deoxygenated sprinkler system water. Where water flow is fairly common with flushing, filling, draining, or pump testing, aerobic bacteria thrive. Layers of aerobic and anaerobic bacteria can result when systems are emptied and refilled. MIC can occur in wet or dry sprinkler systems. Treatment can be difficult and expensive, as MIC may reside in isolated areas of a sprinkler system and can be a recurring problem. If MIC is suspected, a metallurgical/biological test facility may be contacted to test the system for its presence.

Additional information regarding MIC is available on-line or from the National Fire Protection Association, National Fire Sprinkler Association (on-line seminar available), American Fire Sprinkler Association, or the National Association of Corrosion Engineers.

Summary

The materials and components comprising a fire sprinkler system play a critical role in the proper function of the system over the lifetime of the building in which it is installed. The engineer needs to consider more than simply specifying listed or approved materials. Function, space constraints, potential corrosion problems, aesthetics, and future maintenance should all be considered to varying degrees as appropriate to each project. Some projects may call for the use of more expensive materials where certain aesthetic issues are important to the customer, others may demand the most cost-effective solutions, and some may have very specific AHJ requirements. These items are only a few of the many needing the attention of the designing engineer. It is the engineer's job to determine what stakeholders have requirements or desires beyond those of mere code compliance and to incorporate those items into the system's design specifications while still maintaining a design in compliance with applicable codes and standards.

BIBLIOGRAPHY

References Cited

1. FM Global Loss Prevention Data Sheet 3-10, June 1992 (Revised September 2000), Copyright 2003, Factory Mutual Insurance Company, 1151 Boston-Providence Turnpike, PO Box 9102, Norwood, MA.
2. Schultz, G. R., "Water Distribution Systems," in Cote, A. E., ed., *Fire Protection Handbook*, 19th edition, NFPA, Quincy, MA, 2003, p. 10-37.
3. Ductile Iron Pipe Research Association, 245 Riverchase Parkway East, Suite O, Birmingham, AL 35244.
4. *Uni-bell Handbook of PVC Pipe: Design and Construction*. Uni-Bell PVC Pipe Association, 2655 Villa Creek Drive, Suite 155 Dallas, TX 75234, Copyright © 2003.
5. Bsharat, Tariq, "Detection, Treatment and Prevention of Microbiologically Influenced Corrosion in Water Based Fire Protection Systems," June 1998, National Fire Sprinkler Association.
6. Bauer, R. O., "Microbiologically Influenced Corrosion (MIC) in Fire Sprinkler Systems," www.sprinklersolutions.com.

NFPA Codes, Standards and Recommended Practices

NFPA Publications National Fire Protection Association, 1 Batterymarch Park, Quincy, MA 02169-7471

The following is a list of NFPA codes, standards, and recommended practices cited in this chapter. See the latest version of the *NFPA Catalog* for availability of current editions of these documents.

NFPA 13, *Standard for the Installation of Sprinkler Systems*
NFPA 24, *Standard for the Installation of Private Fire Service Mains and Their Appurtenances*

Additional Reading

Automatic Fire Sprinkler Systems Handbook, 9th edition, NFPA, Quincy, MA, 2002.
Clarke, B. H., and Aguilera, A. M., "Microbiologically Influenced Corrosion in Fire Sprinkler Systems," in Cote, A. E., ed., *Fire Protection Handbook*, 19th edition, NFPA, Quincy, MA, 2003, p. 10-403.

6

Evaluation of Water Supply

Michael A. Crowley, PE

Water supplies for major inhabited areas have been around for thousands of years. The early Roman aqueduct systems conveyed water from the mountains to the city of Rome. In the early 1800s, in the United States, cities began developing water works and methods of delivering water throughout the city. These urban water uses included bathing, drinking, and in many instances fire protection, and consisted of wooden mains with wooden plugs. When needed for fire protection, the plugs were removed, water was allowed to fill the hole around the main, and bucket brigades were used to transport water from the main to the fire scene. In some cases, the wooden main would be uncovered and a hole opened in the main that would be closed later with a wooden plug. This method was not very efficient, but did help the cities grow beyond the banks of a river or lake. Figure 6.1 shows a wooden water main arrangement with a wooden plug and hole around the water main.

Over the centuries, water works systems have developed into highly reliable methods of delivering water of sufficient quality and quantity throughout a city or region. The uses for water have expanded to include industrial water usage and direct connections to automatic sprinkler systems. This chapter will discuss the evaluation of the adequacy of water supply systems and their use in automatic sprinkler systems.

Providing Adequate Water Supply

Providing adequate water supply to an automatic sprinkler system can be done in several ways:

- A typical automatic sprinkler system water supply features a direct connection to a municipal water system with acceptable pressure and water flow capabilities.
- When the water supply is not capable of providing adequate pressure but has adequate water flow characteristics, a fire pump can be used to provide the required pressure for the automatic suppression system.

FIGURE 6.1
Typical Wooden Water Main
with Wooden Plug and Hole.

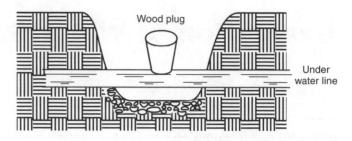

FIGURE 6.1
Typical Wooden Water Main
with Wooden Plug and Hole.

- Other means of providing pressure and flow include pressure tanks, gravity tanks, or the use of pumps with raw water sources such as lakes, streams, or other bodies of water.
- Special requirements are needed to use raw water sources in automatic suppression systems. The selection of the type of water supply is usually determined by local availability.
- Sites requiring fire protection that are remote from a municipal system will probably use a combination of pressure tanks, gravity tanks and pumps, portable water supplies, or raw water sources and pumps.

The cost of providing a reliable means of water supply to a fire protection system will guide the type of acceptable system to be used.

Methods of Determining Water Supply

The two primary methods of determining water supply are water flow tests and mathematical graphing.

Water Flow Tests of Hydrants

When determining the viability of a water supply system, the ability of the system to provide adequate water must be evaluated. Water supply data can be determined using water flow tests of hydrants in accordance with NFPA 291, *Recommended Practice for Fire Flow Testing and Marking of Hydrants*, water authority projections, modeled water supply predictions, and fire department records of past system performance. This chapter will discuss in detail the methods of determining water supply data and estimated system demand for initial evaluations of the available water supply.

Once the water supply is determined, automatic sprinkler system designers must also determine the automatic sprinkler system demand. The demand data will consist of estimating or calculating water flow for fire suppression, estimating or calculating pressure requirements for the proper operation of the automatic suppression system, and applying the required duration of the flow and pressure. In addition to this, if the available water supply is not adequate, methods for determining the sizing of an acceptable water supply, including storage tanks, must be applied.

Evaluating water supplies for fire protection system design is the first step in developing a reliable automatic sprinkler system for the hazards involved.

The proper selection of a water supply, the proper determination of its supply characteristics, and a thorough understanding of the demand of an automatic sprinkler system are key to the successful design of an automatic suppression system.

Piped, underground public water supply systems are the primary water supplies used for water-based fire protection purposes. To determine the water supply available, a water flow test must be conducted. Prior to conducting the test, permission should be obtained in writing from the water authority or operator of the underground water supply system, and a map of the underground system should be obtained to determine the general layout of the water supply system. Information regarding the size and type of underground piping is essential when a detailed evaluation of the water supply is performed. Once the building site or location for the automatic sprinkler system is located on a water supply map, the designer should select at least two hydrants located as close as possible to the property.

Site evaluation should also consider elevations of the fire hydrants, slope of the site, and water discharge areas, since water from a fire hydrant during a water flow test can cause flooding and water damage to the site. Adequate drainage and protection from water flow should be taken into account for each water flow test.

The equipment needed for the water flow testing as recommended by NFPA 291, 2002 edition, Section 4.4, includes the following items, which are ordinarily provided in most commercially-available water flow test kits:

1. A single 200-psi (14-bar) bourdon pressure gauge with 1-psi (0.0689-bar) graduations

2. Pitot tubes

3. Hydrant wrenches

4. 50- or 60-psi (3.5- or 4.0-bar) bourdon pressure gauges with 1-psi (0.0689-bar) graduations and scales with 1-in. (1.6-mm) graduations [one pitot tube, a 50- or 60-psi (3.5- or 4.0-bar) gauge, a hydrant wrench, a scale for each hydrant to be flowed]

5. A special hydrant cap tapped with a hole into which a short length of ¼-in. (6.35-mm) brass pipe is fitted. This pipe is provided with a T connection for the 200-psi (14-bar) gauge and a cock at the end for relieving air pressure

As recommended by NFPA 291, 2002 edition, subsection 4.4.2, all pressure gauges should be calibrated at least every 12 months, or more frequently, depending on use. Figure 6.2 displays several possible testing arrangements.

The fire hydrant closest to the property is designated as the residual hydrant, onto which a pressure gauge is mounted on a hydrant cap (item 5), and used to record the static pressure and residual pressures during the flow portion of the water supply test. The next hydrant (flow hydrant) downstream will be used to obtain the water flow information.

Before beginning a flow test, it is recommended that the residual hydrant be flushed. The water should be allowed to flow clear or relatively clear prior to beginning the test. Once the residual hydrant has been flushed, it can be shut down and the flow test setup begun once the hydrant cap and gauge are in place. This initial step helps to ensure that the attached pressure gauge will

FIGURE 6.2
Suggested Test Layout for Hydrants.

(Source: NFPA 291, 2002, Figure 4.3.4)

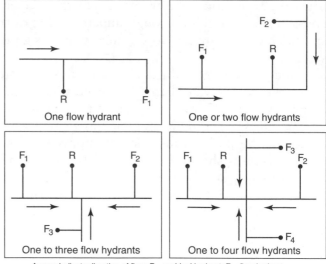

Arrows indicate direction of flow: R – residual hydrant; F – flow hydrant

not become obstructed or partially obstructed by debris or sediment, resulting in inaccurate pressure readings. Figure 6.3 shows the effective flow test point in a water supply flow test.

With these two hydrants selected, the next step will be to record the static pressure of the water supply system. This is done by fitting the hydrant cap and pressure gauge on the residual hydrant (hydrant closest to the suppression system) and slowly opening this hydrant and bleeding out excess air within the hydrant barrel. The bleeding of air can be done by a stopcock on the pressure gauge or by loosely fitting the cap of the pressure gauge to the hydrant butt

FIGURE 6.3
Effective Flow Test Point.

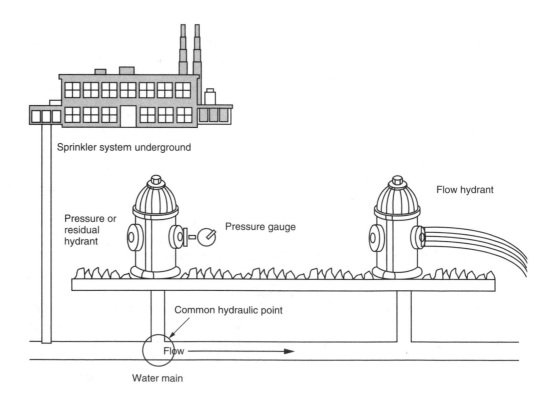

and securing it after the air has emptied from the hydrant barrel. The pressure gauge should record a steady pressure within the piping. This pressure will represent the static pressure of the water supply system with no fire protection water flowing at the time. The static pressure will also represent normal domestic water flow in the pipe if the water supply system is a dual-use system (domestic and fire protection).

The timing of the test should reflect, where possible, a worst-case scenario for the water supply system, that is, the largest domestic demand that would normally occur on the system. The largest domestic demand will be determined hourly and seasonally, depending upon the water supply system. The water authority should be consulted to determine the normal pressure and capacity fluctuations it encounters in the normal operation of the system.

After the static pressure is obtained, the residual hydrant is then used to determine the residual pressure of the water supply system during water flow conditions. When the flow hydrant is opened, and after the water stream has stabilized into a steady, clear flow, the pressure recorded at the residual hydrant will be recorded as the residual pressure at the flow quantity (in gpm), determined at the flowing hydrant.

The next step involves determining how much water is flowing at the flow hydrant. Before the flow hydrant is opened, the water supply system operator should determine that all underground valves are open and that the system will be operating under its normal water supply arrangement, which includes ensuring that all pumps are on automatic control, and verifying that water storage tanks filled to at least the minimum volume.

The hydrant outlet must be inspected to determine its shape and its hydrant coefficient. Figure 6.4 shows typical hydrant coefficients based on the type of hydrant opening at the barrel of the fire hydrant. To determine the appropriate hydrant coefficient, persons conducting the water flow test must, in a no-flow condition, reach into the hydrant opening and feel the back throat of the hydrant opening. If this throat area is smooth and contains a machined curve, the hydrant coefficient would be considered 0.9, as shown in Figure 6.4 (left). If the hydrant throat is square, causing the water to turn at a 90-degree angle as shown in Figure 6.4 (center), the hydrant coefficient is 0.8. If the hydrant opening is threaded into the barrel of the hydrant and the opening extends into the barrel area, the coefficient is considered to be 0.7, as shown in Figure 6.4 (right). The hydrant coefficient is a dimensionless number and is the same in English and SI units.

The diameter of the flow opening is the final piece of information needed to determine water flow from a given hydrant orifice. The diameter of this

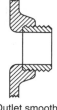  

Outlet smooth
and rounded
(coef. 0.90)

Outlet square
and sharp
(coef. 0.80)

Outlet square and
projecting into barrel
(coef. 0.70)

FIGURE 6.4

Three General Types of Hydrant Outlets and Their Coefficients of Discharge.

(Source: NFPA 291, 2002, Figure 4.7.1)

opening is recorded in inches. The diameter of the opening, the hydrant coefficient, and the pitot tube pressures are used to determine the water flowing from the hydrant at a given pressure (velocity pressure).

The flow hydrant is then opened and the water stream should be stabilized and allowed to flow clear or relatively clear prior to beginning the actual measurement of the water flow. The residual pressure should be checked to verify there is a water flow adequate to drop the residual pressure by 25 percent, as compared to the static pressure. A 25-percent reduction in residual pressure will produce test points on the water supply curve far enough apart to produce an accurate curve. If the static pressure has not dropped adequately, more water flow should be obtained, if possible. A second hydrant outlet on the flow hydrant can be opened or a second flowing hydrant may be needed to obtain an adequate drop from the static to the residual pressure reading. The flow from all flowing outlets must be measured.

The methods of measuring the water flow involve the use of a pitot tube or a special pressure-measuring device for the fire hydrant (see Figures 6.5 and 6.6). Figure 6.5 is a pitot tube and fire hydrant setup after flushing and before the test. The equipment allows for a steady pressure reading and accurate location of the pitot tube. Figure 6.6 is the flow test of a fire hydrant with a fixed pitot tube arrangement.

Both the pitot tube and the pressure-measuring device record the velocity pressure of the water stream leaving the hydrant, which is utilized to determine the flow in gallons per minute (gpm). Using a pitot tube to determine the water flow requires some practice. The pitot tube must be equipped with a pressure gauge to record the velocity pressure in the middle range of the gauge. Pitot tube gauge readings of less than 10 psi or greater than 30 psi

FIGURE 6.5
Pitot Tube and Hydrant After Flushing and Before the Test.

FIGURE 6.6
Flow Test of a Hydrant.

should be avoided. The pitot tube must be inserted in the water flow in an area called the *vena contracta* (see Figure 6.7). The area of the vena contracta is located approximately half the diameter away from the opening and at the center of the hydrant opening, which represents a point where turbulence conditions that exist at the edge of the hydrant outlet are not expected to be present. The pitot tube must be held at a 90-degree angle from the face of the hydrant opening (Figure 6.8). Figure 6.9 shows the proper orientation of a pitot tube relative to a hydrant opening, with D being the diameter of the hydrant opening. The pitot tube should be located near the center of the flow. The person conducting a water flow test must be familiar with the equipment to ensure an accurate reading. Some pitot tube gauges require the bleeding of air within the gauge prior to taking measurements; others do not.

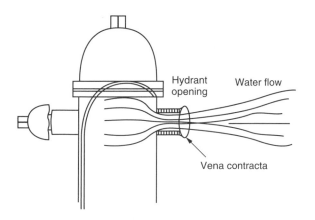

Hydrant opening

Water flow

Vena contracta

FIGURE 6.7
Vena Contracta.

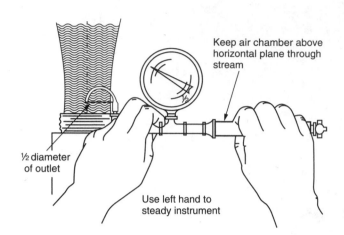

FIGURE 6.8

Proper Application of Pitot Tube to Hydrant Outlet.

(Courtesy of Insurance Services Office, Inc.)

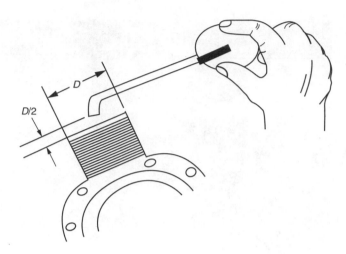

FIGURE 6.9

Proper Orientation of a Pitot Tube.

The pitot tube and gauge record the velocity pressure from the flow hydrant stream(s). This velocity pressure must be converted into a flow (gallons or liters per minute), which requires the hydrant coefficient and diameter of the flow opening recorded during the water flow test. Figure 6.4 shows typical hydrant coefficients, based on the type of hydrant opening from the barrel of the fire hydrant.

$$Q = 29.84 \; cd^2(p)^{1/2} \qquad \text{(6.1a) (English)}$$

The variables in Equation 6.1 are identified as follows:

Q = hydrant flow (in gpm)

c = the hydrant coefficient (dimensionless)

d = the diameter of the hydrant opening (in inches)

p = the velocity pressure measured from the pitot tube (in psi)

29.84 is a constant accounting for the physical constraints in unit conversions

$$Q = 0.666 \; cd^2(p)^{1/2} \qquad \text{(6.1b) (SI)}$$

For the International System of Units (SI), or metric units, the variables in Equation 6.1 are identified as follows:

Q = hydrant flow (in liters per minute)

c = the hydrant coefficient (dimensionless)

d = the diameter of the hydrant opening (in millimeters)

p = the velocity pressure measured from the pitot tube (in bars)

0.666 is a constant accounting for the physical constraints in unit conversions in SI units.

EXAMPLE 6.1 Calculating Hydrant Flow

As an example, the water supply determination for a pitot gauge reading of 14 psi (0.97 bar), a hydrant coefficient of 0.9, and a hydrant opening of 2.5 in. (63.5 mm) would be calculated as follows:

$$d = 2.5 \text{ in. (63.5 mm)}$$
$$p = 14 \text{ psi (0.097 bar)}$$
$$c = 0.9$$

The flow for this example is Q (gpm) $= 29.84 \times 0.9 \times (2.5 \text{ in.})^2 \times (14 \text{ psi})^{1/2}$.

$$Q = 29.84 \times 0.9 \times 6.25 \times 3.74$$
$$= 627.8 \text{ gpm}$$

In the International System of Units (SI), the problem is solved as follows:

$$Q(\text{L/m}) = 0.666 \times 0.9 \times (63.5 \text{ mm})^2 \times (0.97 \text{ bar})^{1/2}$$
$$= 2{,}382.5 \text{ liters per minute}$$

In this water supply test, the flowing hydrant has 627.8 gpm or 2,382.5 lpm flowing.

The next step in determining the water supply data for a site involves the plotting of a water supply curve for the data obtained. This water supply curve is plotted on special logarithmic graph paper (shown in Figure 6.10). Water flow is recorded along the x log axis, which is raised to the $N^{1.85}$. The logarithmic arrangement of the grid lines permits water supplies to be plotted as straight lines.

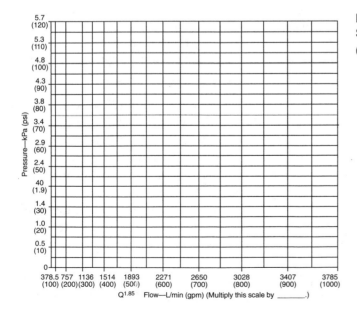

FIGURE 6.10
Sample Graph Sheet.
(Source: NFPA 15, 2001, Figure B.1c)

FIGURE 6.11

$N^{1.85}$ Water Supply Graph for Example 6.2.

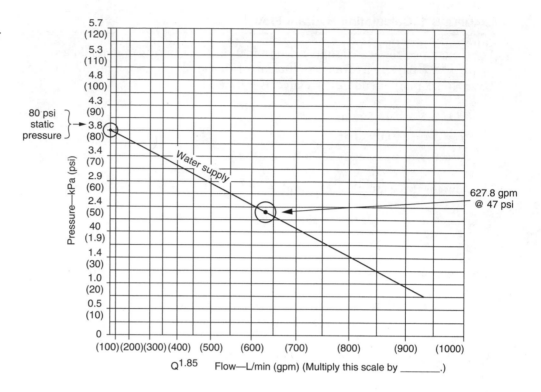

EXAMPLE 6.2 Flow Test Graphing

Assuming a water supply with a static pressure of 80 psi, and a residual pressure of 47 psi at a flow of 627.8 gpm, draw a graph of the water supply. The static condition is recorded on the left-hand side of the graph at the flow point of 0 gpm. The static pressure is recorded as 80 psi on Figure 6.11. The second point on the water supply graph is the actual flowing point recorded in our example test. The residual pressure available at the pressure hydrant was given as 47 psi while flowing at 627.8 gpm from the flowing hydrant, as shown on Figure 6.11.

With those two points recorded on the graph, a straight line connecting them provides the water supply data available at the base of the static hydrant as shown on Figure 6.11. The data provided by the residual hydrant describe the water supply system for determination of an effective water supply.

The water supply curve displays the data used to evaluate the water demand of automatic suppression systems. When the water demand point for an automatic suppression system is plotted above the curve, insufficient water is available from this water supply to meet that given demand. If the water demand point were located below the water supply curve, this would indicate adequate water supply is available for the system. It should be noted that some jurisdictions require a safety factor of 5 to 10 psi between the system demand and the water supply curve to account for seasonal variations in a water supply system. Some jurisdictions require a time log of the static pressures to determine the daily or weekly pressure fluctuations.

The water supply graph can be used to determine pressures available at any given flow along the graph line. There are two methods of determining flow at a given point. One is using the graph. An example would be to determine the pressure available at the 400 gpm flow in Figure 6.11. By using the graph, tracking along the x axis to 400 gpm, going up to the point it intersects the water supply line, and tracing straight across to the y axis, to determine that 66 psi is available at 400 gpm based on this water supply curve.

Mathematical Graphing

The second method, a mathematical method of pressure and flow determination, using a formula shown in Equation 6.2, can be done to determine pressure at a given flow:

$$P = (P_R - P_S)\left(\frac{Q}{Q_R}\right)^{1.85} + P_S \qquad (6.2)$$

where:

$\quad P$ = pressure at any given flow (in psi)

$\quad P_R$ = residual pressure measured in the test (in psi)

$\quad P_S$ = static pressure measured in the test (in psi)

$\quad Q$ = flow at which one wants to know P (in gpm)

$\quad Q_R$ = residual flow measured in the test (in gpm)

EXAMPLE 6.3 Calculating Pressure and Flow by Graphing

For the graph in Figure 6.11 for Example 6.2, the following data would be collected during the test:

$\quad P_S$ = 80 psi

$\quad P_R$ = 47 psi

$\quad Q_R$ = 627.8

The pressure available at 400 gpm is

$$P = (47 - 80) \times (400 \div 627.8)^{1.85} + 80 = 65.7 \text{ psi}$$

This value confirms that the 66 psi value obtained by graphing was reasonably accurate.

Either the graphing or calculation method can be used to determine pressure at any given point along the water supply line, using water flow test data. The water supply curve and calculation methods of determining water flow available are valid by testing up to the actual tested water flow. Water flow and pressure relationships above the actual tested water flow should be used with caution. Water supply systems in some cases do not have a predictable flow and pressure relationship at flows significantly higher than the tested water flow. Other adjustments to the water supply data that must be considered include elevation differences between the water supply test point and the fire protection system location; seasonal adjustments to the water supply availability; peak daily usage information, if available; and water supply pumping information to determine the pumping start sequences or use of additional pumps during peak demand periods.

TABLE 6.1		
Pumper Outlet Coefficients		
Pitot Pressure (Velocity Head)		Coefficient
psi	bar	
2	0.14	0.97
3	0.21	0.92
4	0.28	0.89
5	0.35	0.86
6	0.41	0.84
7 and over	0.48 and over	0.83

Source: NFPA 291, 2002, Table 4.8.2.

Other adjustments to be evaluated are the use of the pumper outlets on the fire hydrants. Using the pumper outlet on a fire hydrant is a method of obtaining adequate test flow or residual pressure drop when only a single flow hydrant is available. These pumper outlets are generally the 4- or 5-in. outlets that connect directly to the fire truck. NFPA 291 recommends using adjusted hydrant coefficients based on the velocity pressure recorded from the pumper outlet. Table 6.1 is a summary of the velocity pressure and coefficient adjustments for the use of the 4- or 5-in. outlets when determining water flow.

Equation 6.3 is the formula to be used when a pumper outlet is the flowing outlet for a water supply test. In Equation 6.3,

$$Q = 29.83 \times cc_p d^2 p^{1/2} \tag{6.3}$$

c_p is the pumper outlet coefficient from NFPA 291. The use of the pumper outlet should be limited to times when the static pressure cannot be dropped by flowing from the normal 2½-in. hydrant connections. The pumper outlet coefficient can make a major difference in the flow calculation, depending upon the velocity pressure. These differences can have significant impact on the water supply data; see Table 6.1.

EXAMPLE 6.4 Calculating Pumper Outlet

A pumper outlet on a fire hydrant is used on a water flow test. The pitot pressure is 4 psi, the pumper outlet is rounded, with a diameter of 4 in. Calculate the flow through the pumper outlet.

$$Q = 29.83 \times cc_p d^2 p^{1/2}$$

where:

$d = 4$ in.

$p = 4$ psi

$c = 0.9$

$c_p = 0.89$ (from Table 6.1)

Thus,

$$Q = 29.83 \times 0.9 \times 0.89 \times 16 \times 2$$
$$= 764.6 \text{ gpm}$$

By not using the pumper outlet coefficient,

$$Q = 29.83 \times 0.9 \times 16 \times 2$$
$$= 859.4 \text{ gpm}$$

This is an additional 94.8 gpm.

These two calculations reflect the adjustments required when using the 4-in. pumper connection to measure velocity pressure with a pitot tube gauge. The actual flow is overcalculated without the pumper correction factor c_p.

Determining the Water Demand

With the water supply determined, only half of the information has been gathered to determine if the water supply is adequate for the system requirements. Although detailed hydraulic calculations can be performed to determine a relatively accurate water demand, the following are estimation techniques that provide a quick-and-easy way to estimate whether the water supply is sufficient. This will also help determine whether fire pumps and storage tanks are required to comply with the suppression system demands. Chapter 10, "Hydraulic Calculations," details the method for precise determination of fire sprinkler system demand. The following section demonstrates how to perform a quick estimate of fire sprinkler system demands.

To determine the estimated water demand, the following information is needed:

- The estimated flow in gpm of the suppression system.
- The estimated pressure needed to produce this above flow.
- The estimated total capacity of the flow and duration needed for the automatic suppression system.
- If the water supply is not adequate, the estimated size of a tank or storage system.
- If the pressure of the existing system is not adequate, the estimated size of the fire pump for increasing system pressure. (See Chapter 12, "Fire Pump Design.")

With this information, an estimate of the automatic suppression system demand can be determined and compared to the available water supply. In determining the flow requirements of the system, the first step is to evaluate the type of automatic sprinkler being used. Dry-pipe sprinkler systems, double-interlock preaction sprinkler systems, and sprinklers installed on excessively sloped ceilings require adjustments that increase the basic operating areas, although the use of quick-response sprinklers allows the reduction of the basic operating areas in some cases and occupancies. Some automatic sprinkler types have minimum required flows (Q_m) that differ from that of the standard

FIGURE 6.12
Density/Area Curves.

(Source: NFPA 13, 2002, Figure
11.2.3.1.5)

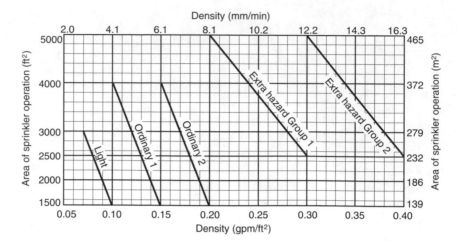

K-5.6 sprinkler. The next step would be to determine the total number of sprinklers to be used in the design area (*N*). The total number of sprinklers can be determined using a density/area method, the room design method, residential sprinkler design, or a specific or special application as dictated by the engineer of record or manufacturer of a special product.

Density/area curves from NFPA 13, *Standard for the Installation of Sprinkler Systems*, are the starting point for one method of calculating flow. Figure 6.12 is a reproduction of the NFPA density/area curves. These curves have been developed over the years and represent the minimum automatic sprinkler discharge density required to protect a given hazard when that discharge density is applied over the selected operating area. This figure lists the discharge densities along the *x* axis and the operating areas along the *y* axis. A typical example of a discharge density for a light-hazard occupancy would be 0.1 gpm per square foot over the most hydraulically remote 1500 ft². This point is represented on the light-hazard curve at the bottom of Figure 6.12 where the light-hazard line intersects 1500 ft² of operating area.

To determine the flow, use Hydraulic Formula No. 1 (see Chapter 10, "Hydraulic Calculations"), shown here as Equation 6.4:

$$Q_m = d \times (\text{area per sprinkler}) \tag{6.4}$$

where:

> Q_m = the minimum required flow for the most demanding sprinkler (gpm)
>
> d = density (gpm per square foot)
>
> area per sprinkler = square foot/sprinkler, the actual or maximum square footage of the most remote or demanding automatic sprinkler

Q_m is the minimum demand in an ideal world that does not consider the frictional resistance of pipe to water flow. The estimate does not take into account friction loss in the pipe, flow adjustments for multiple sprinklers, and increased pressure due to friction loss from the remote point back to the water supply source. An estimated overage factor should be applied to these initial flow calculations. The friction loss factors can vary from 15 percent to more

TABLE 6.2

Hose Stream Demand and Water Supply Duration Requirements for Hydraulically Calculated Systems

Occupancy	Inside Hose (gpm)	Total Combined Inside and Outside Hose (gpm)	Duration (minutes)
Light hazard	0, 50, or 100	100	30
Ordinary hazard	0, 50, or 100	250	60–90
Extra hazard	0, 50, or 100	500	90–120

For SI units, 1 gpm = 3.785 L/min.

Source: NFPA 13, 2002, Table 11.2.3.1.1.

than 25 percent for high friction loss systems. Some government agencies require a 30 percent overage calculation in the preliminary evaluations of water supply. This overage factor is intended to address the increased water flow due to the increased pressure that is related to the friction loss in the system. The specific friction loss of a given system can be vastly different from this estimate. Overage factors in the 25 to 30 percent range are not uncommon during the first evaluation pass for a given sprinkler system.

In addition to the friction overage adjustment, a hose stream allowance must be accounted for in the water flow estimations. Hose stream requirements are taken directly from NFPA 13, 2002 edition, Table 11.2.3.1.1, reproduced here as Table 6.2. The hose stream flow is added at the point of connection of interior hoses, or at the base of the fire sprinkler riser at the pressure required for the sprinklers per NFPA 13. The hose flow is needed for the total water demand.

The overage factors are generally based on experience. Normal systems with rectangular operating areas would be estimated using a 15 percent overage factor. Systems with large-orifice sprinklers, large operating areas, or large elevation changes within the operating area would use a 25 percent overage factor. Equation 6.5 shows the estimated total flow (ETF) based on the minimum flow per sprinkler and the number of sprinklers in the most remote area. Equation 6.5 also shows the estimated adjustments for flow due to friction loss in the remote area.

$$\text{ETF} = Q_m \times N \times F \text{ plus the hose demand} \tag{6.5}$$

where:

ETF = the total estimated flow

Q_m = the minimum required flow for the most demanding sprinkler (gpm per sprinkler)

N = the number of sprinklers in the operating area

F = the friction loss factor (an estimated friction loss of between 1.15 to more than 1.25 as shown on Table 6.3)

See Table 6.3.

TABLE 6.3

Overage Factors for Estimating Sprinkler System Demand

Amount of Friction Loss	Overage Factor
Very low friction loss in system	1.15
Average friction loss in system	1.20
High friction loss in system	1.25 or more

EXAMPLE 6.5 Estimating Sprinkler System Demand

Estimate the sprinkler system demand for a light hazard building with 0.1 gpm over the most remote 1500 ft^2 with sprinklers spaced at 225 ft^2 per sprinkler, where $K = 5.6$.

0.1 gpm per sq ft $\times$ 225 sq ft per sprinkler = 22.5 gpm per sprinkler

Check minimum flow:

$$Q = K\sqrt{P}$$

where:
　$K = 5.6$
　$P = 7.0$ psi

Thus,

$$5.6 \times 7^{1/2} = 14.8 \text{ gpm per sprinkler}$$

This check is done to determine if the minimum pressure required by NFPA 13 at the most remote sprinkler is adequate to provide the minimum water flow per sprinkler to meet the minimum discharge density. Since the 7-psi minimum sprinkler pressure allowed in NFPA 13 will produce only 14.8 gpm, the minimum pressure at the remote sprinkler will exceed 7 psi to produce the minimum discharge density of 22.5 gpm per sprinkler. The 22.5 gpm per sprinkler will provide the 0.1 gpm per square foot density, so the operating area of 1500 ft^2 requires a minimum of seven sprinklers to cover the remote area.

This is a normal, same-elevation operating area with low friction loss. An overage factor of 15 percent (multiply by 1.15) will be used. Estimated flow is

$$\text{ETF} = Q_m \times N \times F \quad \text{(plus the hose demand)}$$
$$= 22.5 \times 7 \times 1.15 = 181.25 \text{ gpm}$$

Per Table 6.2, hose demand for a light hazard building, from NFPA 13 is 100 gpm. The total demand estimate is

$$181.25 + 100 = 281.25 \text{ gpm}$$

This flow estimation formula can be used with sprinkler systems having similar demands at the individual sprinklers. Where there are multiple ceiling elevation changes or different special-type sprinklers, adjustments must be made to this estimation. The total volume of water needed can be determined by multiplying the estimated flow (determined above) by the duration (the time required for the flow).

The duration requirements are from Table 11.2.3.1.1 of NFPA 13, reproduced here as Table 6.2. These durations are based on the hazard being protected. The automatic sprinkler system is designed to contain and control a fire. The durations in NFPA 13 are minimum durations anticipating the intervention of the fire department to complete the suppression process. If there are reasons for a delay in fire fighter response to a site, the duration of water flow should be increased to address this risk. A remote rural site with a volunteer fire department responding within 25 to 30 min may want to extend the duration of the fire supply water flow for light hazard from 30 min to 60 min.

When estimating durations for a tank that serves only fire sprinklers and does not serve hydrants or hose stations within the protected property, the hose stream demand does not need to be added into the estimated flow to determine the volume needed for this system.

Determining the Required System Pressure

With the system water demand determined, the next step is to evaluate the pressure required to deliver the water. The hydraulic formula shown in Equation 6.6 determines the minimum flow in gpm at a given pressure within the system.

$$Q = K \times P^{1/2} \tag{6.6}$$

where:

Q = the minimum flow in gpm

K = the K-factor of the sprinkler

P = the pressure in psi at the sprinkler

Required Minimum Pressure at the Remote Sprinkler (P_s)

The required minimum pressure (P_s) at the hydraulically most remote sprinkler is determined by solving this equation for P

$$P_s = \left(\frac{Q}{K} \right)^2 \tag{6.7}$$

P_s can also equal the minimum pressure needed for special sprinklers as specified by the manufacturer, or 7 psi as specified by NFPA 13 as the minimum sprinkler operating pressure. Q is the Q_m for the sprinkler in the density area as determined earlier when estimating the water demand. K is 5.6 for ½-in. orifice sprinklers, and other K-factors are available sprinkler manufacturers. Using K-factors, the minimum pressure at the sprinkler can be determined. SI units use the same equation but different K-factors for the sprinkler. This SI K-factor is available from the manufacturer.

EXAMPLE 6.6 Pressure Calculation for Sprinkler Given Flow and K-Factor

Using the Q_m from the previous illustration of flow estimation,

$$Q = K \times P^{1/2}$$
$$Q_m = 22.5$$
$$K = 5.6$$

$$P_s = \left(\frac{Q_m}{K}\right)^2$$

$$= \left(\frac{22.5}{5.6}\right)^2 = 16.29 \text{ psi}$$

Elevation Pressure Loss (P_E)

The next pressure calculation, expressed in Equation 6.8, accounts for elevation changes between the most demanding sprinkler area and the water supply.

$$P_E = [(0.433 \text{ psi/ft}) \times h \text{ (in ft)}] \tag{6.8}$$

where:

P_E = the pressure requirements for the system due to changes in elevation

h = the change in height from the water source to the sprinkler elevation

P_E = [(0.433 psi/foot) $\times$ the elevation change (in feet)]

EXAMPLE 6.7 Calculating Elevation Changes

If the water supply entrance in Example 6.6 is at elevation 1 ft and the highest sprinkler and branch line is located at elevation 22 ft, the change in elevation is 21 ft (22 − 1 = 21 ft).

$$P_E = [(0.433 \text{ psi/ft}) \times h \text{ (in ft)}]$$
$$= 0.433 \times 21 = 9.09 \text{ psi}$$

Frictional Pressure Loss (P_F)

The next pressure estimation needed is an estimated friction loss in the piping along the piping path from the sprinkler to the water supply. This pressure estimation is designated P_F. A rule-of-thumb estimate is 0.15 psi per foot of straight pipe plus equivalent lengths for fittings and valves as they are encountered.

EXAMPLE 6.8 Calculating Estimated Friction Loss

In Example 6.6, there is 50 ft of straight steel pipe (Schedule 40) from the remote operating area. There are six 3-in. 90° standard elbows and one gate valve. NFPA 13, 2002 edition, Table 14.4.3.1.1, reproduced here as Table 6.4, provides a list of equivalent lengths of straight pipe for fittings and valves.

$$P_F = [(\text{pipe length}) + (\text{fitting loss}) + (\text{valve loss})] \times 0.15 \text{ psi/ft}$$
$$= [50 + 6(7) + 1] \times 0.15 \text{ psi per foot} = 13.95 \text{ psi} \tag{6.9}$$

Pressure Loss from Special Valves (P_V)

Lastly, P_V is the pressure loss estimation for additional valves and special devices such as backflow preventers. Backflow prevention devices usually have

TABLE 6.4

Equivalent Schedule 40 Steel Pipe Length Chart

Fittings and Valves	Fittings and Valves Expressed in Equivalent Feet of Pipe														
	½ in.	¾ in.	1 in.	1¼ in.	1½ in.	2 in.	2½ in.	3 in.	3½ in.	4 in.	5 in.	6 in.	8 in.	10 in.	12 in.
45° elbow	—	1	1	1	2	2	3	3	3	4	5	7	9	11	13
90° standard elbow	1	2	2	3	4	5	6	7	8	10	12	14	18	22	27
90° long-turn elbow	0.5	1	2	2	2	3	4	5	5	6	8	9	13	16	18
Tee or cross (flow turned 90°)	3	4	5	6	8	10	12	15	17	20	25	30	35	50	60
Butterfly valve	—	—	—	—	—	6	7	10	—	12	9	10	12	19	21
Gate valve	—	—	—	—	—	1	1	1	1	2	2	3	4	5	6
Swing check*	—	—	5	7	9	11	14	16	19	22	27	32	45	55	65

For SI units, 1 in. = 25.4 mm; 1 ft = 0.3048 m.

Note: Information on ½-in. pipe is included in this table only because it is allowed under 8.14.19.3 and 8.14.19.4.

*Due to the variation in design of swing check valves, the pipe equivalents indicated in this table are considered average.

Source: NFPA 13, 2002, Table 14.4.3.1.1.

recommended friction loss factors based on the size and estimated flow through the device. Each manufacturer has a table or graph of friction loss versus flow rate. These should be used to determine the friction loss for each device.

The total estimated pressure demand, EPD, can be determined by summing up the preceding pressure loss estimates. Equation 6.10 is the sum of the pressure adjustments required to estimate the sprinkler system pressure demand. This equation includes minimum operating pressure, elevation changes, friction loss in pipe and special valve friction loss.

$$\text{EPD} = P_s + P_E + P_F + P_V \tag{6.10}$$

The combination of the estimated pressure needed for a system and the estimated flow can be plotted on a water supply test curve to determine if adequate water is available.

EXAMPLE 6.9 Calculating System Demand

A retail store (Ordinary Hazard Group 2) is 150 ft by 100 ft wide and 15 ft in height. The building is a large, open, uncompartmented space with a flat roof/ceiling arrangement. A standard quick-response spray sprinkler is selected with a K-factor of 5.6. The sprinkler density protection requirements are 0.2 gpm per square foot. The maximum spacing per sprinkler is 130 ft^2 per sprinkler.

$$Q_m = 0.2 \text{ gpm per ft}^2 \times 130 \text{ ft}^2 = 26 \text{ gpm per sprinkler}$$

Using the density/area method for determining the total number of sprinklers, and adjusting for the use of quick response, the area for remote sprinkler activation is 1500 ft² × 67.5 percent [adjustment for quick-response sprinklers in a wet-pipe system, NFPA 13, Figure 11.2.3.2.3.1 (shown here as Figure 6.13)], which equals an area of 1013 ft². The total number of sprinklers (N) is 1013 ÷ 130 = 7.8, or 8 sprinklers. An overage of 1.2 is estimated based on regular-sized piping and average friction loss for the system.

From Table 6.2, a hose stream of 250 gpm is required. The estimated flow equals

$$Q_m \times N \times F + \text{hose stream} \tag{6.11}$$
$$26 \times 8 \times 1.2 + 250 = 499.6 \text{ gpm (rounded up to 500 gpm)}$$

With this example, assume that a tank is going to be used for both the sprinklers and hose demand. NFPA 13 requires a 60- to 90-min duration for the flow based upon Table 11.2.3.1.1. This example has used a 90-min duration. This gives a total water volume required of the flow times for the duration, or 500 gpm × 90 min = 45,000 gal. So far there is a demand of 500 gpm and a total volume of water required of 45,000 gal.

The next step in determining system demand is to estimate the pressure required. The minimum operating pressure for a single sprinkler (P_s) is determined by Equation 6.12:

$$P_s = \left(\frac{Q}{K}\right)^2 \tag{6.12}$$

where:

$Q = Q_m = 26$ gpm
$K = 5.6$

Thus,

$$P_s = \left(\frac{26}{5.6}\right)^2 = 22 \text{ psi}$$

Pressure loss due to elevation change (P_E) is

$$0.433 \times \text{the height of the system} \tag{6.13}$$

FIGURE 6.13
Design Area Reduction for Quick-Response Sprinklers.

(Source: NFPA 13, 2002, Figure 11.2.3.2.3.1)

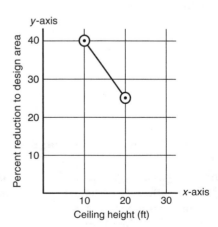

Note: $y = \frac{-3x}{2} + 55$

For ceiling height ≥10 ft and ≤20 ft, $y = \frac{-3x}{2} + 55$

For ceiling height <10 ft, $y = 40$

For ceiling height >20 ft, $y = 0$

For SI units, 1 ft = 0.31 m.

The height of the system = 15 ft.

$$P_E = 0.433 \times 15 = 6.5 \text{ psi}$$

The estimated friction loss in the flow path will assume a simplified layout of the system to determine the minimum piping and estimated number of fittings. For this example, assume a center-fed tree system with 165 ft of piping to the base of the riser, and the addition of at least three tees (30 ft), a check valve (35 ft), and a control valve (10 ft). Estimated friction loss:

$$P_F = (165 + 30 + 35 + 10) \times 0.15 \text{ psi per foot} = 36 \text{ psi}$$

Additional friction loss estimations for special valves and other assemblies (P_V) assume a double check valve backflow assembly with a friction loss of 10 psi at our estimated flow. To complete Equation 6.10:

$$\begin{aligned} \text{EPD} &= P_s + P_E + P_F + P_V \\ &= 22 + 6.5 + 36 + 10 \\ &= 74.5 \text{ psi} \end{aligned}$$

For this example, the demand of the system is 500 gpm at a minimum of 74.5 psi pressure. If the water supply cannot provide 500 gpm at 74.5 psi, the designer could consider changing the standard sprinklers from a K of 5.6 to a K of 8. This would reduce the P_s from 22 psi to 11 psi and reduce the total pressure (EPD) from 75 psi to 63.5 psi.

If after K-factor changes, the estimated demand is still above the supply, a detailed calculation, shown in Chapter 10, "Hydraulic Calculations," where pipe sizes are carefully changed to try to balance the demand to the supply may be needed to demonstrate compliance. When all methods of balancing the demand versus the supply fail, a fire pump would be needed to address any deficiencies in pressure.

If the water supply data indicate that 500 gpm is not available in our example, on-site water storage capable of meeting the volume calculations of 45,000 gal would be required. Water storage systems would need an adequately elevated tank, pressure tank, or a tank and fire pump combination to supply both the flow and pressure if the municipal water supply is not adequate.

Tanks and Other Water Supplies

With the advent of the modern automatic sprinkler, the demand on municipal water supply systems for a reliable flow at a given pressure was increased. Although many public water supply systems have increased their capacities and ability to supply large amounts of water at adequate pressure, there are situations where water storage tanks are required for fire protection purposes. These situations include storage tanks for properties that are not served by a public water supply, or storage tanks installed on upper zones in high-rise buildings with multizoned riser systems, or storage tanks installed as a secondary water supply for redundancy and reliability. Some facilities are high risk or mission-critical and require redundancy to increase the reliability of an automatic suppression system, such as nuclear power plants or critical military installations. Also, storage tanks are used in areas subject to earthquakes or where the reliability of the public water supply system is in question. Public water supply systems use tanks to meet system flow demands when wells cannot produce enough water flow and elevated tanks to provide pressure to the system.

Water supply storage tanks are used in rural settings where there are limited or no public mains, for example, remote industrial sites, residential occupancies, warehousing facilities, or other protected properties in areas with limited or no municipal water supply. Storage tanks are also needed when the public supply cannot provide adequate flow due to insufficient capacity in the public system or the presence of high-hazard facilities supplied by public mains of marginal capacity.

Water storage tanks are divided into two major categories: atmospheric tanks and pressure tanks. Atmospheric tanks work at normal atmospheric pressure and are generally open and vented to the exterior. Pressure tanks are enclosed vessels that use internal pressures to move the water. The size or volume of the tank would be determined in the estimated system demand. Commercial water storage tanks range from 5000 gal to 1,000,000 gal. Residential storage tanks are in the range of 100 to 5000 gal. Table 6.5 presents standard steel tank storage sizes.

Atmospheric Tanks

Atmospheric tanks are subdivided into two categories: gravity tanks and suction tanks. A gravity tank is generally a tank elevated above the ground or a tank located at the highest point in a municipal area. Elevated tanks provide pressure to a system based on the height of the tank. The pressure available from an elevated

TABLE 6.5

Standard Net Capacity Sizes of Steel Tanks

m³	gal
18.93	5,000
37.85	10,000
56.78	15,000
75.70	20,000
94.63	25,000
113.55	30,000
151.40	40,000
189.25	50,000
227.10	60,000
283.88	75,000
378.50	100,000
567.75	150,000
757.00	200,000
1135.50	300,000
1892.50	500,000

Source: Based on NFPA 22, 2003, paragraph 5.3.1.1.

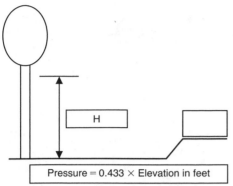

Pressure = 0.433 × Elevation in feet

H is the height of the lowest point of storage, usually the top of the riser into the storage tank.

Example : *H* = 50 ft
Pressure at the base of the riser
(ground) = 0.433 psi/foot × 50 ft
Pressure = 21.65 psi

FIGURE 6.14
Typical Elevated Tank.

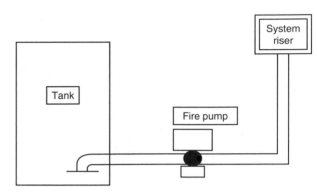

FIGURE 6.15
Typical Suction Tank Arrangement.

atmospheric gravity tank is equal to 0.433 times the elevation of the bottom of the tank. Figure 6.14 illustrates a typical elevated tank. The bottom of the tank, used as the reference point, is a conservative measure to ensure that adequate pressure is available under the worst-case condition. The other atmospheric tank is a suction tank. The suction tank is generally located at grade or below grade and has a fire pump connected to it to provide the flow and pressure for an automatic suppression system. Figure 6.15 is a schematic depiction of a suction tank arrangement. When drawing water from nonpotable water sources at grade, strainers and corrosion-resistant piping materials, as a minimum, may be required. Limits are placed on the size of the tank by the way it is constructed.

Pressure Tanks

Pressure tanks must comply with *ASME Boiler and Pressure Vessel Code* [1]. Pressure tanks require interior inspection every 3 years. Although pressure tanks are still available and are used in designs, other methods of providing water supply, such as atmospheric tanks and fire pumps, are gaining due to their large capacities for flow rate and total water stored. Table 6.6 shows typical sizes of commercially available pressure tanks.

As mentioned, pressure tanks are not in wide use; when they are used, they are generally set up to be approximately two-thirds full of water and one-third full of pressurized air. The air/water ratio can be adjusted to provide the correct pressure for a hydraulically calculated sprinkler system, using the following formula:

$$P_i = \left[\frac{P_F + 15}{A} \right] - 15 \qquad (6.14)$$

where:

P_i = tank pressure (psi)

P_F = pressure required from hydraulic calculations (psi)

A = proportion of air

TABLE 6.6

Typical Dimensions of Horizontal Pressure Tanks of Standard Sizes

Approx. Gross Capacity		Approx. Net Cap ⅔ Full		Inside Diam.		Inside Length		Approx. Wt. of Water ⅔ Full	
gal	L	gal	L	in.	m	ft	m	lbs	kg
3,000	11,355	2,000	7,570	60	1.5	20.2	6.2	16,670	7,568
3,000	11,355	2,000	7,570	66	1.7	17.0	5.2	16,670	7,568
3,000	11,355	2,000	7,570	72	1.8	14.2	4.3	16,670	7,568
4,500	17,033	3,000	11,355	66	1.7	25.4	7.7	25.000	11,350
4,500	17,033	3,000	11,355	72	1.8	21.3	6.5	25,000	11,350
4,500	17,033	3,000	11,355	78	2.0	18.2	5.5	25,000	11,350
6,000	22,710	4,000	15,140	72	1.8	28.2	8.6	33,340	15,136
6,000	22,710	4,000	15,140	78	2.0	24.2	7.4	33,340	15,136
6,000	22,710	4,000	15,140	84	2.1	21.0	6.4	33,340	15,136
7,500	28,388	5,000	18,925	78	2.0	30.3	9.2	41,670	18,918
7,500	28,388	5,000	18,925	84	2.1	26.2	8.0	41,670	18,918
7,500	28,388	5,000	18,925	90	2.3	22.7	6.9	41,670	18,918
9,000	34,065	6,000	22,710	84	2.1	31.4	9.6	50,000	22,700
9,000	34,065	6,000	22,710	90	2.3	27.3	8.3	50,000	22,700
9,000	34,065	6,000	22,710	96	2.4	24.0	7.3	50,000	22,700

Note: 4,5000-gal (17-m^3) gross capacity is the minimum ordinarily accepted for pressure tanks for automatic sprinkler systems.

Source: *Fire Protection Handbook*, NFPA, 2003, Table 10.2.7.

EXAMPLE 6.10 Pressure Tank Calculation

Assuming a calculated pressure of 75 psi and a tank 15% full of air, determine the tank pressure.

$$P_i = \left[\frac{P_F + 15}{A} \right] - 15$$

$$= \left[\frac{75 + 15}{0.15} \right] + 15 = 165 \text{ psi}$$

The stored air pressure is used to push the water out of the tank and into the system as needed. The air pressure would be required to be significantly higher than the minimum pressure demand for the system. Figure 6.16 is a schematic of a typical pressure tank arrangement. Pressure tanks, due to their cost and compressed air requirements, are generally limited in volume. For large-volume water storage tanks, atmospheric tanks might be more cost-effective than pressure tanks.

With the advent of relatively inexpensive special-duty fire pumps (250 gpm and lower), the use of pressurized tanks has diminished greatly throughout the United States. Designers will still be required to evaluate modifications to existing pressure tank systems and may be required to design new pressure tank systems where conditions dictate.

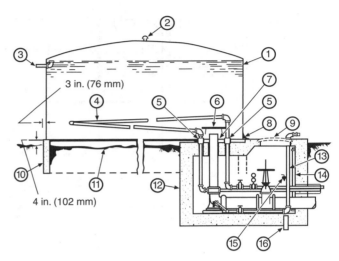

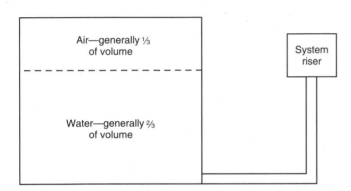

FIGURE 6.16
Typical Pressure Tank Arrangement.

1. Pump suction tank
2. Screened vent
3. Stub overflow pipe
4. Steam coil for heating
5. Extra-heavy couplings welded to tank bottom
6. Vortex plate
7. Watertight lead slip joint
8. Flashing around tank
9. Manhole with cover
10. Concrete ring wall
11. Sand or concrete pad (depending on soil condition)
12. Valve pit
13. Drain pipe
14. Ladder
15. Drain cock
16. Valve pit drain

FIGURE 6.17
Ground-Level Suction Tank Showing the Discharge Pipe Connected to the Bottom of the Tank in a Valve Pit.

(Source: *Fire Protection Handbook*, NFPA, 2003, Figure 10.2.1; courtesy of FM Global)

EXAMPLE 6.11 Estimating Demand for a Pressure Tank

System demand minimums are shown in Table 6.2 (Table 11.2.3.1.1 of NFPA 13). Light-hazard occupancies require a minimum flow duration of 30 min. If the hydraulically calculated flow rate is 200 gpm, the minimum storage volume is

$$30 \text{ min} \times 200 \text{ gpm} = 6000 \text{ gal}$$

Assuming that 6000 gal is $\frac{2}{3}$ of the pressure tank's total volume, the tank capacity is $6000 \div \frac{2}{3} = 9000$ gal.

The closest standard tank size is 9000 gal. The minimum ordinary accepted pressure tank size for an automatic sprinkler system is 4500 gal gross capacity.

Tank Materials

The design of water tanks for private fire protection systems is addressed in NFPA 22. This standard covers allowable materials, tank locations, capacities, requirements for gauges, fill pipes, overflow piping, cleanouts, drains, and heating requirements. Tanks can be designed and built out of many different types of materials, including wood, steel (both welded and bolted), concrete, and embankment-supported rubberized fabric (ESRF). Fiberglass can be used for tank materials in residential applications only. Figures 6.17 through Figure 6.19 show examples of tanks of different materials.

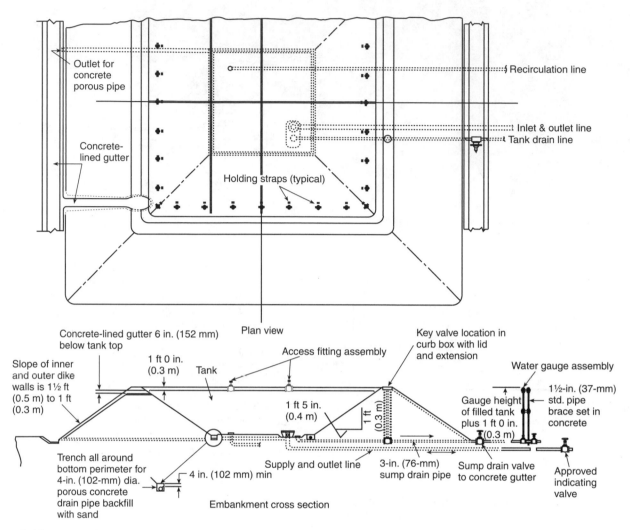

FIGURE 6.18

Installation Details of a Typical Embankment-Supported Fabric (ESF) Tank, Including Fittings.

(Source: *Fire Protection Handbook*, NFPA, 2003, Figure 10.2.7)

Tank Location and Heating

In addition to tank material, the location of the tank should be considered. Water storage tanks should not be located in areas subject to fire exposures. Where possible, these tanks should be at least 20 ft away from combustible structures. If the tanks must be located closer to the structure, fireproof tanks or open sprinklers to protect the tanks from fire exposures should be considered. NFPA 80A, *Standard for the Protection of Buildings from Exterior Fire Exposures*, sets requirements for evaluating fire exposure hazards and protecting exposures if needed. Locating the tank in areas not subject to freezing must be considered in the tank design. If a tank cannot be located within a heated space, then a tank heating system should be considered. For water tanks that are subject to freezing, the water within the tank and piping system must be maintained at or above 42°F, as required by NFPA 22, subsection 15.1.2. This can be done by installing a tank heating system, or by installing the tank below the frost line. The frost line is determined from Figure 6.20. NFPA 22 lists numerous methods

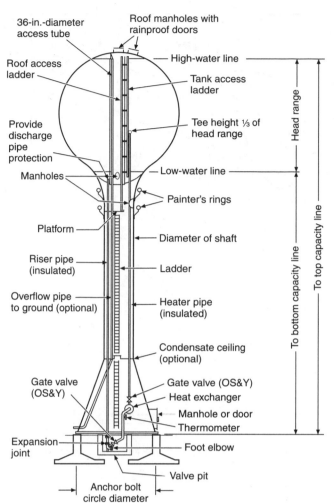

36-in.-diameter access tube

Roof manholes with rainproof doors

Roof access ladder

High-water line

Tank access ladder

Provide discharge pipe protection

Tee height ⅓ of head range

Manholes

Low-water line

Painter's rings

Platform

Diameter of shaft

Riser pipe (insulated)

Ladder

Overflow pipe to ground (optional)

Heater pipe (insulated)

Condensate ceiling (optional)

Gate valve (OS&Y)

Gate valve (OS&Y)

Heat exchanger

Manhole or door

Thermometer

Expansion joint

Foot elbow

Valve pit

Anchor bolt circle diameter

Head range

To bottom capacity line

To top capacity line

FIGURE 6.19
Typical Pedestal Tank.

(Source: *Fire Protection Handbook*, NFPA, 2003, Figure 10.2.3)

of providing heating systems for water storage tanks, including steam, gas-fired, oil-fired, electric and solar heating methods. The intent of these heating systems is to maintain the water temperature above 42 degrees at all times, thus assuring a reliable water supply for fire protection purposes.

Calculating Tank Elevation

An elevated tank can minimize or eliminate the requirements for fire pumps on a site. The elevated tank should be sized to meet the required system demand multiplied by the duration. The duration should be modified to address fire department response and whether the automatic suppression systems are monitored or not. NFPA 13, 2002 edition, paragraph 11.2.3.1.8(10), allows the durations of water to be reduced with an automatic sprinkler system monitored by a remote station or a central station. The duration of flow identified in NFPA 13 is a recommended minimum, and other risk components should be evaluated in determining the duration for an elevated tank. A tank size can be developed based on the flow demand and determined duration.

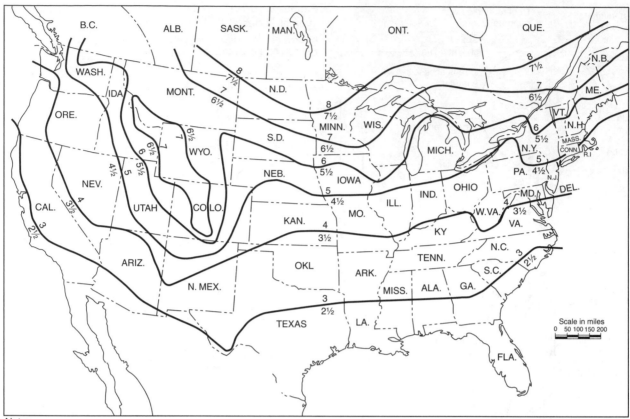

Notes:
1. For SI Units, 1 in. = 25.4 mm; 1 ft = 0.304 m.
2. Where frost penetration is a factor, the depth of cover shown averages 6 in. greater than that usually provided by the municipal water works. Greater depth is needed because of the absence of flow in yard mains.

FIGURE 6.20
Recommended Depth of Cover (in feet) Above Top of Underground Yard Mains.

(Source: NFPA 24, 2002, Figure A.10.4.1)

EXAMPLE 6.12 Calculating Tank Elevation

Flow demand of 500 gpm and duration of 90 min gives a minimum tank size of 45,000 gal. The elevation of the tank is determined by the minimum system pressure requirements. Rearranging Equation 6.13 [P_E = 0.433 psi per foot × H (the height)] to

$$H = \frac{P_E}{0.433} \qquad (6.15)$$

where H = the height necessary to achieve system pressure and the system demand is 500 gpm at 45 psi. This pressure takes into account all of the valves and pipe estimated from the sprinkler system remote area to the base of the elevated tower. The height of the tower required to provide 45 psi is H.

$$H = \frac{45 \text{ psi}}{0.433}$$
$$= 104 \text{ ft}$$

Assuming a 90 minute duration, the elevated tank would require its 45,000 gal of storage to be 104 ft above the sprinkler system connection to the base of the water tower.

Tank Attachments

NFPA 22 requires that water storage tanks be fitted with a number of monitoring and control devices. Water level gauges are needed (subsection 13.8.1 of NFPA 22, 2003 edition). These water level gauges can be a site gauge, an altitude gauge, or closed circuit high- and low-level alarms. The discharge piping from the tank should contain a control valve and check valve to maintain system pressure (subsection 13.2.12 of NFPA 22). The filling of these tanks can be automatic or manual. Manual filling is strongly discouraged. Automatic filling valves will increase the reliability of the overall system Depending on the manual observation of water level and then the manual operation of the valve introduces decreased reliability. During a water flow, the automatic valve will open and fill the tank while it is supplying water to the system, which will extend the duration slightly. Storage tanks must be equipped with overflow piping to allow water to escape the tank if accidentally overfilled (subsection 13.5.3 of NFPA 22). Cleanout access for the tanks and riser drains must be provided for maintenance of the tank (Section 13.6 of NFPA 22). Additional monitoring for water temperature must be provided in areas subject to freezing (subsection 13.8.1 of NFPA 22).

Summary

This chapter covered how to determine the water supply available for an automatic sprinkler system and methods of estimating the demand of a given automatic sprinkler system. Comparing these two measurements can help determine the adequacy of the available water supply system for a given suppression system design. These preliminary calculations of demand should be refined and confirmed as the specifics of the system design evolve.

When the available water supply capacity is inadequate, and some method of on-site water storage is needed, the designer has many options available. Depending upon the economics of the location, there are many materials and methods of construction available for a water storage tank: Elevated tanks can be used to reduce the need for fire pumps, but they increase the complexity of water filling and freeze protection for the system; pressure tanks can be used for automatic sprinkler systems with relatively low flow rates and low total water storage; and ground level tanks can be used to meet the needs for a large volume of water with fire pumps. The main purpose of these water storage tanks is to provide an adequate volume of water for the automatic sprinkler systems.

BIBLIOGRAPHY

Reference Cited

1. *ASME Boiler and Pressure Vessel Code*, American Society of Mechanical Engineers, New York, 2001.

NFPA Codes, Standards, and Recommended Practices

NFPA Publications National Fire Protection Association, 1 Batterymarch Park, Quincy, MA 02169-7471

The following is a list of NFPA codes, standards, and recommended practices cited in this chapter. See the latest version of the *NFPA Catalog* for availability of current editions of these documents.

NFPA 13, *Standard for the Installation of Sprinkler Systems*

NFPA 15, *Standard for Water Spray Fixed Systems for Fire Protection*

NFPA 22, *Standard for Water Tanks for Private Fire Protection*

NFPA 24, *Standard for the Installation of Private Fire Service Mains and Their Appurtenances*

NFPA 80A, *Recommended Practice for Protection of Buildings from Exterior Fire Exposures*

NFPA 291, *Recommended Practice for Fire Flow Testing and Marking of Hydrants*

Additional Reading

Wilcox, W. E., "Fixed Water Storage Facilities for Fire Protection," in Cote, A. E., ed., *Fire Protection Handbook*, 19th edition, NFPA, Quincy, MA, 2003.

Impact of Unique Applications on Design Approaches

7

Brian Foster, PE, and Grant Cherkas

Since the 1990s, the fire sprinkler industry has seen a number of new and innovative developments, some of which were developed based on inadequacies inherent in the use of traditional pendent or upright sprinklers, exemplified by sprinklers under steeply pitched roof structures or ceilings, and protection of concealed combustible spaces. The coverage of unique applications provides the necessary tools to handle situations not common to most sprinkler installations, and the designer must be cognizant of situations where older fire protection technologies and practices may not provide the level of protection required or expected by the stakeholders.

Many of the sprinklers in this chapter are identified as Special Sprinklers in NFPA 13, *Standard for the Installation of Sprinkler Systems*, and may not have complete installation requirements published in NFPA 13 for a given application. The listing organizations, such as Underwriters Laboratories (UL) and FM Global (FM), will approve or list these sprinklers based on an installation guideline developed by the manufacturer. In order to achieve a listed installation, the rules in the manufacturer's data must be followed without deviation.

This chapter will review the use of sprinkler technologies in three broad categories: Special Listed Sprinklers, Special Service Sprinklers, and Other Types of Sprinklers.

Special Listed Sprinklers

Special listed sprinklers include the following:

- Attic sprinklers
- Concealed combustible space sprinklers
- High-pressure sprinklers
- Institutional sprinklers
- Low-pressure water mist sprinklers
- Metal building sprinklers
- Window sprinklers

FIGURE 7.1
Typical Attic Back-to-Back
Sprinkler.

(Courtesy of Tyco Fire & Building
Products)

Attic Sprinklers

Attic sprinklers are listed to be specified for combustible attic spaces. These upright-style sprinklers are designed with a unique deflector design that allows them to provide excellent coverage under a sloped roof. These sprinklers may not be in trussed rafters that have no ceiling below. In other words, they are intended for attic protection only, and cannot be used for occupancy fire control or hazard protection. Figures 7.1, 7.2, and 7.3 display typical attic sprinklers.

The installation of attic sprinklers is limited to roof slopes of 4:12 to 12:12. Roofs having slopes of less than 4:12 can be effectively protected using conventional sprinklers provided that specific NFPA 13 rules are followed. The engineer must be aware of special conditions that apply when the slope exceeds 2:12. For conventional sprinklers, a 30 percent design area increase for sloped roofs is required for slopes greater than a 2:12 pitch. With the 30 percent penalty, attic sprinklers may be a better choice for some installations since the penalty does not apply to these specially listed sprinklers.

Attic sprinklers are installed in a row under the peak of a trussed roof line. Spacing rules for various styles of attic sprinklers vary, and the designer must follow the installation requirements in the manufacturer's data in addition to NFPA 13.

Attic sprinklers come in several configurations. For installation under a peaked roof with a need for protection on both sides, a back-to-back style sprinkler (as shown in Figure 7.1) provides coverage from eave to peak in two directions. The distance from the attic sprinkler to the eave can be up to 30 ft. If the attic is too wide, conventional sprinklers can be used to supplement the coverage area. Conventional sprinklers can also be used to provide supplemental protection in areas of obstruction.

As shown in Figure 7.3, a single-direction attic sprinkler is available for use in single pitched roof structures, for example, a shed style roof. In this situation, the sprinkler is again installed at the peak and water is directed in one direction down the slope toward the eave.

FIGURE 7.2
Typical Attic Hip Sprinkler.

(Courtesy of Tyco Fire & Building Products)

FIGURE 7.3
Typical Attic Single-Direction Sprinkler.

(Courtesy of Tyco Fire & Building Products)

For hip roof structures, a special attic sprinkler is available to protect the hip portion of a combustible trussed roof. See Figure 7.2. Manufacturer's literature provides installation and spacing requirement details that must be followed.

There are some additional considerations related to the use of attic sprinklers. The number of actuated sprinklers in steeply pitched roof structures can be less when using attic sprinklers versus conventional sprinkler protection, since the attic sprinklers are most effectively positioned at the roof peak. Another consideration is the available water supply, since higher pressures are required for attic sprinklers in most situations. In some cases, a poor water supply may dictate the use of conventional sprinkler protection.

Concealed Combustible Space Sprinklers

The concealed combustible space sprinkler (CCS) is most commonly used to protect the floor framing of a multistory building of combustible construction. The concealed space is often formed by the application of a ceiling either directly to the bottom of the wood framing or supported below this framing (i.e., a dropped acoustic tile ceiling). Figures 7.4 and 7.5 illustrate typical concealed combustible space sprinklers.

The use of sprinklers in areas of concealed combustible construction should only be considered when more practical alternatives do not exist, such as the NFPA 13 provision of alternatives to the protection of some concealed combustible spaces, including using blocking and space-filling insulation. One concern that must be kept in mind during the design of a sprinkler system using these devices is how to access the sprinklers once installed. If a hard ceiling is installed, access to the sprinklers may be virtually impossible and would result in the inability to meet the inspection requirements found in NFPA 25, *Standard for the Inspection, Testing, and Maintenance of Water-Based Fire Protection Systems.*

The installation of concealed combustible sprinklers must be in accordance with the manufacturer's listing. Typical limitations include a horizontal roof slope that does not exceed a 2:12 pitch. Any insulation in the space must be "netted" or secured such that when wet, it will not impair the discharge

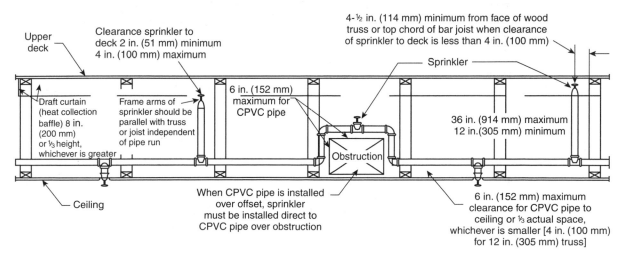

FIGURE 7.4
Wood and Steel Open Truss Construction Using CPVC Pipe (Cross Section View).
(Courtesy of The Viking Corporation)

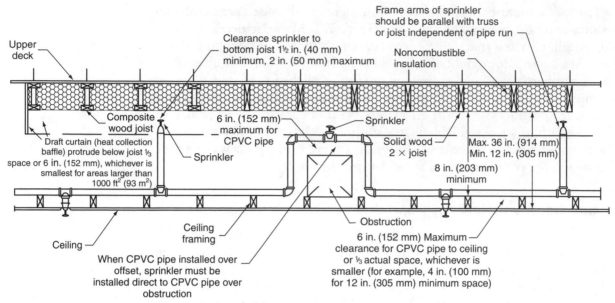

FIGURE 7.5
Solid Wood or Composite Wood Joist Construction with Noncombustible Insulation Filling Upper Deck
Using CPVC Pipe (Cross Section View).

(Courtesy of The Viking Corporation)

pattern of the sprinkler heads. The use of coarse screen wire is typical for this
purpose. The height of the protected space must be within the limits of the listing and there cannot be additional hazards in the space that would exceed a
light hazard occupancy. The listed concealed sprinklers have a minimum spacing of 6 ft, which is similar to conventional sprinklers. This limitation is imposed
to prevent a wetting of ("cold soldering") adjacent sprinklers and preventing
them from activating.

The CCS sprinklers can be piped with listed CPVC plastic fire sprinkler
pipe. Additional sprinklers may be needed to protect offsets or obstructions
when using CPVC pipe to prevent its failure under fire conditions. As of this
writing, only certain CPVC sprinkler piping is listed for use with specific CCS
sprinklers. Any design must include a verification of all the listing criteria
prior to the creation of a specification or sprinkler system layout.

CCS sprinklers will typically be designed for a density of 0.10 gpm/ft^2
(0.41 L/m^2) over the hydraulically most remote 1000 ft^2 (93 m^2) and, again,
will require verification with the product listings.

High-Pressure Sprinklers

High-pressure sprinklers are listed for pressures that exceed 175 psi, and the
implications of their use can result in more cost-effective designs, especially in
tall buildings. In a typical fire protection system, the piping, fittings, and sprinklers are rated at 175 psi (12.065 bar) working pressure. Consider a situation
with a building that features a roof manifold that requires 100 psi (6.89 bar)
on the roof elevation of 190 ft (57.9 m). Assuming an 8 psi (0.55 bar) friction
loss in the standpipe and supply piping back to a fire pump, under these conditions, you would require

$$190 \text{ ft} \times 0.433 \text{ psi/ft} + 8 \text{ psi} + 100 \text{ psi} = 190.27 \text{ psi} (13.12 \text{ bar})$$

where:

$$0.433 \text{ psi/ft} = \text{pressure attributed to elevation}$$

$$100 \text{ psi (6.89 bar)} = \text{required pressure at the roof manifold under full standpipe flow conditions}$$

Since 175 psi (12.065 bar) is the typical maximum working pressure for fire sprinklers and sprinkler components, the lowest 35 ft (10.67 m) of equipment would be "overpressurized." Options would include the use of pressure-reducing control valves on the floor risers up to the fourth floor. It would also require a drain riser large enough to test the pressure-reducing valve at the full sprinkler system demand. An alternative would be to use listed sprinklers rated above 190 psi (13.12 bar) and to use fittings and valves listed for these higher pressures. This would eliminate extensive testing and maintenance on the system and improve the overall reliability with only modest cost considerations. Most of the sprinkler manufacturers have a line of listed high-pressure sprinklers suitable for working pressures up to 250 psi (17.24 bar) and hydrostatically testable to 300 psi (20.68 bar).

Institutional Sprinklers

Institutional sprinklers are designed to be "tamper-resistant" sprinklers. They are also designed to prevent the attachment of heavy loads to the device prior to failure, an important feature in correctional or mental health facilities where there is a possibility of attempted suicide by hanging. They are also designed to eliminate the removal of pieces of metal from the sprinkler that could be fashioned into weapons.

NFPA 13 specifies that the use of institutional sprinklers in light hazard occupancies requires a quick-response listed sprinkler. These sprinklers are available in standard and extended coverage listings. Figure 7.6 illustrates a typical institutional sprinkler.

Low-Pressure Water Mist Sprinklers

Low-pressure "water mist" sprinklers (see Figure 7.7) produce water droplets associated with a water mist system with a percentage of the droplet sizes exceeding 430 microns. These sprinklers are provided in a threaded base format either with an open orifice or with a closed orifice and frangible bulb element. The closed sprinklers are listed for the protection of light and ordinary hazard occupancies as

FIGURE 7.6
Tyco Max Institutional Sprinkler.

(Courtesy of Tyco Fire & Building Products)

FIGURE 7.7
AM24 Low-Pressure Water Mist Sprinkler.

(Courtesy of Tyco Fire & Building Products)

defined by NFPA 13. These sprinklers have a minimum operating pressure in excess of 100 psi (20.68 bar). The original benefit of these sprinklers in light and ordinary hazard occupancies was their potential to reduce the fire protection system water flow needs. However, with the advent of quick-response sprinklers and the design area reductions allowed by NFPA 13 since the 1996 edition, applications for these water mist sprinklers are limited.

The open style low-pressure water mist sprinkler is used with water mist systems in total flooding applications, primarily for the protection of turbines and machine spaces. Applications and requirements for water mist systems can be found in NFPA 750, *Standard on Water Mist Fire Protection Systems*. The manufacturer provides installation and design requirements in addition to the information found in NFPA 750. The designer needs to ensure that all the applicable conditions can be met prior to specifying or designing a system with water mist sprinklers.

Metal Building Sprinklers

Metal building sprinklers are truly a unique application for typical preengineered metal buildings. These buildings are identified by the use of metal "Z" purlins supported on a rigid frame (honch columns). Figure 7.8 illustrates a typical metal building sprinkler.

In typical metal building sprinkler installations, two or more rows of sprinklers, oriented in a sidewall configuration, are run up the slope perpendicular to the purlins and hung from the rigid frame. The spacing of these lines using conventional sprinklers is based on area-density factors, limits in NFPA 13, or a special listing of the sprinkler. Often, conventional sprinklers are obstructed by items such as bridging, lighting, and mechanical systems. In recent years, the metal building industry has engineered these lightweight metal structures with smaller and smaller structural safety margins, which often restricts the amount of weight that can be supported from the purlins. Metal building sprinklers result in lower loads on the lightweight building structural elements (purlins), since the branch-line piping is supported from the rigid frame. The space around the rigid frame is usually not heavily congested with other building equipment or constructed auxiliary features. The metal building sprinkler works well in this situation since the piping for these systems is run parallel and adjacent to the rigid frame.

One additional feature of metal building sprinklers is that they are listed for extended coverage. At the time of publication, the sprinklers available were listed for up to 175 ft² (16.3 m²) coverage. The systems may be used in wet, dry, or preaction-type sprinkler systems.

There are significant limitations that must be evaluated before electing to specify or design a system with these sprinklers. These limitations include the following:

- Building height restrictions
- A minimum design density of 0.21 gpm/ft² (0.83 L/m²)
- Minimum sprinkler spacing to prevent "cold soldering" and skipping during activation
- A limitation of storage height
- Deflector to roof deck clearances per the sprinkler's listing
- Compliance with obstruction rules in NFPA 13 for sidewall sprinklers

FIGURE 7.8
Model MB14-EC Sprinkler.
(Courtesy of Reliable Automatic Sprinkler Co., Inc.)

- A roof pitch not exceeding 2 in 12
- The bay spacing (dimension between rigid frames)

As with all specially listed sprinklers, the manufacturer's product data sheet must be reviewed for limitations specific to that sprinkler.

Window Sprinklers

Window sprinklers, which have been on the market for a number of years, are a valuable tool in exposure protection and can also be used to give ordinary glass a fire rating. A window sprinkler has a specially configured deflector that results in a directional water spray over the entire area of a protected glass surface. The window sprinkler is available in a closed-head style with a fusible link/frangible bulb operating element similar to a normal fire sprinkler. They are also available in an open-orifice style usable in deluge applications.

The use of these sprinklers has limitations. They may be used only for fixed glass windows. Intuitively, this makes sense because the effectiveness of the window sprinkler is based on a uniform cooling of the glass surface. This prevents the glass and frame from expanding or distorting, which would result in a glass failure and the ultimate loss of the fire-rated membrane. Reliance on the window sprinkler for fire resistance needs to take into account exposures such as falling or bursting occupancy elements that could be reasonably expected to impair the system operation. Under those circumstances, wired glass or fire-rated high-strength glazing may be a better alternative.

Another consideration with the installation of closed-head window sprinklers is the configuration of mullions (pane dividers) that could interrupt the vertical sheeting of the flow. The presence of a horizontal mullion would preclude the use of these sprinklers. Another requirement is the maintenance of a 2-in. separation between the glass and combustibles such as blinds or curtains. Modern horizontal or vertical blinds are likely to encroach on the minimum space separation and disrupt the water coverage by the window sprinkler.

In other applications involving open-head deluge window sprinklers for exposure protection only, multiple levels of window sprinklers could be arranged to provide protection above and below exposures. Window sprinklers have been effectively used on glass curtain walls of atriums and retail store fronts.

The installation rules for window sprinklers are given in the manufacturer's literature, as this is Special Sprinkler classification.

One final consideration in the use of window sprinklers is the aesthetics of such an installation. Unless the engineer can work with the architect on an acceptable level of concealment, project owners may find the use of these sprinklers aesthetically objectionable.

Special Service Sprinklers

Special service sprinklers include the following:

- Corrosion-resistant sprinklers
- Directional spray nozzles
- High-temperature sprinklers
- Intermediate-level sprinklers

FIGURE 7.9
Central Model A Upright
Wax-Coated Corrosion-
Resistant Sprinkler.

(Source: *Automatic Sprinkler Systems
Handbook*, NFPA, 2002, Exhibit 3.33)

Corrosion-Resistant Sprinklers

Corrosion-resistant sprinklers are designed to be resistant to exposure damage. This is very important in some installations, including near corrosive chemical processes, in tunnels, in salt air (coastal), and in high-humidity locations. There are various means of protection available and their selection should be based on the degree and type of exposure hazard. Figure 7.9 shows a corrosion-resistant sprinkler.

The original corrosion-resistant sprinklers were lead coated, and these sprinklers are still commonly available. An improvement over the plain lead coating is beeswax over lead. This combination provides a very corrosion-resistant sprinkler, but it is not very aesthetically pleasing. For extremely corrosive conditions, stainless steel sprinklers are available in limited styles. These sprinklers are typically reserved for highly corrosive areas where the use of wax over lead sprinklers might not be adequate or where ambient temperatures are too high for the wax. Another common protection is a baked-on polyester paint. These finishes provide a reasonable degree of protection in mildly corrosive conditions provided that the finish is not damaged during the installation process. Failure to use the manufacturer's installation wrenches is a typical cause of damage to the coating.

Directional Spray Nozzles

Unlike the standard spray sprinklers, which are designed to apply water spray over a given square footage of floor area, directional spray nozzles are designed to project a water spray over the surface area of a hazard. These nozzles come in many shapes and sizes and may have a fusible/frangible element or may have an open orifice. Figure 7.10 illustrates a directional spray nozzle. The use of these nozzles is applicable when a high-velocity directed spray is needed. Hazards such as large utility company transformers and flammable gas storage tanks and spot exposure protection have been common applications for these types of devices, designed in accordance with NFPA 13, *Standard for the Installation of Sprinkler Systems*, or NFPA 15, *Standard for Water Spray Fixed Systems for Fire Protection*.

The available coverage angles of directional spray nozzles range from about 65 degrees to 180 degrees. They are available in various orifice sizes to provide flexibility in the design densities possible. Open-head directional spray nozzles are used with deluge-type systems. Exterior applications must take into consideration bees, hornets, and other insects that will attempt to build nests in the open nozzles and piping. A number of the available open-nozzle spray devices have dust caps that are easily dislodged during a system activation yet provide protection from insects. Periodic inspections of open-sprinkler systems need to be made to ensure that all discharge nozzles are open and functional. As with any deluge-type system with dry piping, use of galvanized pipe and fittings is prudent to increase the system lifespan.

High-Temperature Sprinklers

The selection of sprinklers based on the application requires the designer to understand the environment in which the sprinkler will be installed. Use of too low a temperature rating increases the probability of a premature activation, possibly as a result of a nonfire event. Use of too low of a temperature rating in storage applications can result in a fire opening far too many sprinklers

FIGURE 7.10
Grinnell Automatic
Protectospray™ Nozzle.

(Courtesy of Tyco Fire and Building
Products)

with a resulting degradation of the ability of the water supply to maintain adequate flows and pressures. NFPA 13 provides guidance in the selection of sprinklers with respect to exposure temperatures; see Tables 7.1, 7.2, and 7.3. For an illustration of high-temperature sprinklers, see Figures 7.11 and 7.12.

Some applications require the use of extremely-high-temperature fusible elements to prevent activation under normal conditions. Process ovens and baking equipment using conveyors are examples of applications for high-temperature sprinkler protection. High-temperature sprinklers are available in activation temperatures of up to 650°F (343.33°C). The temperature selection criteria should provide at least a 50°F (10°C) margin above the maximum temperature of the protected environment. One must be cautious of localized hot spots, especially in occupancies such as heated ovens, automotive paint baking operations, and heated bakery equipment. The area around the heat source may have localized higher temperatures that require additional consideration.

FIGURE 7.11
Pendent-Style High-Temperature Quarzoid Sprinkler.

(Courtesy of Tyco Fire & Building Products)

TABLE 7.1

Temperature Ratings of Sprinklers Based on Distance from Heat Sources

Type of Heat Condition	Ordinary Degree Rating	Intermediate Degree Rating	High Degree Rating
(1) Heating ducts			
(a) Above	More than 2 ft 6 in.	2 ft 6 in. or less	
(b) Side and below	More than 1 ft 0 in.	1 ft 0 in. or less	
(c) Diffuser	Any distance except as shown under Intermediate Degree Rating column	*Downward discharge:* Cylinder with 1 ft 0 in. radius from edge extending 1 ft 0 in. below and 2 ft 6 in. above *Horizontal discharge:* Semicylinder with 2 ft 6 in. radius in direction of flow extending 1 ft 0 in. below and 2 ft 6 in. above	
(2) Unit heater			
(a) Horizontal discharge		*Discharge side:* 7 ft 0 in. to 20 ft 0 in. radius pie-shaped cylinder *(see Figure 8.3.2.5)* extending 7 ft 0 in. above and 2 ft 0 in. below heater; also 7 ft 0 in. radius cylinder more than 7 ft 0 in. above unit heater	7 ft 0 in. radius cylinder extending 7 ft 0 in. above and 2 ft 0 in. below unit heater
(b) Vertical downward discharge *(for sprinklers below unit heater, see Figure 8.3.2.5)*		7 ft 0 in. radius cylinder extending upward from an elevation 7 ft 0 in. above unit heater	7 ft 0 in. radius cylinder extending from the top of the unit heater to an elevation 7 ft 0 in. above unit heater
(3) Steam mains (uncovered)			
(a) Above	More than 2 ft 6 in.	2 ft 6 in. or less	
(b) Side and below	More than 1 ft 0 in.	1 ft 0 in. or less	
(c) Blowoff valve	More than 7 ft 0 in.		7 ft 0 in. or less

For SI units, 1 in. = 25.4 mm; 1 ft = 0.3048 m.

Source: NFPA 13, 2002, Table 8.3.2.5(a).

FIGURE 7.12
Upright-Style High-Temperature Quarzoid Sprinkler.

(Courtesy of Tyco Fire & Building Products)

TABLE 7.2

Ratings of Sprinklers in Specified Locations

Location	Ordinary Degree Rating	Intermediate Degree Rating	High Degree Rating
Skylights		Glass or plastic	
Attics	Ventilated	Unventilated	
Peaked roof: metal or thin boards, concealed or not concealed, insulated or uninsulated	Ventilated	Unventilated	
Flat roof: metal, not concealed	Ventilated or unventilated	Note: For uninsulated roof, climate and insulated or uninsulated occupancy can necessitate intermediate sprinklers. Check on job.	
Flat roof: metal, concealed, insulated or uninsulated	Ventilated	Unventilated	
Show windows	Ventilated	Unventilated	

Note: A check of job condition by means of thermometers might be necessary.

Source: NFPA 13, 2002, Table 8.3.2.5(b).

TABLE 7.3

Ratings of Sprinklers in Specified Residential Areas

Heat Source	Minimum Distance from Edge of Source to Ordinary-Temperature Sprinkler		Minimum Distance from Edge of Source to Intermediate-Temperature Sprinkler	
	in.	mm	in.	mm
Side of open or recessed fireplace	36	914	12	305
Front of recessed fireplace	60	1524	36	914
Coal- or wood-burning stove	42	1067	12	305
Kitchen range	18	457	9	229
Wall oven	18	457	9	229
Hot air flues	18	457	9	229
Uninsulated heat ducts	18	457	9	229
Uninsulated hot water pipes	12	305	6	152
Side of ceiling- or wall-mounted hot air diffusers	24	607	12	305
Front of wall-mounted hot air diffusers	36	914	18	457
Hot water heater or furnace	6	152	3	76
Light fixture: 0 W–250 W	6	152	3	76
250 W–499 W	12	305	6	152

Source: NFPA 13, 2002, Table 8.3.2.5(c).

Intermediate-Level Sprinklers

Intermediate-level sprinklers are standard spray sprinklers with a flat metal plate either attached to the deflector or installed around the threaded base of the sprinkler (see Figure 7.13). The sprinklers with the plate attached to the deflector are to be installed in an upright position. The sprinklers with a plate near the base are designed to be installed in a pendent position.

The typical application of intermediate sprinklers is within racks. When sprinklers are installed vertically, the sprinklers at lower elevations are subject to being wetted by falling water from the sprinklers above, resulting in a cooling of the fusible or frangible element. This condition is commonly referred to as "cold soldering." The circular metal plate on the intermediate-level sprinkler prevents the water falling from above impacting the heat-sensitive element on the sprinkler. This results in faster activation of the sprinkler with less activation skipping during the earlier stages of a fire.

Intermediate-level sprinklers come in various orifice sizes, in upright style, pendent style, with or without protective metal cages, and with standard or quick-response elements.

NFPA 13 provides design requirements for in-rack sprinkler system design and requirements. In selecting in-rack intermediate-level sprinklers, the designer must be cognizant of the potential for physical damage to the sprinklers, typically located in the rack flue spaces. Consideration has to be given to using either upright or pendent-style intermediate-level sprinklers. With the upright style, relatively little impact force to the water shielding plate is necessary to warp the sprinkler frame and potentially cause a leak or activation. On the other hand, a pendent-style intermediate-level sprinkler installed so that its deflector clears the bottom of a horizontal rack member may lend itself to becoming an easy "target." An evaluation of the racks and operations needs to be made before deciding on the style of sprinkler. In practice, the use of metal cages on the intermediate-level sprinklers in racks would be almost universally recommended.

Another application of the intermediate-level sprinkler is as a pilot sprinkler on a deluge or preaction-type sprinkler system. Some designers have mistakenly likened the water shield to a "heat collector" when used as a pilot sprinkler. To be effective as a pilot sprinkler, the installation rules in NFPA for standard sprinklers should be observed. For some high-challenge fires that can be expected to produce extreme heat and radiant energy, the sprinklers could be installed in open areas in close proximity to the potential anticipated fire. Examples include large utility transformers with thousands of gallons of combustible oil or a compressed flammable gas storage cylinder.

FIGURE 7.13
Upright-Style, Intermediate-Level TY-B Sprinkler.

(Courtesy of Tyco Fire & Building Products)

Other Types of Sprinklers

Other types of sprinklers include the following:

- Dry pendent and dry sidewall sprinklers
- Foam-water sprinklers
- Old-style sprinklers
- Vertical sidewall sprinklers

FIGURE 7.14
Viking Model E Dry Pendent
Sprinkler.

(Source: *Automatic Sprinkler Systems
Handbook*, NFPA, 2002, Exhibit 3.34)

Dry Pendent, Dry Upright, and Dry Sidewall Sprinklers

Dry sprinklers come in several configurations that allow installations in not only the normal upright and pendent positions but also in a sidewall orientation; see Figure 7.14. These sprinklers have a number of applications, including protecting coolers and freezers from a heated space outside the chilled area, protection of outside loading docks, and protection of attic spaces. The sprinklers come with a variety of orifice sizes, temperature ratings, and finishes (including corrosion-resistant). Dry pendent sprinklers are also used for the protection of canopies, as seen in Figure 7.15.

The dry sprinkler consists of a threaded barrel with a plug near the threaded connection and a sprinkler installed at the other end. The barrel is filled with pressurized nitrogen to prevent water from entering and filling the barrel. When the sprinkler fuses, the internal pressure is released and the plug holding water back is dislodged. The barrel then fills and the sprinkler operates as would any normal sprinkler. These sprinklers are good for a single use and must be replaced after an activation.

Dry sprinklers must be installed to ensure that colder temperatures cannot provide exposure back to the water-filled pipe through convection, or negatively affect heat loss through conduction. The penetration between the colder area and the warmer space must be well sealed to prevent air movement between the spaces. Gaskets, silicone caulk, and other arrangements, including rubber "boots," have been used successfully. Dry sprinklers are ordered cut to a specific length. Manufacturers typically make a variety of standard lengths and will also custom make special-order lengths. The sprinkler K-factor for the heads changes with the length of the barrel, so the length of the device can affect the hydraulic calculations.

There are installation considerations with dry sprinklers. They cannot be piped to a 1-in. elbow. Insertion of the dry sprinkler into an elbow can dislodge the seated plug on the threaded connection end of the assembly. Use of a tee will eliminate this potential problem. The length of the dry sprinkler barrel must be increased for extremely low-temperature freezers and in areas with low winter temperatures. For situations where there are temperatures at or near freezing, a 1-ft-long barrel may be adequate to protect against heat loss conduction of the warm temperature from the wet portion of the system. With lower anticipated temperatures, longer barrels may be required. The product manufacturer should be consulted to determine the minimum needed lengths.

FIGURE 7.15
Dry-Pendent Sprinklers for
Protection of Covered
Platforms, Loading Docks,
and Similar Areas.

(Source: NFPA 13, 2002, Figure
A.8.14.7)

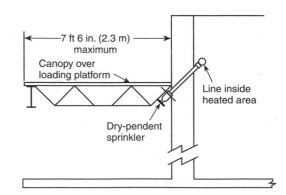

Air-Aspirating Foam-Water Sprinklers

Air-aspirating foam-water sprinklers are of an open head design for use in deluge foam-water sprinkler systems. A typical 1 to 6 percent foam-water solution is discharged from these sprinklers to distribute low expansion foam. Standard spray sprinklers can be used to aspirate foam-water solutions, but the air-aspirated foam-water sprinkler is able to generate a thicker foam blanket that will increase the burn back time, especially when protecting flammable liquid pool fires. Although these sprinklers are not common, they may be more practical than a high-expansion foam system due to the lower costs of the foam-water sprinkler systems.

The design of sprinkler systems using foam water as the extinguishing agent is covered by NFPA 16, *Standard for the Installation of Foam-Water Sprinkler and Foam-Water Spray Systems*. For an illustration of foam water sprinklers, see Figure 7.16.

Old-Style Sprinklers

The old-style sprinkler was used to protect sprinklered occupancies until the early 1950s. Its replacement, the standard spray sprinkler, is still in use today. The old-style sprinkler is still manufactured since it achieves a different discharge pattern than a modern spray sprinkler. In fact, outside the United

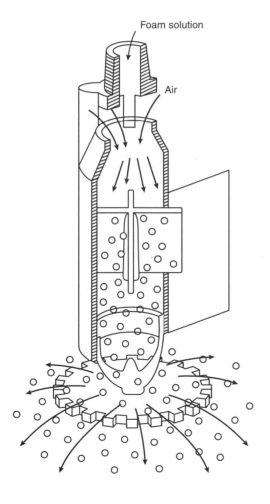

Foam solution

Air

FIGURE 7.16
Air-Aspirating Foam Sprinkler.

(Source: *Fire Protection Handbook*, NFPA, 2003, Figure 11.5.30)

States, the old-style sprinkler is still widely used under combustible ceilings, piers, and wharves, and its use may be appropriate in the U.S. as well.

An old-style sprinkler in both its pendent and upright versions will disperse about 40 percent of the discharged water up and 60 percent of the flow down. The exact percentage of discharge by direction varies by the specific sprinkler. Although this distribution was determined to be generally less effective for protection of commodities at the floor than that of the modern spray sprinkler, which directs nearly all water down, certain applications may need this up-and-down water distribution. The dual-direction spray reduces the capability of the sprinkler to spray horizontally and necessitates a reduced spacing.

In situations where wetting of a combustible ceiling as well as floor protection would be desired, an old-style sprinkler might be appropriate. The designer should take into account the reduction in delivered density directed at a fire on the floor in the spray pattern. Compensation for this reduced water may be desirable from a design standpoint. Studies have shown that the modern spray sprinkler is considered as being capable of providing good cooling below combustible ceilings. In cases where accumulations of lint or sawdust might promote a fast-burning fire across a ceiling, the use of old-style sprinklers could make good engineering sense.

In an older building that contains existing old-style sprinklers, it would not be appropriate to mix them with new spray sprinklers in any one fire area. Due to the likely advanced age of old-style sprinklers, the need to update the system with new spray sprinklers should be evaluated, keeping in mind the original performance objective of ceiling and floor protection.

Upright and Pendent Sidewall Sprinklers

Upright and pendent sidewall sprinklers are designed for installation in the upright or pendent position and work like a standard sidewall sprinkler. Water delivery from this type of sidewall sprinkler is in a 180-degree arc in the direction of projection. These sprinklers are installed against a wall/ceiling interface. The deflector distance below the ceiling needs to be within the listed distance of the specific sprinkler.

The upright or pendent sidewall sprinkler is often used as a retrofit sprinkler with exposed piping running down a hallway or narrow room. Use of a sidewall in an upright position provides better physical protection for the head than a standard sidewall sprinkler, since the pipe is located below the sprinkler.

Summary

In response to industry needs and the variety of field conditions, sprinkler manufacturers have developed numerous solutions beyond the standard upright and pendent sprinklers. The number of types and configurations grew exponentially through the 1980s and 1990s, and we now have useful tools for unique situations. It is important for a sprinkler system designer to keep current with not only the NFPA codes and standards, but also with changes in technology.

BIBLIOGRAPHY

NFPA Codes, Standards, and Recommended Practices

NFPA Publications National Fire Protection Association, 1 Batterymarch Park, Quincy, MA 02169-7471

The following is a list of NFPA codes, standards, and recommended practices cited in this chapter. See the latest version of the *NFPA Catalog* for availability of current editions of these documents.

NFPA 13, *Standard for the Installation of Sprinkler Systems*
NFPA 15, *Standard for Water Spray Fixed Systems for Fire Protection*
NFPA 16, *Standard for the Installation of Foam-Water Sprinkler and Foam-Water Spray Systems*
NFPA 25, *Standard for the Inspection, Testing, and Maintenance of Water-Based Fire Protection Systems*
NFPA 750, *Standard on Water Mist Fire Protection Systems*

Design Implementation

8

Overview of Sprinkler System Layout

Al Moore, PE

The intent of this chapter is to make the designer aware of sprinkler system zoning and layout information that should be included in the engineering contract documentation. Many design considerations for installing sprinklers inside a building may or may not be indicated in the NFPA standards but may need to be included in the engineering documentation. This chapter also points out items that sprinkler contractors consider when determining the layout of mains, branchlines, and sprinklers with regard to the building structure and architecture. The designer will be better able to prepare engineering documents through an understanding of the principles involved with developing the sprinkler contractor's shop drawing layout. Fire protection engineering is a discipline that requires the designer to understand and coordinate with the work of all other trades. Civil, architectural, structural, mechanical, and electrical designs all have an impact on sprinkler system design. Sprinkler systems cannot be designed without basic knowledge of these disciplines.

SFPE Position Statement

In 1998, the Society of Fire Protection Engineers (SFPE) developed a position statement entitled "The Engineer and the Technician: Designing Fire Protection Systems" [1]. The position statement describes the roles and responsibilities of engineers and technicians when designing and laying out fire protection systems, including sprinkler systems. A discussion of the position statement can be found in Chapter 19, "Professional Issues in Fire Protection Engineering Design." In discussing sprinkler system layout, the position statement notes that the engineer or the technician develops the layout of the system based upon the design concept.

Review of Applicable Information

During the early design portion of the project, the designer should review available information, such as preliminary contract documents and information from meetings with the building owner to get a feel for the use of each space within the building and for the use of areas around the outside of the building. The intent of this phase of the design process is to determine what sprinkler system design criteria are appropriate for each area of the building and whether additional or alternative fire protection analyses may be required. During the review of the available information, the designer should be making determinations relative to the occupancy classifications for each area and room of the building. It is especially important to identify areas that are proposed to contain a higher fuel load or be considered hazardous areas, such as paint spray booths, kitchen equipment, hazardous storage rooms, high bay storage, hazardous production areas, and fuel oil tank storage rooms.

Owner Information

The owner or personnel who will use the building can provide some of the most useful information with regard to determining what the appropriate sprinkler system design criteria will be. The owner knows what items will be stored in each area, including the types of materials, equipment, and processes that will be used. Preparing a list of pertinent questions relative to the use of the building will help the designer understand the situation better. The questions will vary greatly depending on the use of the building and should cover the types of materials, equipment, processes, and storage involved. It is especially important to develop an understanding of potential future changes in use that the building may encounter when tenancy of the building changes. A designer should consider an approach that permits flexibility for future changes in the building that are foreseeable or probable so that the sprinkler system will not require replacement or expensive upgrades when changes occur.

Owner's Existing Facilities. Another resource may be an on-site review of another building of similar construction and proposed use. The owner of a building to be built may have other existing buildings that can be visited. Even though the building is not likely to be identical, it may provide the designer with a better feel for the fire hazards that exist.

Material Safety Data Sheets. The owner may be able to supply Material Safety Data Sheets (MSDS) for each material proposed for use in the building. Material Safety Data Sheets contain specifications for a material that provide information about its composition, physical and chemical properties, health and safety hazards, emergency response, and waste disposal. They also contain information about appropriate fire protection. For example, an MSDS for isopropyl alcohol might indicate that an alcohol-type foam, carbon dioxide, or dry chemical should be used as the extinguishing media. The Internet is a good source for MSDS information on a variety of materials and chemicals. By typing in the name of the chemical followed by MSDS in an Internet browser, a number of sources for MSDSs can usually be found.

Preliminary Construction Documents

The designer needs to gather and review information about the building and systems to be installed in the building. This should include a review of the architectural documents and contract documentation from other engineering disciplines, including civil, structural, mechanical, and electrical. As additional information becomes available, the designer should continue to review it for changes as the construction documents progress to completion. It is not unusual for a building to change dramatically during the design process due to budget constraints or changes in the owner's vision of the use of the building.

Applicable Codes and Standards

The designer must correctly identify and apply the codes and standards adopted by the jurisdiction where the building is being built. These include but are not limited to national, state, and local code requirements, city policies, and insurance underwriter requirements. It is also important to know the applicable edition of the building and fire codes that have been adopted by the jurisdiction and if there are any local amendments to the adopted codes. Building codes generally adopt certain editions of standards from standards-developing organizations (SDOs) such as ANSI, ASTM, and NFPA. The possibility exists that not all volumes of a particular standard have been adopted. A local government may have its own set of building and fire codes and not use a nationally published building code.

Even in cases where NFPA 13, *Standard for the Installation of Sprinkler Systems*, is adopted, the NFPA 13 requirements may not apply if other provisions are specified in the building and fire codes. An example is the difference between NFPA 13D, *Standard for the Installation of Sprinkler Systems in One- and Two-Family Dwellings and Manufactured Homes*, and NFPA *101®*, *Life Safety Code®*, on the requirements for eliminating sprinklers in closets (sprinklers are required in closets in excess of 24 ft² minimum for NFPA 13D, and 12 ft² for NFPA *101*). The building code or fire code requirements will usually take precedence over information in an adopted standard.

Also, it is important to review whether the local government adopted the latest edition of NFPA 13 or a previous edition. Many designers and sprinkler contractors will simply use the latest edition of NFPA 13. This can cause problems if the requirements in the adopted edition of NFPA 13 are different from the latest edition. The designer should indicate which editions of national, state, and local codes, standards, city policies, and insurance underwriter requirements apply to the project. Generally, this should be stated in the project specifications.

Fire Protection Resources

There are many books—such as NFPA's *Fire Protection Handbook* [2], *The SFPE Handbook of Fire Protection Engineering* [3], the *Automatic Sprinkler Systems Handbook* [4], and NFPA 13—that provide requirements or guidance for sprinkler system design criteria once a hazard has been identified. Insurance company guidelines from FM Global or Industrial Risk Insurers, such as *FM Global Loss Prevention Guidelines* [5], and FM Global data sheets have a substantial amount of information that is a good resource whether a building is insured by FM Global or not. Insurance companies may also have fire loss data for a variety of buildings. There are also several reference books published by NFPA to help identify hazardous materials, such as *Building and Fire Code Classification of Hazardous Materials* [6] and *Fire Protection Guide to*

Hazardous Materials [7]. The Internet is also a valuable source for information. FIRESPRINKLER.org, NFSA.org, and SFPE.org, and the web site of your local SFPE chapter are a few sites that include several links to other fire protection related sites.

At all times, the designer needs to keep in mind that the applicable codes and standards indicate minimum requirements. Additional fire protection can and should be specified if the designer determines that a particular hazard is not adequately protected.

Items To Be Included on Sprinkler System Contract Documents

Sprinkler Systems Design Criteria

Designers should perform a hazard analysis of the building and determine the appropriate sprinkler system design criteria for each area and room of the building. The design criteria should indicate the sprinkler system occupancy classification; what type of sprinkler system is to be used for each area such as wet, dry, deluge, or preaction; design density and design area of operation to be calculated; maximum spacing per sprinkler; and what type of sprinkler is to be used, including type, style, response, finish, and K-factor. The size and location of each sprinkler zone should also be indicated.

Fire Protection System Infrastructure

Designers should show the location of pertinent fire protection system infrastructure items on the drawings. This includes the location of the fire protection system underground piping and lead-ins, water tanks, fire pumps, standpipes and standpipe distribution mains, fire department connections, sprinkler system control valves, drain provisions, and other information necessary to describe the fire protection system water supply to the sprinklers. Special alternative suppression and local application systems such as kitchen wet chemical, CO_2, and clean agent fire suppression systems should also be indicated on the documents. The water supply source, including available pressures and flow rates, should be indicated on the drawings.

Sprinkler System Zoning

Sprinkler zones should be determined and indicated on the designer's drawings. A sprinkler zone consists of a defined area protected by a single sprinkler system. One way to indicate sprinkler system zone boundaries is to show a heavy dashed line around the area of the building to be included in each sprinkler zone. See Figure 8.1 for an example of how sprinkler zones can be indicated. Sprinkler zones need to be carefully coordinated with HVAC smoke control zones, fire alarm zones, and building fire separations. The sprinkler zones may or may not need to match the exact boundaries of other system zones depending on the interaction between systems and the zone sizes.

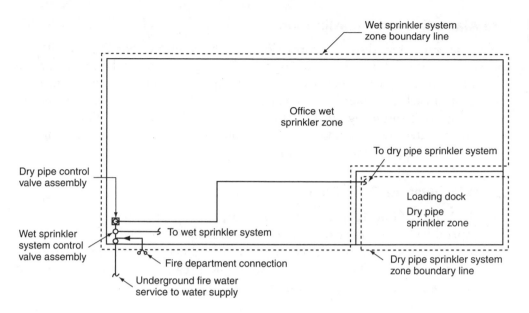

FIGURE 8.1
Sprinkler System Zones.

Labeling of Sprinkler Zones

Each sprinkler zone should be labeled by a letter or number, such as Zone A or Zone B, or labeled by the primary use of the area, such as Office, Production, or Storage. A sprinkler zone is designed such that each sprinkler zone is controlled by a separate control valve and water flow switch. Each sprinkler control valve that can shut off the water supply to a sprinkler zone at the riser assembly is required to have a sign or graphic indicator to clearly indicate the zone the control valve is serving; see Figure 8.2, which shows two wet sprinkler system control valve assemblies with graphic diagrams indicating the area of the building that each sprinkler system serves.

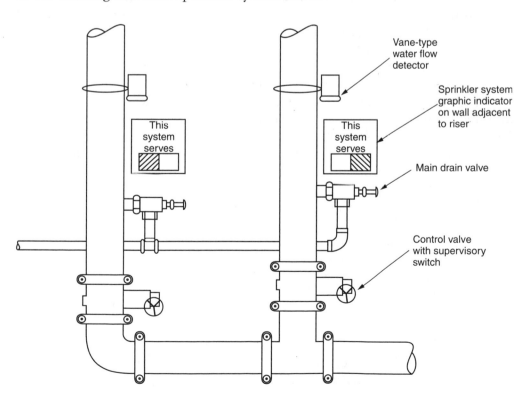

FIGURE 8.2
Sprinkler System Risers with Sprinkler System Graphic Indicators.

Fire Alarm System Coordination

The fire alarm system display information must be carefully coordinated with the sprinkler zones established by the engineer. A diagram should be placed at the main fire alarm control panel and also at the sprinkler control valve indicating the extent of each sprinkler zone. The text description of a sprinkler zone alarm, supervisory or trouble signal that is displayed on the main fire alarm control panel must match the description of the sprinkler zone indicated on the diagram located at the sprinkler system shutoff valve.

Sprinkler System Zone Size

Per NFPA 13, 52,000 ft^2 is the maximum area allowed to be controlled by a single sprinkler system riser on any one floor for light and ordinary hazard occupancies. In hydraulically calculated extra hazard occupancies and high-piled storage occupancies, 40,000 ft^2 is the maximum area allowed to be controlled by a single sprinkler system riser on any one floor.

Zoning Strategies. With up to 52,000 ft^2 allowed per floor, a single sprinkler system riser could protect a two-story or higher light/ordinary hazard occupancy building with up to 52,000 ft^2 on each of the floors. Depending on the use of the building, the designer may recommend that the sprinkler system be divided into two or more zones per floor, even though it is not required by code. When sprinkler contractors are bidding the project, and no direction is given in the bid documents as to the number of zones to install, they will choose the least-expensive method so they have a chance of being awarded the job. The least-expensive method of designating the entire building as one sprinkler zone may not be in the best interest of the owner. The building may have a use that is primarily for small office tenants, and each floor of the building may be divided into many parts. Each time remodeling occurs in one of the tenant areas, the entire building would have to be drained and refilled to make the sprinkler modifications.

If a building is to be protected with a wet sprinkler system, there are a few points to consider with regard to the zoning of the system inside the building. A wet sprinkler system is defined by NFPA 13 as a sprinkler system employing automatic sprinklers attached to a piping system containing water and connected to a water supply so that water discharges immediately from sprinklers opened by heat from a fire. First, if the building has a single wet sprinkler system serving multiple floors, the draining and refilling of the entire building sprinkler system may cause more corrosion in the piping over time when compared to draining only a single floor or a portion of the floor, and second, the entire building is left without fire protection when modifications are made to one floor. Third, there is a waste of water when refilling the entire sprinkler system each time in lieu of a floor or portion of a floor. In the case of a two-story building, the owner may not be willing to pay the premium for the additional valves and monitoring devices associated with additional zones. But what if the building is three stories, four stories, or higher? The owner may find the convenience of only having to drain individual floors an advantage; see Figure 8.3, which shows a schematic of a building sprinkler riser with a separate sprinkler system control valve assembly on each floor. Building codes are likely to require separate sprinkler zones with sprinkler system control

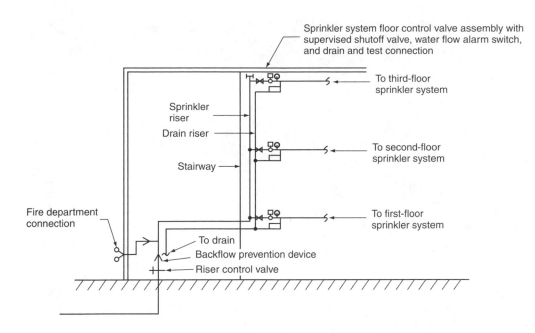

FIGURE 8.3
Wet Sprinkler Riser Diagram with Isolation Valves at Each Floor.

valves on each floor of the building if the building is considered a high-rise. The building code should be reviewed by the designer to determine when separate floor control valves are required. A wet supply riser with a sprinkler control valve assembly on each floor will provide more flexibility for draining and refilling of sprinkler zones without impacting other areas of the building. The designer should review the design and pricing for each option with the owner.

Mezzanines. Mezzanines are common in warehouse buildings and other occupancies with high ceiling heights. The floor area occupied by mezzanines is not to be included in the allowable square footage per floor per subsection 8.2.2 in NFPA 13, 2002 edition. Figures 8.4 and 8.5 show mezzanine areas.

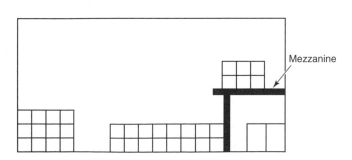

FIGURE 8.4
Mezzanine Used for Storage.

(Source: *Automatic Sprinkler Systems Handbook*, NFPA, 2002, Exhibit 8.3)

FIGURE 8.5
Mezzanine Storage Area in Warehouse.

(Source: *Automatic Sprinkler Systems Handbook*, NFPA, 2002, Exhibit 8.4)

FIGURE 8.6
Interstitial Space with Catwalk.

Interstitial Spaces. An interstitial space is a space above a floor area that usually has a floor or barrier above the suspended ceiling. They are like a separate floor, not used for occupancy but exclusively for installation of, and access to, mechanical and electrical equipment. An interstitial space may include catwalks or other walking platforms to allow access to complicated mechanical and electrical equipment and systems located above the suspended ceiling. Interstitial spaces are found in occupancies such as hospitals and clean rooms and in specialized production such as medical device manufacturing. The photograph in Figure 8.6 was taken from the interstitial space above the ceiling of a medical device company production area.

The requirement for sprinklers in an interstitial space must be reviewed with regard to whether the construction and occupancy of the space is noncombustible. Where a significant amount of electrical and mechanical equipment with motors and control panels are placed in these spaces, the local AHJ may consider them an area similar to a mechanical room and require sprinkler protection. Per Section 13.23 of NFPA 13, 2002 edition, sprinklers are required in interstitial spaces and plenums located above clean rooms. It does not address whether sprinklers protecting the interstitial space are to be on a separate sprinkler system zone.

Smoke Control Zones. Another consideration when determining sprinkler zones is the coordination of sprinkler zones with smoke-control zones. A smoke-control zone is defined by NFPA 92A, *Recommended Practice for Smoke-Control Systems*, as a space within a building enclosed by smoke barriers, including the top and bottom, which is part of a zoned smoke-control system. A smoke-control system is defined by NFPA 92A as an engineered system that uses mechanical fans to produce pressure differences across smoke barriers to inhibit smoke movement. This is an important consideration where it is a requirement that sprinkler system water flow activate mechanical smoke exhaust

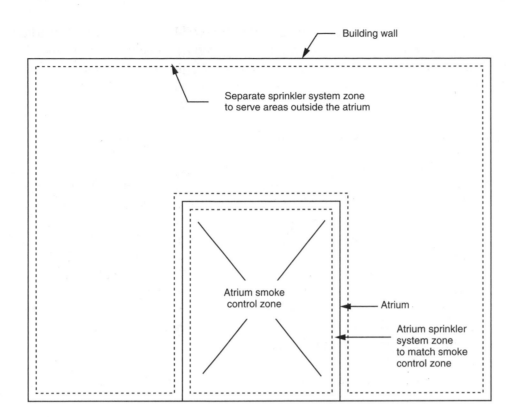

FIGURE 8.7
Atrium Sprinkler Zones
Coordinated with Atrium
Smoke Control Zone.

upon a signal from the sprinkler system water flow alarm switch. If the two systems are improperly aligned, a sprinkler activating outside the smoke-control zone would erroneously activate the mechanical smoke exhaust system.

Atrium Zones. Atriums may need to have a separate sprinkler zone to activate the atrium smoke exhaust system. Figure 8.7 shows sprinkler zones coordinated with the atrium smoke zone.

Smoke Compartments. Smoke compartments are created in some building occupancies to limit the spread of smoke in the building. This is especially important in buildings where the occupants may not be able to evacuate themselves, such as in a hospital or prison. Where an automatic smoke-control system is not being used in conjunction with the smoke compartments, there may not be any advantage to making the sprinkler system zones match the smoke compartment boundaries. Smoke compartments are usually smaller than the allowable size limit for sprinkler zones. Matching the sprinkler zones to the smoke compartments will generally require more sprinkler zones and sprinkler control valve assemblies.

Dry Pipe Systems. Dry pipe sprinkler systems are defined by NFPA 13 as sprinkler systems employing automatic sprinklers that are attached to a piping system containing air or nitrogen under pressure, the release of which (as from the opening of a sprinkler) permits the water pressure to open a valve known as a dry pipe valve, and the water then flows into the piping system and out the opened sprinklers.

   Dry pipe sprinkler system zones may protect building areas up to the same size as wet sprinkler systems; however, dry pipe systems have additional criteria with regard to volumetric capacity and the time required to deliver water to the inspector's test connection once the test connection valve is opened fully. The time it takes to trip the valve, push the air out of the piping, and deliver water to the inspector's test connection can have a significant impact on the size of a dry pipe sprinkler system zone.

Dry System Water Delivery. Depending on the volumetric capacity of the piping in a system, and whether a quick-opening device is used, a system may be required to deliver water to an inspector's test connection within 60 sec. Per NFPA 13, the 60-sec limit does not apply to dry systems with capacities of 500 gal or less, or to dry systems with capacities of 750 gal or less if equipped with a quick-opening device such as an accelerator. Figure 8.8 shows a dry-pipe valve with an accelerator. NFPA 13, Table A.7.2.3 (reproduced here as Table 8.1), indicates volumetric capacities for 1 ft of pipe for common sizes of Schedule 10 and Schedule 40 piping. Where dry pipe systems exceed 750-gal capacity, they must be able to deliver water to the system's test connection in not more than 60 sec, starting at normal air pressure on the system and at the time of a fully opened inspection test connection. If a system does not exceed the 750-gal limit and includes a quick-opening device, the delivery time to the inspector's test connection may take up to 3 min, which is considered acceptable by NFPA 13. See *Automatic Sprinkler Systems Handbook*, commentary to section 7.2.3.3, which also cautions the designer on the risks of delaying water delivery to such an extent.

FIGURE 8.8
Dry Pipe Valve and
Associated Components.

(Source: *Automatic Sprinkler Systems Handbook*, NFPA, 2002, Exhibit 7.3)

TABLE 8.1

Capacity of 1 ft of Pipe (Based on Actual Internal Pipe Diameter)

Nominal Pipe Diameter (in.)	Pipe		Nominal Pipe Diameter (in.)	Pipe	
	Schedule 40 (gal)	Schedule 10 (gal)		Schedule 40 (gal)	Schedule 10 (gal)
¾	0.028		3	0.383	0.433
1	0.045	0.049	3½	0.513	0.576
1¼	0.078	0.085	4	0.660	0.740
1½	0.106	0.115	5	1.040	1.144
2	0.174	0.190	6	1.501	1.649[b]
2½	0.248	0.283	8	2.66[a]	2.776[c]

For SI units, 1 in. = 25.4 mm; 1 ft = 0.3048 m; 1 gal = 3.785 L.
[a]Schedule 30.
[b]0.134 wall pipe.
[c]0.188 wall pipe.

Source: NFPA 13, 2002, Table A.7.2.3.

NFPA 13 permits a procedure for calculating the water delivery time using a listed calculation program. A listed calculation program can predict the trip and transit time (maximum delivery time) for a sprinkler system based upon the anticipated minimum number of the most remote sprinklers initially opened in the early stages of fire growth. See Table 8.2, reproduced from Table 11.2.3.9.1 of NFPA 13. Where the dry pipe system water delivery time is calculated by the method prescribed in NFPA 13, the piping volume limitations may be exceeded.

Dry Pipe System Zone Sizes. Even with fire pump pressure supplying a dry pipe system, a dry sprinkler zone can be restricted to an area of much less than 52,000 ft². It is not unusual to see dry pipe sprinkler zones of 25,000 ft² or less. The predominant factor affecting the size of a dry pipe sprinkler zone is the volumetric capacity of the dry pipe system, which will impact the amount

TABLE 8.2

Dry System Water Delivery

Hazard	Number of Most Remote Sprinklers Initially Open	Maximum Time of Water Delivery
Residential	1	15 seconds
Light	1	60 seconds
Ordinary I	2	50 seconds
Ordinary II	2	50 seconds
Extra I	4	45 seconds
Extra II	4	45 seconds
High piled	4	40 seconds

Source: NFPA 13, 2002, Table 11.2.3.9.1.

of time it takes to deliver water to the inspector's test connection, because the pressurized air in the piping system must escape before water can flow to the remote point of the system. The volumetric capacity of the piping may have to be reduced by limiting the size of the sprinkler zone. Designers should be witnessing dry pipe trip tests to get a feel for what sprinkler zone size is appropriate for various dry pipe system configurations.

The size of a dry pipe sprinkler zone will be greatly affected by the water pressure available, the system piping configuration, the air pressure carried in the system piping, and the particular dry pipe valve utilized. Dry pipe valves with a lower maintained air pressure requirement, such as 10 to 22 psi, will allow the water to reach the inspector's test connection more quickly than a standard dry pipe valve with 10 to 50 psi maintained air pressure. This is because air is compressed in the piping and takes longer to be expelled. Some dry-valve manufacturers do not make low pressure differential valves, so the designer will need to specify this feature. The amount of air pressure required for a dry pipe system varies for each dry pipe valve manufacturer and can be found in the printed manufacturer catalog information for each valve.

Preaction Sprinkler Systems. A preaction sprinkler system is defined by NFPA 13 as a sprinkler system employing automatic sprinklers attached to a piping system containing air that might or might not be under pressure, with a supplemental detection system installed in the same areas as the sprinklers. There are three types of preaction sprinkler systems that are further defined by NFPA 13 as follows: a single interlock system, which admits water to sprinkler piping upon operation of detection devices; a noninterlock system, which admits water to sprinkler piping upon operation of detection devices or automatic sprinklers; a double interlock system, which admits water to sprinkler piping upon operation of both detection devices and automatic sprinklers. Figure 8.9 shows a double interlock preaction valve assembly.

The size of preaction sprinkler system zones depends on the type of preaction system used. Preaction sprinkler systems are limited to not more than 1000 sprinklers being controlled by any one preaction valve. The sprinkler zone size for single and noninterlock preaction systems are not restricted by pipe volume capacity as is a dry system. Single and noninterlock systems are expected to operate more quickly than dry pipe and double interlock systems. Since they do not rely on piping that is pressurized with air, they will fill with water in less time when activated.

Double interlock preaction systems with piping pressurized by air are restricted to piping volumes that are not more than 750 gal. Where the system exceeds a 750-gal capacity, it must be able to deliver water to the system test connection in not more than 60 sec, starting at normal air pressure on the system and at the time an inspection test connection is fully opened. The size of a preaction sprinkler system zone will be greatly affected by the water pressure available, the system piping configuration, the air pressure carried in the system piping, and the particular valve being installed.

Data Centers. Sprinkler systems serving data center and other computer areas are required by NFPA 75, *Standard for the Protection of Information Technology Equipment*, to be a separate sprinkler zone from other areas of the building. Data centers often use preaction sprinkler systems to protect areas with high-value computer equipment.

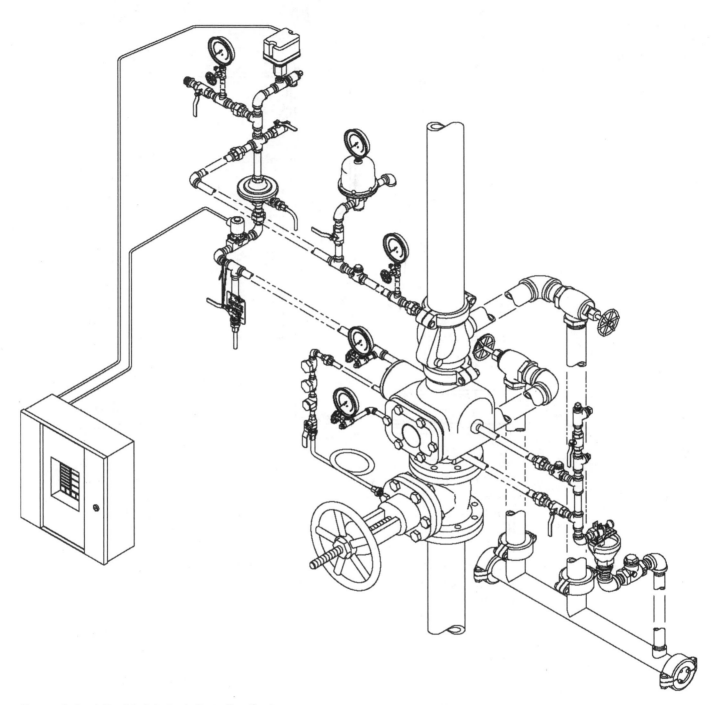

FIGURE 8.9 A Double Interlock Preaction System. (Courtesy of Reliable Automatic Sprinkler Company)

Malls. The model building codes may require that mall retail tenant areas be supplied by a separate sprinkler control valve from the mall common spaces.

Boundary Considerations

A common problem can occur when two sprinkler system zones are adjacent to each other. It is difficult for maintenance personnel to know where one sprinkler system stops and the other system starts if there is not a partition or

Figure 8.10
Sprinkler Zones with
Potentially Confusing
Sprinkler System Boundaries.

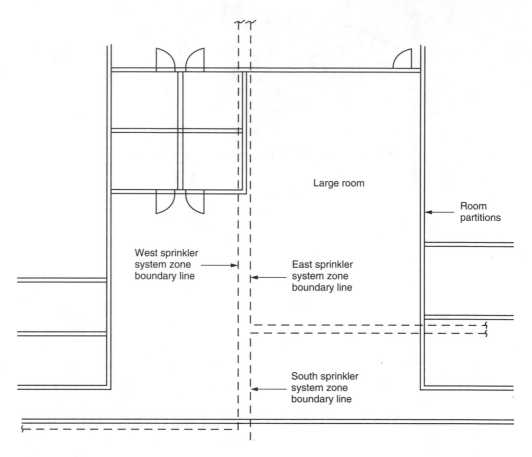

some other easily discernible boundary between the systems. Figure 8.10 indicates sprinkler system boundaries that may cause an accident when field revisions are made to the sprinkler piping in the large room. A contractor may drain down one sprinkler zone to revise the location of sprinklers for some remodeling. If the remodeling is taking place near where two sprinkler zones meet, the contractor may drain one sprinkler system but remove a sprinkler to start remodeling on the other "live" sprinkler zone. This has obvious consequences and happens periodically in buildings where the boundaries of the sprinkler system zones are not known. This is one reason to specify that sprinkler system zone maps be placed at each sprinkler control valve assembly indicating the areas protected by each zone.

Performance-Based Design (PBD) Considerations

Sprinkler zones may need to be coordinated with PBD requirements. For example, a separate sprinkler control zone may be required for "window sprinklers" if they are used as a part of a fire-resistive rating at a glass partition. Also see Chapter 17, "The Performance-Based Design Process and Automatic Sprinkler System Design."

Structural Considerations

The type of structure and the location of structural members will most certainly impact the location and orientation of the sprinkler mains and branchlines. The sprinkler system zone orientation should consider the installation direction of sprinkler mains and branchlines.

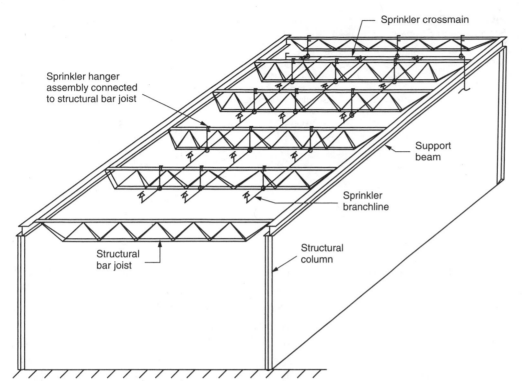

Sprinkler crossmain

Sprinkler hanger
assembly connected
to structural bar joist

Support
beam

Sprinkler
branchline

Structural
bar joist

Structural
column

FIGURE 8.11
Sprinkler Branchlines
Installed Perpendicular to
Structural Bar Joist.

Beam and Bar Joist Construction. When the structure is comprised of steel beams and bar joists, the branchlines should run perpendicular to the bar joists. Mains can be placed directly under a bar joist or hung between two bar joists with a "trapeze" hanger. A top beam clamp is normally used to hang each branchline from the top of the bar joist. Figure 8.11 shows the sprinkler branchlines perpendicular to the bar joist. Branchline piping installed parallel to the bar joist limits the location of the branchline to the location of the bar joist or requires the use of trapeze hangers. Sprinkler locations are expected to be symmetrically located in a fairly precise manner, but bar joists may not be installed precisely where shown on the structural drawings and may not be run precisely parallel to each other. Trying to align sprinkler branchlines parallel to bar joists will make it difficult to prefabricate the sprinkler system piping and may make installation significantly more expensive due to the additional piping required to do the same job.

The structural engineer should be made aware of where large sprinkler mains are to be installed. The structural engineer may need to change the size of structural members (especially with beam and bar joist construction), or specialized hangers that distribute load among more than one structural member may be required where several large sprinkler mains are to be installed in the same area.

Vertical Pipe Penetrations. Sprinkler system standpipes and sprinkler risers that extend from floor to floor typically are located within a stairway. When the designer indicates these risers on the contract documents, the risers need to be coordinated with the structural beams so that the piping does not penetrate steel beams. Figure 8.12 shows a stairway with a beam at the landing where a standpipe might be placed. Also note the required egress clearance in the stairway.

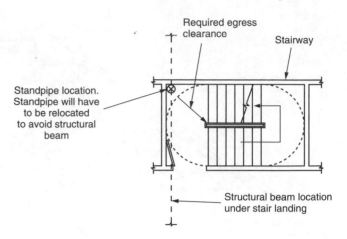

FIGURE 8.12
Standpipe in Stairway
Conflicting with Structural
Member.

The location of the risers is not an item that the contractor should work out in the field. If the sprinkler riser is moved into the stairway further to avoid a beam, the riser may now be inside the required egress clearance of the stairway. These are items that should be resolved prior to contract documents being issued. The riser may need to be located in a shaft with cabinets in the stairway walls or in some other arrangement. If the riser location has to be changed in the field it may have to happen on the day the concrete is being poured, which means decisions may have to be made without a thorough review of the situation. It may also mean a costly change order to the project. If the structure is comprised of posttension concrete members, it is critical that sleeves be set in the concrete prior to the posttension members being installed. This should be noted on the drawings, which should also state that coring of structural members is not allowed without approval from the structural engineer.

Coordination Among Building Trades

The fire protection designer has a responsibility to perform a portion of the coordination that occurs among building trades. The extent of that coordination will change with each type of project, but in general the designer should be trying to help the contractor understand special conditions and areas where conflict might exist between trades. This section indicates several examples of these special conditions and areas. The designer needs to think through a project similar to the way the contractor does when laying out the sprinkler system mains and branchlines. Typically, ductwork and electrical shop drawings are being prepared at the same time as the sprinkler system shop drawings. Therefore, the sprinkler contractor may rely on the designer's layout of ductwork and piping to determine the preliminary location of mains and branchlines.

Preliminary Layout

Many sprinkler contractors will contact the mechanical and electrical contractors just prior to setting main and branchline locations to coordinate the placement of major runs of ductwork, cable trays, and large plumbing and electrical piping with sprinkler piping. The contractor will lay out the sprinklers in such a way as to minimize the number of sprinklers and the amount of piping required to install the system in accordance with the design documents and NFPA 13. But the system layout can not indicate sprinklers in the same location as lights and HVAC diffusers. Where suspended ceilings are to be installed, the architectural reflected ceiling plan and HVAC and lighting layout will have an impact on the location of branchlines.

Lighting

Sprinkler water discharge is affected by obstructions located below a suspended ceiling, such as soffits, cable trays, ductwork, or lighting. It is common to see pendent-hung, chain-suspended, or surface-mounted lighting that can be an obstruction to sprinklers, and the designer must make the contractor aware of these potential obstructions on the drawings. The amount of sprinklers required will most likely increase when the lights create an obstruction to the sprinkler spray pattern. This is also an opportunity for the designer to discuss the lighting with the design team and determine how it affects the overall cost of the project. If pendent-hung lights installed close to a suspended ceiling will obstruct the sprinklers, maybe the lights can be lowered in a way not to obstruct the sprinklers. See Figure 8.13.

FIGURE 8.13
Pendent-Hung Lighting Installed Close to Suspended Ceiling and Automatic Fire Sprinklers.

Special Situations

Storage Warehouses. In areas where high-piled or rack storage are to be located, consider specifying that the sprinkler piping mains be located above the bottom of bar joists to protect the system and provide adequate clearance for product loading into the rack units.

Congested Areas. In congested areas of a building where there is marginally adequate space to install sprinkler piping with other building systems, the designer should verify that all trades can fit in the space. If sprinkler piping is required to avoid such areas, this should be noted on the drawings.

Overhead Doors. Overhead garage doors that store horizontally will cause an obstruction to the sprinklers installed at the ceiling. NFPA 13, paragraph 8.4.2(3), indicates that sidewall sprinklers can protect the area below overhead doors. Sidewall sprinklers are to be installed under the door in a location such that when the overhead door is up, sprinkler protection is still provided for the area under the door. See Figure 8.14.

Code Clearances. Building codes require minimum clearances above the finished floor for standard vehicles and handicap vehicles in parking garages. These clearance heights should be verified with the architect. Depending on the structural height available, the minimum clearances required may conflict with the desired location of sprinkler piping. Sprinkler piping may have to be installed in a manner that may cost significantly more if the contractor is not made aware of the clearance requirements.

FIGURE 8.14
Overhead Door Obstructs Sprinkler Discharge When in the Open Position.

(Source: *Automatic Sprinkler Systems Handbook*, NFPA, 2002, Exhibit 8.9)

Building Section

A section of the building indicating the infrastructure of the sprinkler and standpipe systems is a useful tool for the contractor and the designer, especially for high-rise buildings. The diagram can include feed mains, standpipes, drain risers, fire department valves, standpipe shutoff valves, and sprinkler control valves. The diagram can also indicate where pressure-regulating valves are required, where high-pressure standpipes are used, and where pressures are reduced at each sprinkler system floor control valve assembly. Figure 8.3 is a riser diagram for a three-story building.

Mechanical Plenums

Some mechanical rooms have unheated plenums where fresh air enters through louvers into mechanical units. If these plenums are accessible via a small door or access panel such that storage could occur in the plenum, the local AHJ may require that these areas be sprinklered. During an inspection of a new building, a code official found a gas barbeque grill in one of these plenums. In another case, where the mechanical room was near the ground floor, a small lawn mower and gas can were stored in an air plenum. Granted these items should not be in these plenums for air quality reasons, but the point is that these areas can end up being used for storage regardless of intent. The designer needs to consider how a space could be used regardless of the intended use. Plenum areas may also be subject to freezing temperatures, so the designer should indicate whether a dry pipe system, or more simply, dry pendent-type sprinklers are to be installed.

Typical Sprinkler System Layouts

The designer must understand the sprinkler system main and branchline configurations available to the contractor. This will also help with the determination of sprinkler zones. The type of sprinkler zone configuration chosen by the designer will affect the type of system that the contractor will install. Designers are encouraged to provide piping details on sprinkler contract documents. Designers should include piping details to demonstrate the intent of the system, to provide results of coordination, to confirm placement of piping to match aesthetic concerns, and to demonstrate zoning intent. There are three primary types of sprinkler configurations: tree, loop, and gridded. There are also systems that will combine two or more of these configurations.

Tree System

A "tree" system is the simplest layout using the least amount of sprinkler main piping to supply the sprinkler branchlines. The layout is structured very much like the trunk of a tree, with branches hanging off each side. Because it uses the least amount of main piping, the tree system is usually the least expensive. See Figures 8.15(a) and 8.15(b) for general diagrams of a tree system.

Figure 8.15(a) and (b)
Tree Sprinkler Systems.

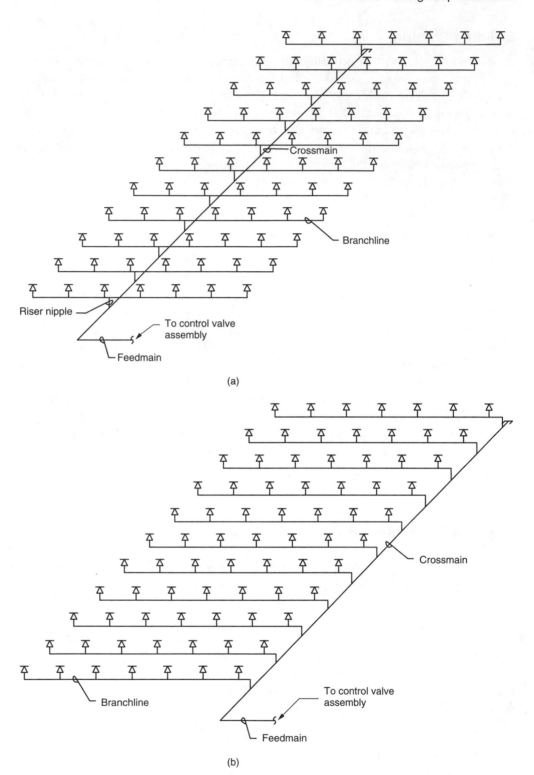

(a)

(b)

Looped System

A looped system allows water that flows through the system to split and feed the branchlines from a two-direction "loop." Supplying water to the branchlines with two loop mains can reduce friction loss in the system. Looped systems provide enhanced distribution of water to the branchlines, with a subsequent reduction in pipe diameter. Looped systems are commonly used in buildings with a

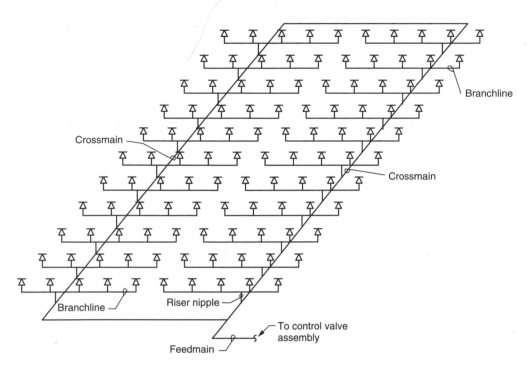

FIGURE 8.16
Looped Sprinkler System.

central elevator core, such as a high-rise. The sprinkler riser for a looped system is the same as for a tree system. After the main feed pipe leaves the riser control valve and reaches the system branchlines, the feed main splits and runs in two directions to form a large loop. Branchlines are connected onto the loop mains in the same manner as in a tree system. See Figure 8.16 for a general diagram of a looped system.

Gridded System

A gridded system is comprised of at least two crossmains, with many branchlines connecting the two mains. The main nearest to the feed main is called the "near main" or "primary main," and the main farthest from the feed main is called the "far main" or "secondary main." Gridded piping configurations are not permitted for the installation of dry pipe and preaction sprinkler systems per NFPA 13. Gridded systems are often used for warehouse wet sprinkler systems because they are efficient at delivering water to large rectangular areas. See Figure 8.17 for a general diagram of a gridded system.

Placement of Mains and Branchlines

The primary consideration with regard to the placement of mains and branchlines is the spacing of the sprinklers. In general, it will take less piping to lay out a system that uses fewer branchlines with more sprinklers on a branchline than to use more branchlines with fewer sprinklers. It follows that it is less expensive to install larger branchlines with more sprinklers than to install many smaller branchlines with fewer sprinklers. See Figures 8.18(a) and 8.18(b). For additional information, Chapter 9, "Sprinkler System Spacing," includes extensive information with regard to the spacing of sprinklers.

The spacing of sprinklers is also affected by obstructions, which vary from floor-mounted obstructions such as high office partitions, demountable walls,

FIGURE 8.17
Gridded Sprinkler System.

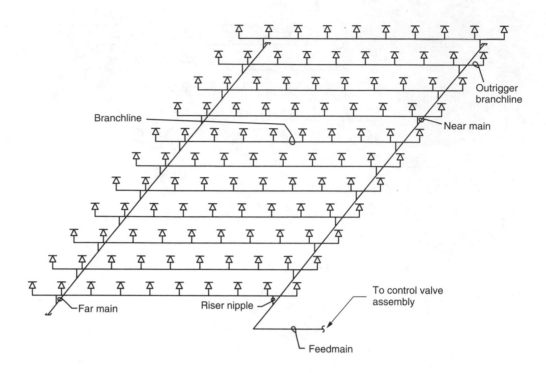

file cabinets, and large art sculptures to ceiling-mounted obstructions such as soffits, lighting, and mechanical and electrical system piping and ductwork. Sloped ceilings will also change the spacing of the sprinklers. With all these considerations for the placement of sprinklers, the placement of branchlines and mains is a custom layout for each and every branchline. Even in warehouses, where one might assume that the layout of sprinklers and branchlines would be consistent from end to end, the layout is impacted by structural members, lights, roof drain piping, and exhaust vents. Add in the variety of standard and extended coverage sprinklers, and it can take a considerable amount of time to lay out the most cost-effective plan.

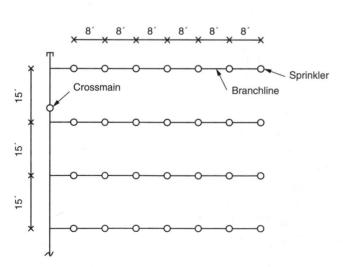

FIGURE 8.18(A)
Layout with Less Piping to Be Installed.

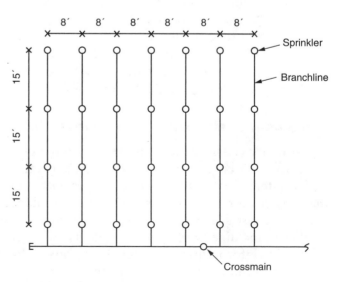

FIGURE 8.18(B)
Layout with More Piping to Be Installed.

Hanger Assemblies

The type of hanger used to support sprinkler piping depends on the type of structure. NFPA 13 has an entire chapter on hanging, bracing, and restraint of system piping. This NFPA chapter covers a wide variety of hanger assemblies for use with concrete, steel, and wood structures, including seismic bracing. At a minimum, the designer's documents should indicate that the sprinkler piping hanger assemblies and structural attachments are to comply with NFPA 13. It is up to the designer to decide if specific types of hanger assemblies are to be used for the project. One of the most common mistakes found in field installations is the use of hangers that are not listed for the way the contractor has installed them.

Requirements for the location of hangers on piping, including the maximum hanger spacing for each size of pipe and the support of vertical piping, are contained in NFPA 13. A thorough review of the NFPA 13 section on hangers will give the designer an understanding of the selection available. The distance between hangers is dependent on the deformation or sagging that would be expected between hangers, so hangers are more closely spaced for smaller-diameter pipe as opposed to larger-diameter pipe and are more closely spaced for plastic pipe than for steel pipe. The maximum distance allowed between hangers is indicated in Table 9.2.2.1 in NFPA 13, reproduced here as Table 8.3. The distance between hangers varies by pipe type and size.

Support Criteria

According to NFPA 13, hanger assemblies shall be designed to support 5 times the weight of the water-filled pipe plus 250 lb at each point of piping support. The building structure must support the added load of the water-filled pipe plus a minimum of 250 lb applied at the point of hanging. The load represents the weight of a sprinkler pipefitter and associated work equipment who grasps the pipe during a mishap.

TABLE 8.3

Maximum Distance Between Hangers (ft-in.)

	Nominal Pipe Size (in.)											
	¾	1	1¼	1½	2	2½	3	3½	4	5	6	8
Steel pipe except threaded lightwall	N/A	12-0	12-0	15-0	15-0	15-0	15-0	15-0	15-0	15-0	15-0	15-0
Threaded lightwall steel pipe	N/A	12-0	12-0	12-0	12-0	12-0	12-0	N/A	N/A	N/A	N/A	N/A
Copper tube	8-0	8-0	10-0	10-0	12-0	12-0	12-0	15-0	15-0	15-0	15-0	15-0
CPVC	5-6	6-0	6-6	7-0	8-0	9-0	10-0	N/A	N/A	N/A	N/A	N/A
Polybutylene (IPS)	N/A	3-9	4-7	5-0	5-11	N/A	N/A	N/A	N/A	N/A	N/A	N/A
Polybutylene (CTS)	2-11	3-4	3-11	4-5	5-5	N/A	N/A	N/A	N/A	N/A	N/A	N/A
Ductile iron pipe	N/A	N/A	N/A	N/A	N/A	N/A	15-0	N/A	15-0	N/A	15-0	15-0

For SI units, 1 in. = 25.4 mm; 1 ft = 0.3048 m.
Note: IPS iron—pipe size; CTS—copper tube size.

Source: NFPA 13, 2002, Table 9.2.2.1.

Engineered Hanger Support

Hanger supports designed by a licensed structural engineer are acceptable per NFPA 13 if five criteria set forth in paragraph 9.1.1.2 of NFPA 13 are met. Those five criteria are as follows:

1. Hangers shall be designed to support five times the weight of the water-filled pipe plus 250 lb (114 kg) at each point of piping support.
2. These points of support shall be adequate to support the system.
3. The spacing between hangers shall not exceed the value given for the type of pipe as indicated in Table 9.2.2.1.
4. Hanger components shall be ferrous.
5. Detailed calculations shall be submitted, when required by the reviewing authority, showing stresses developed in hangers, piping, and fittings and safety factors allowed.

It is not common to have hanger supports designed by a licensed structural engineer because there is a large number of listed hanger assemblies available to fit almost any installation.

Trapeze Hangers

The installation of ductwork, accommodation of sprinkler mains running parallel to two bar joists, or requirements for distribution of sprinkler piping load among more than one structural member may require special trapeze hanger assemblies to provide proper support. See Exhibit 9.2 from NFPA 13, reproduced here as Figure 8.19.

Pressure Over 100 psi

Where maximum pressure in the sprinkler system piping is over 100 psi, NFPA 13, paragraph 9.2.3.4.3, has a few additional requirements with regard to hanging sprinkler piping. The maximum unsupported length of piping between the end sprinkler in a pendent position or drop nipple and the last hanger on the branchline cannot be greater than 12 in. for steel pipe or 6 in.

FIGURE 8.19
Trapeze Installation.

(Source: *Automatic Sprinkler Systems Handbook*, NFPA, 2002, Exhibit 9.2)

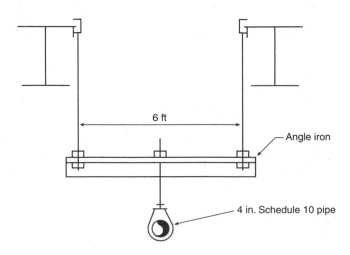

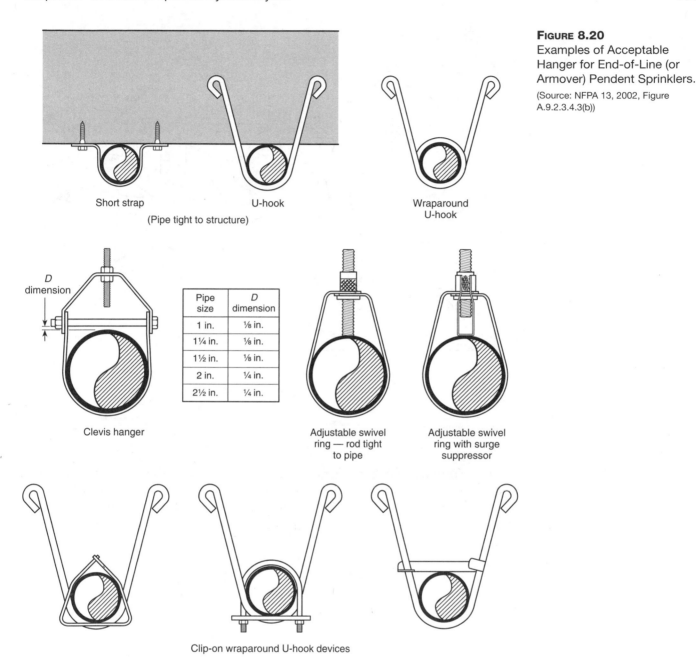

FIGURE 8.20
Examples of Acceptable
Hanger for End-of-Line (or
Armover) Pendent Sprinklers.
(Source: NFPA 13, 2002, Figure
A.9.2.3.4.3(b))

Short strap

U-hook

(Pipe tight to structure)

Wraparound
U-hook

D
dimension

Pipe size	*D* dimension
1 in.	⅛ in.
1¼ in.	⅛ in.
1½ in.	⅛ in.
2 in.	¼ in.
2½ in.	¼ in.

Clevis hanger

Adjustable swivel
ring — rod tight
to pipe

Adjustable swivel
ring with surge
suppressor

Clip-on wraparound U-hook devices

for copper pipe. The hanger at the end sprinkler must also be installed with a hanger that restricts upward movement. See Figure A.9.2.3.4.3(b) from NFPA 13, reproduced here as Figure 8.20. If the maximum pressure expected in the sprinkler system will be over 100 psi, the designer should note this in the portion of the contract documents in which the hanger assemblies are specified.

Seismic Bracing

NFPA 13 has significantly increased the level of detail with respect to seismic design in recent editions. Additional information, if needed, can be obtained by consulting the American Society of Civil Engineers publication, ASCE 7, *Minimum Design Loads for Buildings and Other Structures* [8].

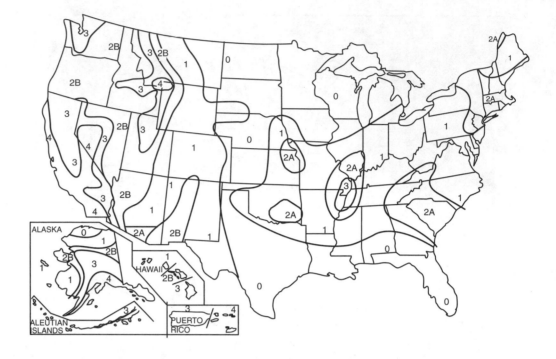

Determining whether seismic bracing is required is the job of the designer, and the requirements should be indicated in the specifications. NFPA 13 has many hanger details for bracing against earthquake forces. The model building codes indicate where seismic protection is required; however, the designer should also review insurance underwriter requirements that may indicate seismic protection requirements. NFPA 13 has examples of seismic zone maps (see Figures A.9.3.1(a) and A.9.3.1(b) from NFPA 13, reproduced here as Figures 8.21 and 8.22) but does not indicate geographically where seismic protection is required.

Drainage

Sprinkler piping has to be installed such that all portions of the piping system can be drained, per NFPA 13. In general, systems that drain back to the riser control valve work the best. This way there is a single location to drain from when maintenance, repairs, or remodeling have to occur. If a system drains back to the riser, the system mains will have few elevation changes in the piping except where the pipe is fed from the sprinkler riser. This is better from both a hydraulic and cost standpoint. It is undesirable but not uncommon to have sprinkler systems with auxiliary drains. This is due to conflicts with structure, ductwork, or ceilings that do not allow the entire system piping to drain back to the riser control valve.

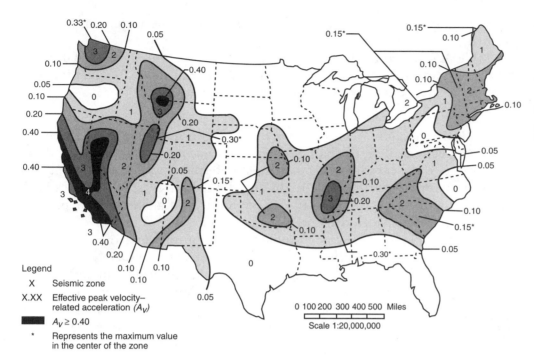

FIGURE 8.22
Map of Seismic Zones and Effective Peak Velocity-Relayed Acceleration (A_v) for the Contiguous 48 States; Linear Interpolation Between Contours Is Acceptable.
(Source: NFPA 13, 2002, Figure A.9.3.1(b))

Legend

X   Seismic zone

X.XX   Effective peak velocity–
related acceleration (A_v)

▮▮▮ $A_v \geq 0.40$

*   Represents the maximum value
in the center of the zone

0 100 200 300 400 500 Miles

Scale 1:20,000,000

Wet Systems and Preaction Systems Not in Areas Subject to Freezing

In wet sprinkler systems and in preaction systems not located in areas subject to freezing, the piping is allowed to be installed level. NFPA 13 has requirements for auxiliary drain valves to be installed where there are trapped sections of pipe. Systems should be designed to facilitate natural drainage and minimize manual drainage of trapped portions of pipe.

Less than 5 Gal. Where the total capacity of the water trapped in sprinkler piping is less than 5 gal, there are three options prescribed in NFPA 13. The 5-gal capacity rule represents the size of the bucket that most sprinkler and maintenance personnel use to drain trapped portions of sprinkler systems. The three options are: an auxiliary drain consisting of a nipple and cap or a plug not less than ½ in. in size, the removal of a single pendent sprinkler, or easily separated flexible couplings such as grooved couplings.

5 Gal up to 50 Gal. Where the total capacity of the water trapped in sprinkler piping is more than 5 gal but less than 50 gal, the auxiliary drain must include a valve at least ¾ in. in size and a plug or a nipple and cap. Designers could specify that a capped ¾-in. hose fitting be installed at the auxiliary drain. That way, maintenance personnel can hook on a garden hose if they want to drain the water to a suitable location.

Over 50 Gal. Where the capacity of trapped sections of piping exceeds 50 gal, an auxiliary drain valve not less than 1 in. in size must be installed. The drain must be piped to an accessible location where the water can be easily drained.

Table A.7.2.3 from NFPA 13 (reproduced previously as Table 8.1) displays the capacity in gallons for 1 ft of Schedule 40 and Schedule 10 piping for sizes up to 8 in. There is a conversion factor given for liters also.

Drain Discharge. NFPA 13 requires that system, main drain, or sectional drain connections shall discharge outside or to a drain connection. A drain discharging to the exterior of a building is preferred from the standpoint that the drain cannot overflow inside the building. However, in climates with freezing conditions, this may cause an ice hazard on a sidewalk or parking area. So what is considered a suitable drain connection? A janitor's floor sink or slop sink will most likely not handle the amount of water that a sprinkler system main drain will discharge with the drain valve completely open. Discharging the main drain pipe to a sump with a suitably sized pump or gravity discharge is another option. In other cases, a sprinkler drain may discharge into an open-fixture waste connection that is tied into the buildings sanitary sewer. The connection must use an air gap, and sprinkler drains cannot be hard-connected to sanitary sewer or storm drain fixtures. In general, at least a 4-in. drain pipe is necessary from the open-fixture waste connection to the sanitary sewer.

The designer must realize that sprinkler system main drains are not only used to gravity drain the water from the system for maintenance, repairs, or remodeling but are also used to verify the residual water pressure at the building from the water supply source. This test is to be performed annually. This involves completely opening the drain valve under pressure until the water pressure stabilizes to determine if the water supply pressure has degraded. Pressure-regulating valves must be tested at full flow rate, which, depending on the valve size, can be a very considerable amount of water flowing under pressure. Where pressure-regulating valves are to be installed, such as in a high-rise situation, the designer needs to verify the flow rate and determine a suitable drain discharge location.

Dry-Pipe and Preaction Systems Installed in Areas Subject to Freezing

It is more difficult to coordinate the location of mains and branchlines for dry-pipe and preaction systems installed in freezing areas because the piping has to be sloped to drain. Mains are to be sloped a minimum of ¼ in. per 10 ft and branchlines are to be sloped a minimum of ½ in. per 10 ft. Sloped piping is more difficult to coordinate since it takes up more space than piping installed level. This is very important, as any trapped water can result in a frozen section of piping that can break. Freezing water can exert pressures within the system piping that will split pipe and high-pressure-rated fittings readily.

Less than 5 Gal. Where the total capacity of the water trapped in sprinkler piping is less than 5 gal, the auxiliary drain must include a valve at least ¾ in. in size and a plug or a nipple and cap.

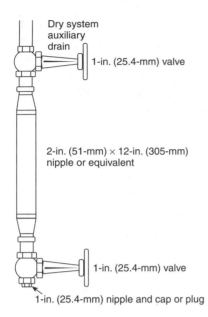

Dry system
auxiliary
drain

1-in. (25.4-mm) valve

2-in. (51-mm) × 12-in. (305-mm)
nipple or equivalent

1-in. (25.4-mm) valve

1-in. (25.4-mm) nipple and cap or plug

FIGURE 8.23
Dry System Auxiliary Drain.
(Source: NFPA 13, 2002,
Figure 8.15.2.5.3.3)

5 Gal or More. Where the total capacity of the water trapped in sprinkler piping is more than 5 gal, the auxiliary drain must consist of what is commonly called a condensate drum drip, drum drip, or condensate drain. A drum drip consists of two 1-in. valves and a 2-in. by 12-in. nipple. See Figure 8.15.2.5.3.3 from NFPA 13, reproduced here as Figure 8.23.

Drum drip auxiliary drains collect condensation that occurs inside the piping during temperature fluctuations and allow for moisture to be drained from system low points prior to freezing temperatures. The valves can be operated such that the drum drip can be drained without losing a significant amount of air pressure from the system. The location of the drum drips should be coordinated to minimize the number of drum drips required. The drum drips must be accessible for maintenance since NFPA 25, *Standard for the Inspection, Testing, and Maintenance of Water-Based Fire Protection Systems*, specifies periodic maintenance schedules for sprinkler systems and components.

Sprinkler Placement

Sprinklers may be considered aesthetically unpleasing and architects may exert pressure on sprinkler designers to arrange sprinklers in symmetric patterns on a suspended ceiling. However, due to sprinkler spacing rules, pressure requirements, and hanger constraints, sprinkler spacing flexibility can be limited. Designers must balance aesthetic concerns with requirements that dictate placement and spacing of sprinklers on reflected ceiling plans, which include locations of soffits, lights, HVAC diffusers, speakers, exit signs, fire alarm devices, and architectural features. There are several factors that the designer should indicate on the contract documents with regard to sprinkler placement.

Lay-in Acoustical Tile Ceiling

When a project has a lay-in acoustical suspended ceiling, the designer should specify whether sprinklers are to be centered in suspended ceiling tiles or be kept a certain distance away from the ceiling grid tees. Centering sprinklers adds expense to a project, since more sprinklers are likely to be required to create centered patterns and since extra fittings are required to install sprinklers in centers of tiles in a prefabricated sprinkler system. Sprinkler escutcheons that overlap a ceiling grid may not be acceptable to the owner from an aesthetic viewpoint. A symmetrical pattern for sprinklers is desirable from an aesthetic viewpoint but is not a code requirement. Sprinkler contractors can easily keep sprinklers in a symmetrical pattern and maintain a certain range of distance away from the ceiling grid tees without having to add additional fittings or "swing joints" that may be necessary to accomplish a center-of-tile installation. For instance, by specifying that the edge of the sprinkler escutcheon be kept at least 3 in. away from the grid tees, the cost of the system will be reduced when compared to sprinklers specified to be in the center of tiles. The installation is also likely to save sprinklers because the contractors will plan to space the sprinklers a little farther apart and will not be restricted to the 2-ft increments typically associated with installing sprinklers in the center of the tiles. The designer should discuss the cost implications of centered versus noncentered sprinklers with the owner. The designer may consider using a concealed sprinkler (as opposed to a semirecessed or pendent) rather than centering the sprinklers. See Figure 8.24 for an example of centered heads.

If the ceiling is plaster or gypsum board and the owner has high expectations for the aesthetics of the space, the designer may want to specify that the sprinklers be aligned with lights, speakers, and other features and devices mounted on the ceiling. In some cases, this may mean more sprinklers than are required by code for coverage. The designer needs to be clear in the directions to the contractor so the contractor will provide additional monies in the bid to install the additional sprinklers and fittings needed to position sprinklers in

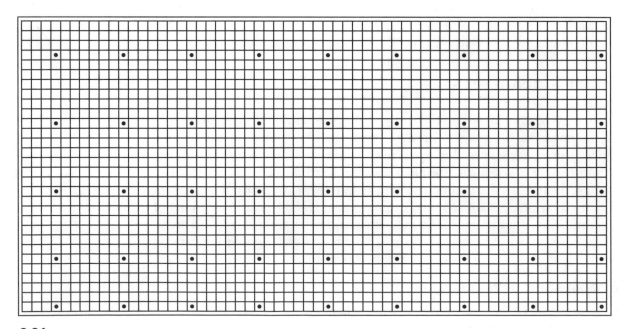

Figure 8.24
2-ft × 2-ft Grid with Sprinkler Centered on 14-ft × 14-ft.

specific locations while still complying with required maximum coverage per sprinkler. A detailed sprinkler layout could be included as a part of the contract documents if there is a complex ceiling in a highly visible area.

Exposed Structure

Where sprinklers are to be installed and an exposed structure exists, there are a few considerations that the designer needs to incorporate into the documents. The contractor must space sprinklers to avoid obstruction by structural members, ductwork, lighting, piping, and partitions.

Shell Space. If the area to be sprinklered is a "shell" space such as an office tenant area that is intended for speculative leasing, the suspended ceilings may not be installed initially. The designer should indicate that the sprinklers be located in the upright position at the exposed structure with 1-in. by ½-in. bushings installed to facilitate 1-in. pipe drops to a future suspended ceiling.

Layout of Sprinklers by Architects

Occasionally, an architect will layout sprinklers for a project to coordinate with the reflected ceiling plan. A qualified sprinkler designer should develop the sprinkler designs and determine sprinkler system performance objectives and criteria.

Small Rooms

There is an exception in NFPA 13 for light hazard occupancies that permits sprinklers to be located up to 9 ft off one wall in "small" rooms, provided that the spacing in all other directions complies with all other spacing rules for light hazard occupancy. See NFPA 13, paragraph 8.6.3.2.4. This is in lieu of the 7-ft 6-in. distance off a wall typically used in light hazard occupancies. The "small room rule," as it is commonly called, is only allowed in rooms less than 800 ft² in area. An important note with regard to using this rule is that if the door to the room does not have a lintel at least 8 in. deep, then the rule cannot be applied. See the definition of small rooms in NFPA 13, subsection 3.3.20. In one case, a contractor was not aware that the doors to tenant offices did not have lintels and that each door opening extended up to the suspended ceiling with only the depth of the door jamb for a lintel. The contractor used the small room rule inappropriately and had to go back and add sprinklers and piping in order to comply with NFPA 13. So what can the designer do to prevent this from happening even though the contractor should have checked the door height? The designer could put a note in the contract documents alerting the contractor to the fact that the small room rule may not be used on this particular project and explain why.

Special Ceiling Finishes and Features

There are many items that impact a sprinkler contractor's bid for a project. The contractors are able to take each of these items into account if they are aware of them. The designer can help reduce conflicts and change orders in the future if special conditions are pointed out that the contractor may not expect to see and may not find in the details of the architect's documents. One example is the installation of cable trays located just below the suspended ceiling in a computer room. The sprinkler contractor may not review the electrical contract documents to find that these cable trays cause obstructions and therefore increase the number of required sprinklers. Sprinklers are required to be coordinated with soffits and ceiling elevation changes. If the ceiling furr down is installed with stone, wood, or other unique surface, determinations must be made with respect to which contractor will be required to provide the hole in the stone ceiling for the sprinkler and whether the sprinkler cover plate needs to match the color of a soffit made of specialized materials. If there is a large piece of artwork hanging from the ceiling, sprinklers may need to be installed under the artwork if it creates an obstruction.

Historic Buildings. Where buildings are of a historic nature and existing finishes cannot be disturbed, placement of sprinklers and piping must be carefully detailed on the contract documents.

Skylights. NFPA 13, subsection 8.5.7, states that sprinklers are not required in skylights and similar ceiling pockets not exceeding 32 ft^2 (3 m^2) in area, regardless of hazard classification, that are separated by at least 10 ft (3 m) horizontally from any other skylights or unprotected ceiling pocket. If sprinklers are required to be installed in a skylight, the skylight window structural framing members may not be designed strong enough to support sprinkler piping, and the designer should contact the structural engineer and coordinate load-bearing capabilities. The location of the piping and the manner in which the piping is to be supported to comply with NFPA 13 are items that may need to be indicated in the contract documents.

Ceiling Pockets

Ceiling pockets are formed in a suspended ceiling when there is a cavity that extends up above the lower suspended ceiling (see Figure 8.25). The designer should indicate where ceiling pockets exist and whether sprinklers are required in the ceiling pockets. NFPA 13 requires sprinklers to be installed in ceiling pockets in subsections 8.6.7 and 8.8.7 except where all of the following conditions are met:

(1) The total volume of the unprotected ceiling pocket does not exceed 1000 ft^3.
(2) The depth of the unprotected pocket does not exceed 36 in.
(3) The entire floor under the unprotected ceiling pocket is protected by the sprinklers at the lower ceiling elevation.
(4) Each unprotected ceiling pocket is separated from any adjacent unprotected ceiling pocket by a minimum 10 ft horizontal distance.
(5) The unprotected ceiling pocket is constructed of noncombustible or limited combustible construction.
(6) Skylights not exceeding 32 ft^2 shall be permitted to have a plastic cover.
(7) Quick response sprinklers are utilized throughout the compartment.

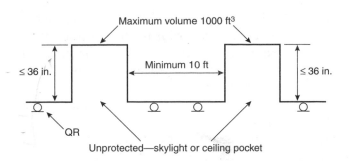

FIGURE 8.25
Unprotected Ceiling Pockets.

(Source: *Automatic Sprinkler Systems Handbook*, NFPA, 2002, Exhibit 8.21)

Vestibules. Buildings in freezing climates may have a vestibule at the building entrance to help keep cold air from entering the building. The vestibule is a part of the building and, according to building code requirements for a completely sprinklered building, should be sprinklered. There are a couple of concerns here. One is the obvious need for heat adequate to keep sprinkler piping from freezing. Even with a unit heater located below the ceiling in the vestibule, the cavity above the ceiling may not receive enough heat to keep the pipe from freezing. Another consideration is that many times these vestibules are very much like a glass box sitting inside a larger building lobby area. The methodology for routing the pipe to the sprinklers in the vestibule is an item that should be determined early in the project.

Hydraulic Considerations. Hydraulic calculations require that each sprinkler within a design area is required to flow water at the minimum required pressure for the occupancy protected. If the standpipe or sprinkler riser to which the system is attached is located centrally to the sprinkler zone, the amount of feed-main piping required for that system will be less, and the size of the piping will be smaller, than if the zone is supplied at the far end of the zone. Figure 8.26 illustrates a sprinkler zone fed from a central stairway and a sprinkler zone fed from a remote stairway. The system fed from the central stairway has obvious hydraulic advantages.

Insurance Underwriter Considerations

Insurance underwriters may have many specific recommendations or requirements that need to be implemented in the sprinkler system design. The underwriter may require larger design areas and higher sprinkler system design densities than NFPA standards if it is determined that a higher level of protection

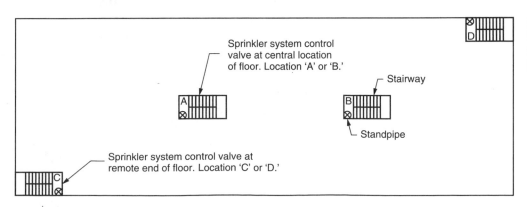

FIGURE 8.26
Sprinkler System Supply from Central (Preferred) and Remote Standpipe.

is necessary. It is important that the designer coordinate the fire protection system scope and design criteria with the underwriter prior to the contract documents being issued for bid. Underwriter recommendations can be broad in scope and include pipe types, design criteria, and sequence of operation. One of the most common recommendations from an insurance underwriter is to provide a higher sprinkler system design density and/or larger design area for various occupancy classifications. This can have a significant impact on the hydraulic calculations. If the owner has directed that the underwriter's recommendations and design criteria be implemented, they must be specified in the designer's contract documents. If the designer creates a coordinated set of contract documents, the contractor would find it unnecessary to contact the insurance underwriter to complete shop drawings. All the pertinent information should already be included in the contract documents.

Obstructions

Sprinkler contractors need to space sprinklers such that the spray pattern of the sprinklers is not obstructed. NFPA 13 is a significant source of information with regard to the placement of sprinklers to avoid a variety of conflicts and obstructions. Some of the obstructions include light fixtures, electrical cable trays, structural members, pipes, ducts, ceiling support systems, ceiling fans, toilet room partitions, signs hung from the ceiling, sloped ceilings, soffits, changes in ceiling elevation, floor-mounted cabinets or equipment, partial height partitions, and artwork. When the designer is aware of conditions that may create obstructions to sprinkler spray patterns, they should be noted on the documents. The best way for the designer to convey the resolution of known obstructions to the contractor is to detail the obstructions, their juxtaposition to sprinklers, and the resolution of the conflict with sprinkler spray patterns on the contract drawings. This enhances coordination between requirements on the contract drawings and details shown on the shop drawings, and the contractor will not miss the conflict on the shop drawings.

Heat and Calculation of Attic Temperature

The designer must be diligent to ensure that adequate heat is provided for wet-pipe sprinkler systems in regions where freezing temperatures are anticipated. Areas subject to freezing include entrance vestibules, mechanical room fresh-air plenums, loading docks, emergency generator rooms, rooms with only an exterior door such as for lawn equipment storage, loading dock or canopy ceiling spaces, stairways located on an exterior wall of the building, attics, and crawl spaces.

A word of caution is in order relative to the specification of insulation as a freeze-protection methodology. Designers should carefully perform heat transfer calculations to demonstrate that an insulation proposal of a given R-rating will be effective for a specific situation. Further, a detail of the proposal needs to be shown on the plan that shows the method of application of the insulation to the protection scenario and the method of attachment of the insulation for a permanent installation.

NFPA 13 requires that wet piping serving a sprinkler system be maintained at or above 40°F. Attics are one of the areas where wet pipes are installed and where wet pipes have the potential to freeze. An example for determining the attic temperature is given in the ASHRAE *Fundamentals Handbook* [9], as follows:

$$t_a = \frac{A_c U_c t_c + t_o (60\rho c_p A_c V_c + A_r U_r + A_w U_w + A_g U_g)}{A_c (U_c + 60\rho c_p V_c) + A_r U_r + A_w U_w + A_g U_g} \tag{8.1}$$

where:

ρc_p = air density times specific heat = 0.018 Btu/ft^3 × °F for standard air

t_a = attic temperature, °F

t_c = indoor temperature near top floor ceiling, °F

t_o = outdoor temperature, °F

A_c = area of ceiling, ft^2

A_r = area of roof, ft^2

A_w = area of net vertical attic wall surface, ft^2

A_g = area of attic glass, ft^2

U_c = heat transfer coefficient of ceiling, Btu/hr × ft^2 × °F, based on surface conductance of 2.2 Btu/hr × ft^2 × °F (upper surface); 2.2 = reciprocal of one-half the air space resistance

U_r = heat transfer coefficient of roof, Btu/hr × ft^2 × °F based on surface conductance of 2.2 Btu/hr × ft^2 × °F (upper surface); 2.2 = reciprocal of one-half the air space resistance

U_w = heat transfer coefficient of vertical wall surface, Btu/hr × ft^2 × °F

U_g = heat transfer coefficient of glass, Btu/hr × ft^2 × °F

V_c = rate of introduction of outside air into the attic space by ventilation per square foot of ceiling area, cfm/ft^2

Note that V_c, the rate of introduction of outside air to the attic, is a significant factor in the determination of the attic temperature. Attics are commonly supplied with continuous roof vents, vents at the eaves, and vents along the walls to provide fresh air to the attic and to permit warm air to escape during the summer months, thus relieving pressure that could deform or damage ceilings. Designers should be aware that insulation may not be capable of protecting wet piping from this temperature differential and that insulation can be displaced by wind or maintenance activity. It is possible to eliminate wet piping in the attic by installing vertical risers within heated walls to sidewall sprinklers at various locations or by installing piping in soffits in the heated area below the ceiling.

Automated Equipment

The designer should note where special equipment is located in the building, and indicate where special sprinkler protection is required or coordination is required to accommodate moving equipment, such as product conveyors, crane rails, automated rack retrieval systems, and automated filing systems.

Special Systems and Considerations

There are additional areas where the designer should note specific requirements on the documents according to the local code requirements. Such considerations include elevator equipment rooms and hoistways, electrical rooms, transformer vaults, stairways and shafts, kitchen hoods and exhaust ducts, prepiped systems, interface with other systems, exposed pipe and fittings, interlocks, paint booths, building service chutes, library stack rooms, escalators, water curtains and draft stops, coolers and freezers, and concealed spaces.

Elevator Equipment Rooms and Hoistways. The code requirements for elevators vary widely, and local AHJs often have specific policies and requirements for these specialized building spaces. Some municipalities use one of the building codes; others may use the ASME/ANSI A17.1, *Safety Code for Elevators and Escalators* [10]. A review of the applicable codes with respect to elevator protection must be performed. There may be different code requirements, depending on whether the elevator is hydraulically operated or electrically operated. In most cases, the elevator equipment must be shut down prior to or upon activation of the automatic sprinkler system. A detection system that will shut down the elevator equipment upon activation of a smoke or heat detector may be required. In other cases, the automatic sprinkler system may be a preaction system. There are many different arrangements that may be acceptable. A code review must be performed and then a proposed sprinkler and elevator power-down system should be discussed with the local building and fire AHJ and the elevator inspector.

The choices available for elevator control include the following:

■ Shut down the elevator immediately on receipt of a detection signal from a detector in an elevator lobby, elevator machine room, or elevator shaft.

■ Permit the elevator to shut down at the nearest floor on receipt of a detection signal from a detector in an elevator lobby, elevator machine room, or elevator shaft.

■ Permit the elevator to descend to the ground floor for use by the fire service on receipt of a detection signal from a detector in an elevator lobby, elevator machine room, or elevator shaft.

■ Permit the elevator movement to be controlled by the combination of the actuation of a smoke or heat detection system and the actuation of a sprinkler system water flow switch.

The choices may be mandated by one or more of the codes or authorities having jurisdiction, and conflicting requirements may result when more than one requirement applies.

Part of the concern relative to elevator movement in the presence of the activation of a water flow signal is the issue of damage to the elevator, motor control, or braking system by fire and damage to these components by water discharge from sprinklers. In dealing with the subject of fire damage, determinations must be made relative to what circumstance would constitute damage that would involve catastrophic failure of the elevator, such as a fire in the elevator machine room or elevator hoistway. It should be noted that elevator pits can collect debris and may collect leaking combustible hydraulic fluid over a period of use, and thus may be the source of a significant fire.

NFPA 13 permits sprinklers to be eliminated from the tops of elevator shafts in cases where the hoistway is noncombustible and the elevator car enclosure materials meet the requirements of ASME/ANSI A17.1, *Safety Code for Elevators and Escalators* [10]. Sidewall sprinklers are required at the bottom of elevator hoistways not more than 2 ft above the bottom of the floor of the pit.

Electrical Rooms. Like the requirements for elevator equipment rooms and hoistways, electrical room requirements will vary by jurisdiction. For a building to be completely sprinklered, the electrical rooms, including electrical switchgear, electrical rooms, and closets, should be protected with sprinklers. NFPA 13, paragraph 8.14.10.3, permits electrical rooms to not be sprinklered when the room is dedicated to electrical equipment only and dry-type equipment is used; the room is 2-hour rated including all penetrations in the room walls, floor, and ceiling; and combustible storage is not allowed in the room. In the absence of a sprinkler system in an electrical room, another type of suppression system should be considered. Some jurisdictions do not want electrical rooms to be sprinklered, others require a preaction system, others may permit the installation of a wet sprinkler system, some may require a special hazard suppression system be installed, and others still may require only smoke detection in the room. Where wet pipe sprinkler systems are installed in electrical rooms, a water flow switch or heat detector may be interlocked with the electrical room control to shut down electrical equipment in advance of water flow. There are specific requirements in NFPA 70, *National Electrical Code*® with regard to where piping may be located when near electrical equipment. The requirements for sprinklers in rooms with electrical equipment should be reviewed with the appropriate building and fire AHJ and the electrical inspector.

Transformer Vaults. Electrical transformer vaults are generally located adjacent to the building but do not have door openings into the building and may be protected by water spray systems in accordance with NFPA 15, *Standard for Water Spray Fixed Systems for Fire Protection*. When transformers are installed inside buildings, either a wet pipe sprinkler system, a water spray system, or a special hazard suppression system should be installed. Some jurisdictions have policies that restrict the use of sprinklers in transformer vaults.

Stairways. Considering a high-rise office building, NFPA 14, *Standard for the Installation of Standpipe and Hose Systems*, requires that standpipes (standpipes or standpipe valves) be located on the intermediate landings of the stairways. Locating standpipes at the intermediate landings is good for the installation of fire department valves but does not work so well with regard to the location for floor control valve assemblies. In some jurisdictions, the AHJ might still want the standpipes located at the main floor entrance landings. The location of the standpipes should be verified with the AHJ prior to indicating standpipes, fire department valves, and floor control valves in the stairways.

Per NFPA 13, subsection 8.14.3, in noncombustible stairways, sprinklers are required at the top of a stairway shaft and under the first landing above the bottom of the shaft. Where the area below a landing or stairway is used for storage, sprinklers are required beneath the landings and stairways. This is usually under the lowest landing of the stairway. Sprinklers at the top of the stairway shaft should protect the entire area of the top of the shaft, including the top landing, and the highest intermediate landing and stairway surfaces. In combustible stairways, sprinklers are required beneath all stairways.

Paint Booths. Paint booths should be called out on the designer's documents along with specifications for indicating shutoff valves and automatic shutdown of the paint delivery piping system. NFPA 13, section 13.4, "Spray Application Using Flammable and Combustible Materials," has requirements for spray areas and mixing areas. It requires an Extra Hazard Group 2 design density for these occupancies.

Building Service Chutes. Building service chutes are to be protected with sprinklers. The service chute may have preinstalled sprinklers located at each door opening to a floor. Typically, building service chutes are supplied by a separate riser at the lowest level of the chute. The riser includes a sprinkler control valve assembly with a supervised shutoff valve, water flow switch, and drain/test connection. The design area for a building service chute requires only the most hydraulically remote three sprinklers be calculated.

Library Stack Rooms. Library stack rooms may require sprinklers in each aisle space if less than 18 in. is maintained between the top of the book shelves, or top of books on the top shelf, and the sprinkler deflectors.

Mobile Shelving. Where high-density mobile shelving is to be installed, the designer should review the sprinkler requirements in NFPA 909, *Code for the Protection of Cultural Resource Properties—Museums, Libraries, and Places of Worship*, and NFPA 232, *Standard for the Protection of Records*. Mobile shelving is defined as a system of records storage, also known as track files, compaction files, or movable files, in which sections or rows of shelves are manually or electrically moved on tracks to provide access aisles. Mobile shelving is usually a type of open-shelf file equipment, available with clearance in excess of 18 in. and also less than 18 in.

Escalators. Escalators over 4 ft wide must have sprinklers installed on the underside to protect the floor area directly under the escalator. An escalator creates an obstruction much like a large duct such that sprinkler protection from above cannot reach the floor area under the escalator.

Water Curtains and Draft Stops. NFPA 13, subsection 8.14.4, has criteria for vertical openings similar to the opening between floors for an escalator. See Figure A.8.14.4 from NFPA 13, reproduced here as Figure 8.27. Sprinklers spaced on 6-ft centers in combination with an 18-in. draft stop should be called out on the contract drawings along with the hydraulic calculation requirements that accompany the use of a water curtain. Per NFPA 13, paragraph 11.2.3.8, water curtain sprinklers are to discharge a minimum of 3 gpm per lineal foot of water curtain with a minimum flow of 15 gpm. A separate design area that includes the water curtain and adjacent sprinklers should be calculated. The designer must also coordinate the location of the draft stop with the architect. Some building codes have requirements for 12-in.-deep draft stops that are different than the requirements for 18-in. draft stops in NFPA 13. The final requirements for the draft stop should be reviewed with the architect and AHJ.

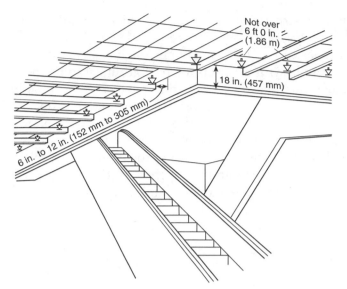

FIGURE 8.27
Sprinklers Around Escalators.
(Source: NFPA 13, 2002,
Figure A.8.14.4)

Concealed Spaces. Concealed spaces requiring sprinkler protection are addressed in NFPA 13. *The Automatic Sprinkler Systems Handbook* [4] has additional commentary to substantiate NFPA 13 requirements for sprinklers in concealed spaces. It states that fire protection is required in concealed spaces when any of three conditions apply: (1) Construction or finish materials are of a combustible nature; (2) the area is used for the storage of combustible materials; (3) the area can contain combustibles associated with building system features such as computer wiring or large quantities of nonmetallic piping. Sprinklers have typically not been used to protect computer room underfloor areas. A clean agent system should be considered.

Coolers and Freezers. Small coolers and freezers that might be found in a retail food store are usually comprised of a self-contained refrigeration box with a top that may be lower than the suspended ceiling in the area where the cooler/freezer is to be located. The coolers/freezers can be protected with dry-pendent-type sprinklers connected to a wet sprinkler system branchline. The designer's drawings should show the location of sprinklers within the cooler/freezer requiring sprinkler protection and also show sprinkler protection needed above the cooler/freezer in addition to the sprinklers inside the cooler/freezer. The penetration for a sprinkler inside a cooler or freezer must be sealed to minimize condensation in the piping. Sprinkler protection may be required above the cooler if the top of the cooler does not extend to the suspended ceiling.

Summary

It has not always been clear what information the engineer or designer should provide on the sprinkler system contract documents. The SFPE Position Statement provides clear direction with regard to this question, and this chapter provides an overview of many of the sprinkler system layout considerations that are to be included in the contract documents. Sprinkler system layout and coordination questions that can be answered early in the project design will benefit the entire design and construction team since sprinkler systems are interrelated to many other building systems.

The preparation of sprinkler contract documents by engineers and designers has been performed by people with a variety of backgrounds and experience. The engineer or designer must specify systems that have been thoroughly analyzed and determined to be code-compliant, cost-effective, and appropriate to the current and future use of a building. Careful planning and coordination with the owner, architect, reviewing authorities, and other engineering disciplines must be accomplished in the design stage of the project. It is prudent to employ an individual who is experienced and competent in the field of fire protection to prepare the sprinkler system documents and specifications.

BIBLIOGRAPHY

References Cited

1. SFPE Position Statement, "The Engineer and the Technician: Designing Fire Protection Systems," Society of Fire Protection Engineers, Bethesda, MD, 1998 [available on-line at http://www.sfpe.org/sfpe/design-fpe.htm].
2. Cote, A. E., ed., *Fire Protection Handbook*, 19th edition, NFPA, Quincy, MA, 2003.
3. DiNenno, P. J., ed., *The SFPE Handbook of Fire Protection Engineering*, 3rd edition, NFPA, Quincy, MA, and SFPE, Bethesda, MD, 2002.
4. *Automatic Sprinkler Systems Handbook*, 9th edition, NFPA, Quincy, MA, 2002.
5. *FM Global Loss Prevention Guidelines*, FM Global, Norwood, MA.
6. Spencer, A. B., et al., *Building and Fire Code Classification of Hazardous Materials*, NFPA, Quincy, MA, 2004.
7. *Fire Protection Guide to Hazardous Materials*, 13th edition, NFPA, Quincy, MA, 2002.
8. ASCE 7, *Minimum Design Loads for Buildings and Other Structures*, American Society of Civil Engineers, Reston, VA, 1998.
9. *ASHRAE Fundamentals Handbook*, ASHRAE, Atlanta, GA, 2001, p. 28-8.
10. ASME/ANSI A17.1, *Safety Code for Elevators and Escalators*, American Society of Mechanical Engineers, New York, 2000.

NFPA Codes, Standards, and Recommended Practices

NFPA Publications National Fire Protection Association, 1 Batterymarch Park, Quincy, MA 02169-7471

The following is a list of NFPA codes, standards, and recommended practices cited in this chapter. See the latest version of the *NFPA Catalog* for availability of current editions of these documents.

NFPA 13, *Standard for the Installation of Sprinkler Systems*
NFPA 13D, *Standard for the Installation of Sprinkler Systems in One- and Two-Family Dwellings and Manufactured Homes*
NFPA 14, *Standard for the Installation of Standpipe and Hose Systems*
NFPA 15, *Standard for Water Spray Fixed Systems for Fire Protection*
NFPA 70, *National Electrical Code*®
NFPA 75, *Standard for the Protection of Information Technology Equipment*
NFPA 92A, *Recommended Practice for Smoke-Control Systems*
NFPA 101®, *Life Safety Code*®

NFPA 232, *Standard for the Protection of Records*

NFPA 909, *Code for the Protection of Cultural Resource Properties—Museums, Libraries, and Places of Worship*

Additional Readings

The Engineer and the Technician: Designing Fire Protection Systems, The Society of Fire Protection Engineers, Bethesda, MD, 1998.

Engineering Technician and Technology Certification Programs, 10th edition, National Institute for Certification in Engineering Technologies, Alexandria, VA, 1995.

9

Automatic Sprinkler System Spacing

Robert M. Gagnon, PE, SET, FSFPE

The spacing of automatic sprinklers is a fundamental component of sprinkler system design and is a driving force with respect to the rules that govern the design and installation of sprinkler systems. Sprinkler spacing decisions are based upon an understanding of the physical properties of the sprinkler, the geometry and materials used in the building construction, the nature of the hazard to be protected, the water supply characteristics, and the obstructions to the water discharged from the sprinkler.

In practice, the design of a sprinkler system is an iterative process that requires an engineer to relate all of the fundamental elements of sprinkler spacing to determine a sprinkler and piping arrangement that meets the requirements of NFPA 13, *Standard for the Installation of Sprinkler Systems*, and the authority having jurisdiction. Sprinklers are positioned to ensure that they adequately perform in both detecting the fire in a timely fashion and discharging a water spray of sufficient density to control the fire. The success criteria for automatic sprinklers are determined through laboratory and field testing at recognized testing facilities and through their empirical performance in actual installations, which has been responsible for an impressive record of achievement in life safety and property protection for over a century.

Sprinkler spacing is a fundamental design parameter of sprinkler system installation. Building construction, occupancy or hazard of the protected space, water supply characteristics, and obstructions all correlate to the spacing and area of coverage of the sprinkler. Hydraulically calculated pipe sizing is normally an affected attribute of the sprinkler system and is determined after the sprinkler spacing and other design attributes are fixed and finalized. For most applications, building construction, obstructions, and water supply characteristics are less variable than the designer's choice in sprinkler spacing and therefore tend to drive the final sprinkler spacing requirements.

It should be stressed that in almost all cases there are multiple spacing combinations that produce sprinkler system design solutions for a specific room. Increasing the sprinkler spacing reduces the number of sprinklers but usually increases the required minimum operating pressure and reduces the frictional pressure capable of being lost in the piping network feeding the sprinkler system. The result can be a trade-off between pipe size and number of sprinklers, which gives the designer the flexibility of numerous code-compliant designs for a given room.

Sprinkler Spacing Overview

Sprinklers are spaced within a room using either the prescriptive rules specified in NFPA 13 or in accordance with laboratory-tested sprinkler spacing criteria given in the manufacturers' literature. There are different spacing rules for standard sprinklers, sidewall sprinklers, and extended coverage sprinklers. The general intent of sprinkler spacing requirements is to ensure that water spray of sufficient density is distributed without obstruction to all areas of the hazard, which is usually defined as the floor area.

The construction features of a building directly affect the spacing of sprinklers. The presence of beams, partitions, sloped roofs, and mechanical features, such as ducts and lights, may create an interference with sprinkler spray and could require that the sprinklers be spaced around or under the obstruction. In addition, building construction materials affect sprinkler system spacing. For example, combustible construction requires different sprinkler spacing criteria than noncombustible construction, and obstructed construction requires different sprinkler spacing criteria than unobstructed construction. In addition, sprinkler spacing is affected by roof slope and geometry.

In the early 1950s, there were essentially two types of sprinklers, upright and pendent. A growing trend in the sprinkler manufacturing industry is for sprinkler manufacturers to invent, test, and produce specialized sprinklers for suppression of specific hazards and submit them for testing and approval by a recognized listing agency. Many of the sprinklers produced in the last 30 years are primarily intended to reduce the number of sprinklers in a given space and the labor associated with their installation. Examples of such sprinklers are extended coverage sprinklers (which are permitted to be spaced farther apart than standard coverage sprinklers), early suppression fast response (ESFR) sprinklers (which may eliminate the need for in-rack sprinklers), and attic sprinklers (which can permit sprinklers to be installed only at the peak of a combustible roof). Once these sprinklers are tested and listed, they are permitted to be spaced in accordance to specific criteria published by the manufacturer. Spacing rules for all extended coverage sprinklers and ESFR sprinklers are based upon NFPA 13 as required by the listing and are no longer based upon the manufacturer's information, but must comply with NFPA 13 as indicated in the appropriate manufacturer's data.

In some cases, notably residential sprinklers, which are primarily designed for life safety, performance characteristics, such as upper room gas temperature and activation time, are part of the listing criteria.

The occupancy or hazard of the space to be protected influences the spacing of the sprinkler system. High hazard spaces, such as those associated with hazardous storage occupancies, require sprinklers to be spaced more closely, with higher water densities and hence higher water flow per sprinkler, than lower hazard spaces, such as light hazard occupancies, including offices.

Sprinkler Spacing Requirements

Sprinkler spacing is measured between the centerline of sprinklers and the centerlines of adjacent branchlines, and between sprinklers and lines along the slope of the ceiling or roof. The maximum distance from the walls shall not exceed half the maximum allowable distance between sprinklers. The designer must use manufacturer's spacing criteria for positioning some sprinklers.

The minimum distance between sprinklers and walls shall comply with each specific sprinkler type and shall never be less than 4 in. The minimum distance between sprinklers shall comply with each specific sprinkler type and shall never be less than 6 ft without baffles (barriers that prevent wetting of adjacent sprinklers).

In a small room of light hazard occupancy classification having unobstructed construction, and floor areas not exceeding 800 ft^2, that are enclosed by walls and ceilings having a minimum 8-in. lintel on all openings, the protection area of coverage for each sprinkler in the small room shall be the area of the room divided by the number of sprinklers in the room, as described in NFPA 13, 2002 edition, paragraph 3.3.20.

Specially listed sprinklers are permitted to be spaced in accordance with their listing. With standard or extended coverage upright or pendent sprinklers, per NFPA 13, paragraph 8.6.3.2.1, ". . . where walls are angled or irregular, the maximum horizontal distance between a sprinkler and any point of floor area protected by that sprinkler shall not exceed 0.75 times the allowable distance permitted between sprinklers, provided the maximum perpendicular distance is not exceeded."

Spacing of Standard Coverage Sprinklers

A summary of steps to be followed for automatic sprinkler system spacing are to determine the following, in order:

- Branchline orientation
- Sprinkler system configuration
- Occupancy classification
- Protection area of coverage
- Maximum area permitted to be protected by a single sprinkler system
- Area protected by each sprinkler (A)
- Maximum area permitted to be protected by each sprinkler (A_{max})
- Maximum permissible spacing between branchlines (L_{max})
- Maximum permissible spacing between sprinklers on branchlines (S_{max})
- Number of branchlines in a bay or room
- Actual distance between branchlines (L)
- Maximum permissible distance between sprinklers (S_{max})
- Minimum permissible number of sprinklers on each branchline
- Actual distance between each sprinkler on branchlines (S_{actual})
- Actual area of sprinkler coverage
- Effect of obstructions on sprinkler spacing
- Crossmains and feedmain placement

Branchline Orientation

Sprinkler branchlines are hung from the roof structure above the sprinkler piping. In modern construction practice, bar joists are lowered into place by cranes, may not always be installed in the precise locations shown on the structural drawing, and may not always be installed straight and true, and a joist installed in one bay may not necessarily line up with a joist in an adjacent bay. For this reason, sprinkler branchlines are always oriented to be perpendicular to bar joists, as shown in Figure 9.1. Branchlines also are run perpendicular to purlins,

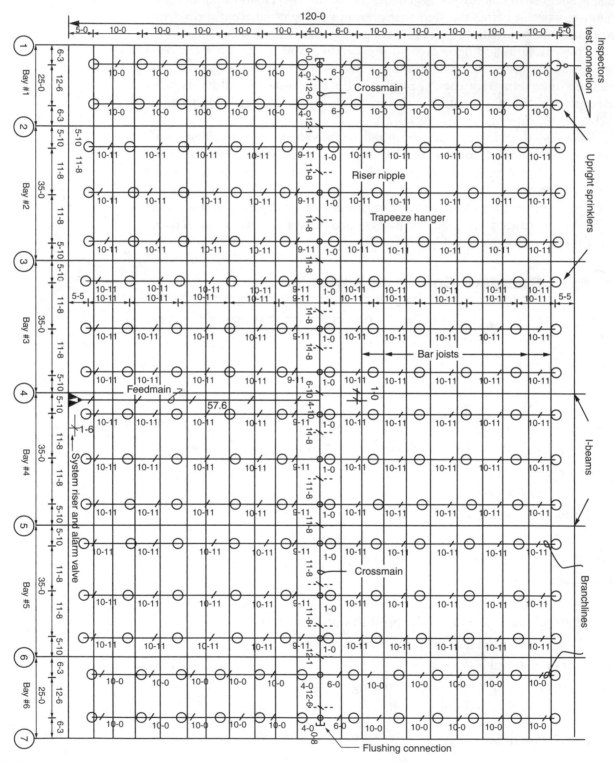

FIGURE 9.1
Typical Layout for a Wet Pipe Center-Fed Tree Sprinkler System.

FIGURE 9.2
Bar Joist Construction.

(Source: *Automatic Sprinkler Systems Handbook*, NFPA, 2002, Exhibit 3.37)

wood joists, or concrete tees (as shown in Figure 9.3) in a building. Crossmains and feedmains connect the branchlines to the source of water supply, as shown in Figure 9.1. By orienting sprinkler branchlines to be perpendicular to bar joists, NFPA hanging requirements are more efficiently enabled and met, and spacing of branchlines is more flexible. Figure 9.2 is a photo of sprinkler piping running perpendicular to bar joists. Figure 9.3 shows typical concrete tee construction.

Branchlines attached to smooth ceilings, such as unobstructed flat concrete ceilings capable of supporting sprinkler pipes, may be oriented in any direction that the designer selects. Branchline logic in this case would therefore be primarily a function of system configuration, such as a tree system, looped system, or grid system, discussed in the next section.

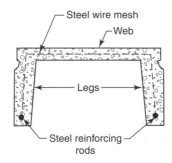

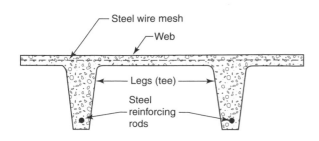

FIGURE 9.3
Typical Concrete Tee Construction.

(Source: NFPA 13, 2002, Figure A.3.7.1(a))

Sprinkler System Configuration

Sprinkler system configurations commonly employed for the layout and installation of sprinkler systems are the tree, the grid, and the loop, as discussed in Chapter 8, "Overview of Sprinkler System Layout." A designer selects sprinkler system configurations primarily by considering building geometry and the potential for hydraulic or economic advantage.

On a tree system, long branchlines should, if possible, be fed so that the crossmain is equidistant from the ends of the branchlines, or alternatively, the branchlines may be supplied by more than one crossmain. A drawing showing a typical center-fed wet pipe tree system is shown in Figure 9.1. On hydraulically calculated sprinkler systems, there is no limitation on the number of sprinklers that are permitted to be supplied on each side of a center-fed crossmain, but the greater the number of sprinklers, the greater the friction loss in each branchline, increasing the probability for uneconomically large branchline pipe sizes.

A designer may opt to use an end-fed tree to supply water to branchlines installed on a flat roof. As another example, where a building features a peaked roof, a designer can opt to configure the system as an end-fed tree system, where branchlines slope up to the roof peak, with crossmains positioned at the eaves of the roof, as shown on Figure 9.4.

A gridded system may be designed only for wet pipe systems and are primarily used to provide a hydraulic advantage for systems with numerous branchlines of significant length. The hydraulic advantage is accomplished when a grid provides water flow to a flowing sprinkler from more than one direction within the grid, decreasing the total friction loss in the piping system and permitting the use of smaller branchline sizes.

Gridded systems are not permitted to be used on dry pipe systems, because such systems would require water to push air out of the system from more than one direction, in advance of water delivery to the sprinkler, which is

Figure 9.4
Sprinklers at Pitched
Roofs; Branchlines Run
Up the Slope.

(Source: NFPA 13, 2002,
Figure 8.6.4.1.3.1(b))

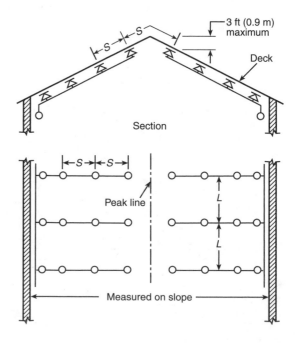

likely to retard water delivery to the open sprinkler and may result in trapped air in the grid.

As a rule of thumb, it is recommended that grid systems feature at least eight branchlines and at least ten sprinklers per branchline. Grid systems configured in this manner are more likely to provide hydraulic and economic advantage for the system.

Looped sprinkler systems permit water flow to crossmains and branchlines from more than one direction and may provide a hydraulic and economic advantage over the standard center-fed tree system.

Occupancy Classification

Occupancy classification, as it relates to the design of automatic sprinkler systems, reflects the level of severity of a fire that is expected to occur in a building space and is directly associated with the fire load density of commodities that are expected to be found in that building space. The fire load density is a value that quantifies the properties related to the potential for and severity of combustion, such as the quantity, arrangement, combustibility, and rate of heat release of the commodities expected to be found in a building space. Combustible commodities are grouped into classifications (light hazard, ordinary hazard, and extra hazard occupancy) related to their potential for severity of combustion, for ease of reference and use by the designer. Further discussion of occupancy classifications is found in Chapter 4, "Hazard and Commodity Classification."

A building may feature numerous rooms with differing uses and occupancy classifications. An example of this is a school that features classrooms of light hazard occupancy but also features mechanical and storage rooms of ordinary hazard occupancy. Further, occupancy classification may even differ within a room, such as a storage warehouse, where storage is arranged into areas of differing storage commodity classifications. Another example is a munitions manufacturing room where increased hazard may be found at specific points within the room, such as nitroglycerin kettles or solid munitions extruding or cutting machines.

A newly constructed building may be in use for a century or more, and the intended function for that building is likely to change over the course of its life span. An example of this is a school that is converted into a storage or manufacturing facility. It is vitally important that the design engineer perform a careful survey and a design reanalysis each time a building undergoes a change in occupancy.

Protection Area of Coverage

The selection of an occupancy classification directly affects the spacing of sprinklers permitted in a building. For standard coverage sprinklers, as discussed next, sprinklers installed for a light hazard occupancy are permitted to cover more square footage than sprinklers in ordinary hazard and extra hazard occupancies. The spacing of sprinklers reflects the perceived potential hazard in a space, and spaces of higher occupancy classifications, such as extra hazard, are expected to have sprinklers spaced more closely to permit a more rapid actuation of the sprinkler and a greater flow and coverage of the protected space.

According to NFPA 13, paragraph 8.5.2.1.1:

The protection area of coverage per sprinkler (A_s) shall be determined as follows:

(1) Along branch lines as follows:

    (a) Determine distance between sprinklers (or to wall or obstruction in the case of the end sprinkler on the branch line) upstream and downstream.

    (b) Choose the larger of either twice the distance to the wall or the distance to the next sprinkler.

    (c) This dimension will be defined as S.

(2) Between branch lines as follows:

    (a) Determine perpendicular distance to the sprinkler on the adjacent branch line (or to a wall or obstruction in the case of the last branch line) on each side of the branch line on which the subject sprinkler is positioned.

    (b) Choose the larger of either twice the distance to the wall or obstruction or the distance to the next sprinkler.

    (c) This dimension will be defined as L.

Hazardous occupancies are spaced as follows:

- Light hazard occupancies are spaced in accordance with Table 9.1.
- Ordinary hazard occupancies are spaced in accordance with Table 9.2.

TABLE 9.1

Protection Areas and Maximum Spacing (Standard Spray Upright/Standard Spray Pendent) for Light Hazard

Construction Type	System Type	Protection Area		Spacing (maximum)	
		ft²	m²	ft	m
Noncombustible obstructed and unobstructed and combustible unobstructed with members 3 ft or more on center	Pipe schedule	200	18.6	15	4.6
Noncombustible obstructed and unobstructed and combustible unobstructed with members 3 ft or more on center	Hydraulically calculated	225	20.9	15	4.6
Combustible obstructed with members 3 ft or more on center	All	168	15.6	15	4.6
Combustible obstructed or unobstructed with members less than 3 ft on center	All	130	12.1	15	4.6
Unoccupied attics having combustible wood joist or wood truss construction with members less than 3 ft on center with slopes having a pitch of 4 in 12 or greater	All	120	11.1	8* × 15 (minimum psi) 10* × 12 (minimum 20 psi)	2.4* × 4.6 (minimum 0.48 bar) 3* × 3.7 (minimum 1.34 bar)

*The smaller dimension shall be measured perpendicular to the slope.

Source: NFPA 13, 2002, Table 8.6.2.2.1(a).

TABLE 9.2

Protection Areas and Maximum Spacing (Standard Spray Upright/Standard Spray Pendent) for Ordinary Hazard

Construction Type	System Type	Protection Area		Spacing (maximum)	
		ft²	m²	ft	m
All	All	130	12.1	15	4.6

Source: NFPA 13, 2002, Table 8.6.2.2.1(b).

- Extra hazard occupancies are spaced in accordance with Table 9.3.
- High-piled storage occupancies are spaced in accordance with Table 9.4.

See Tables 9.1 through 9.4.

Maximum Area Permitted to Be Protected by a Single Sprinkler System

Associated with the occupancy classifications of light hazard, ordinary hazard, and extra hazard, automatic sprinkler systems are constrained by maximum floor area limitations for each occupancy. For example, NFPA 13 requires that light hazard and ordinary hazard sprinkler systems have a maximum system area limitation of 52,000 ft². For extra hazard occupancies, a maximum system area limitation of 40,000 ft² applies.

TABLE 9.3

Protection Areas and Maximum Spacing (Standard Spray Upright/Standard Spray Pendent) for Extra Hazard

Construction Type	System Type	Protection Area		Spacing (maximum)	
		ft²	m²	ft	m
All	Pipe schedule	90	8.4	12	3.7
				[In buildings with storage bays 25 ft (7.6 m) wide, 12 ft 6 in. (3.8 m) shall be permitted]	
All	Hydraulically calculated with density ≥0.25	100	9.3	12	3.7
				[In buildings with storage bays 25 ft (7.6 m) wide, 12 ft 6 in. (3.8 m) shall be permitted]	
All	Hydraulically calculated with density <0.25	130	12.1	15	4.6

Source: NFPA 13, 2002, Table 8.6.2.2.1(c).

TABLE 9.4

Protection Areas and Maximum Spacing (Standard Spray Upright/Standard Spray Pendent) for High-Piled Storage

Construction Type	System Type	Protection Area		Spacing (maximum)	
		ft²	m²	ft	m
All	Hydraulically calculated with density ≥0.25	100	9.3	12	3.7
				[In buildings with storage bays 25 ft (7.6 m) wide, 12 ft 6 in. (3.8 m) shall be permitted]	
All	Hydraulically calculated with density <0.25	130	12.1	15	4.6

Source: NFPA 13, 2002, Table 8.6.2.2.1(d).

In accordance with NFPA 13, paragraph 8.2.3, in cases when a single sprinkler system protects combined occupancies, such as extra hazard, high-piled storage, or storage covered by standards other than NFPA 13, in addition to light or ordinary hazard areas, the extra hazard or storage area coverage shall not exceed the floor area specified for that hazard and the total area coverage shall not exceed 52,000 ft².

Actual Area Protected by Each Sprinkler (*A*)

NFPA 13 uses the formula shown in Equation 9.1 to determine the area protected by a single sprinkler and the spacing between sprinklers.

$$A = S \times L \tag{9.1}$$

where:

A = area protected by the sprinkler, in square feet

S = distance between sprinklers on the branchlines, in feet

L = distance between branchlines, in feet

Maximum Area Permitted to Be Protected by Each Sprinkler (*A*$_{max}$)

NFPA 13 imposes a maximum square footage per sprinkler, A_{max}, as described by the following:

- A_{max} shall not exceed 225 ft² for a hydraulically calculated sprinkler system in a light hazard occupancy.
- A_{max} shall not exceed 200 ft² for a pipe schedule sprinkler system in a light hazard occupancy.
- A_{max} shall not exceed 168 ft² for a light hazard sprinkler system installed in a building of combustible construction, such as a wood frame building.
- A_{max} shall not exceed 130 ft² for a sprinkler system in an ordinary hazard occupancy.

- A_{max} shall not exceed 100 ft^2 for a sprinkler system in an extra hazard occupancy.

- Extended coverage sprinklers are permitted to be used in occupancies for which they are specifically listed and may be spaced up to an A_{max} of 400 ft^2.

Maximum Permissible Spacing Between Branchlines (L_{max})
Maximum Permissible Spacing Between Sprinklers on Branchlines (S_{max})

NFPA 13 requires that the maximum permissible spacing between branchlines (L_{max}) and the maximum permissible spacing between sprinklers on the branchlines (S_{max}) not exceed 15 ft for light- or ordinary-hazard occupancies. NFPA 13 also requires that the maximum permissible spacing between branchlines (L_{max}) and the maximum permissible spacing between sprinklers on the branchlines (S_{max}) not exceed 12 ft for extra hazard occupancies. The required spacing is related to the perceived rate of fire growth, where, for example, sprinklers protecting an ordinary hazard occupancy would be expected to be more closely spaced than sprinklers protecting a light hazard occupancy, to permit a more rapid actuation time for the sprinkler and to enhance water delivery to the more severe hazard.

Each sprinkler has a maximum area of coverage that is based upon the individual tested performance of the sprinkler, or limitations imposed by NFPA 13, or criteria established by the manufacturer, to ensure an acceptable detection and actuation time. Detection of the fire by the heat-responsive element on the sprinkler, which leads to activation of the sprinkler, needs to be rapid enough to ensure that sprinkler activation occurs early in the history of the fire, where the probability for control of the fire is maximized, and in advance of a fire growth that could overwhelm the ability of the sprinkler to control the fire and significantly in advance of room flashover for small rooms.

Number of Branchlines in a Bay or Room

Automatic sprinklers have a maximum permissible separation distance, S_{max} and L_{max}. This maximum distance is based upon the physical limitations of distributing an effective water spray pattern at a given pressure and on the need to match the performance of the sprinkler with the expected commodity as it relates to detecting and actuating in the presence of an expected fire.

In exposed steel construction, such as would be found in most warehouses, the roof structure is supported by horizontal beams, which are supported from the floor with vertical columns. For the purpose of performing a drawing of a building, the line of beams supported by columns are called column lines. Column lines are shown in Figure 9.1 as the lines identified by the circled numbers 1 through 7. The distance or space between column lines is called a bay. Figure 9.1 shows a building with six bays, identified as bay #1 through bay #6, and designers are required to space sprinklers and branchlines within each bay. The spacing of branchlines within each bay is represented by the L dimension, or dimension between each branchline within the bay.

For rooms with smooth ceilings, the dimension between branchlines may be determined in a manner similar to a bay, making adjustments for ceiling-mounted interferences to sprinkler discharge, such as lighting fixtures, speakers, smoke detectors, or supply and return duct diffusers. When ceiling-mounted interferences exist, the location of each sprinkler must be coordinated with the lighting and HVAC (heating, ventilating and air conditioning) obstructions shown on a reflected ceiling plan.

Within a bay, the designer determines the number of branchlines in each bay by dividing the lineal dimension representing the width of the bay by NFPA 13 requirements representing the maximum permissible distance between sprinklers (L_{max}) for the occupancy, as shown in Equation 9.2. The number of branchlines must be a whole number, so the value obtained must be rounded up to the nearest whole number, as shown in Equation 9.2 and Example 9.1.

$$\text{Number of branchlines} = \frac{\text{Width of bay}}{L_{max}} \qquad (9.2)$$

EXAMPLE 9.1 Branchline Spacing (L_{max}) in Bays #1 and #6 in Figure 9.1

An ordinary hazard sprinkler system is proposed for bays #1 and #6, which are 25 ft-0 in. wide, as shown in Figure 9.1. NFPA 13 requires a maximum spacing between branchlines (L_{max}) of 15 ft-0 in. for standard coverage sprinklers. The bay would therefore require no fewer than two branchlines per bay in bays #1 and #6:

$$\text{Number of branchlines} = \frac{\text{Width of bay}}{L_{max}}$$

$$= \frac{25\,\text{ft}}{15\,\text{ft}}$$

$$= 1.7 \text{ branchlines, rounded up to}$$
$$2 \text{ branchlines (see Figure 9.1)}$$

EXAMPLE 9.2 Branchline Spacing (L_{max}) in Bays #2, #3, #4, and #5 in Figure 9.1

An ordinary hazard sprinkler system is proposed for bays #2, #3, #4, and #5, which are each 35 ft-0 in. wide, as shown in Figure 9.1. NFPA 13 requires a maximum spacing between branchlines (L_{max}) of 15 ft-0 in. for standard coverage sprinklers installed in an ordinary hazard occupancy. The bay would therefore require no fewer than three branchlines, per bay, in bays #2, #3, #4, and #5.

$$\text{Number of branchlines} = \frac{\text{Width of bay}}{L_{max}}$$

$$= \frac{35\,\text{ft}}{15\,\text{ft}}$$

$$= 2.3 \text{ branchlines, rounded up to}$$
$$3 \text{ branchlines (see Figure 9.1)}$$

Actual Distance Between Branchlines (L)

The spacing between branchlines within a bay is determined by dividing the width of the bay by the number of branchlines in the bay, as shown in Equation 9.3.

$$L = \frac{\text{Total width of bay}}{\text{Number of branchlines in the bay}} \qquad (9.3)$$

For the following examples, foot-and-inch calculations are used. It is sometimes convenient or necessary to convert decimal feet to foot-and-inch values, and a foot-and-inch calculator may be helpful in this regard.

Example 9.3 Branchline Spacing (L) in Bays #1 and #6 in Figure 9.1

Using the result for number of branchlines per bay that was obtained in Example 9.1, the distance between branchlines in bays #1 and #6, the L dimension, is determined using Equation 9.3 as follows:

$$L = \frac{\text{Total width of bay}}{\text{Number of branchlines in the bay}}$$

$$= \frac{25 \text{ ft}}{2 \text{ branchlines}}$$

$$= 12.5 \text{ ft, or 12 ft-6 in., as shown in Figure 9.1}$$

Note that L in this example is less than L_{max} (15 feet), so the spacing is acceptable. 12 ft-6 in. is, therefore, the L dimension, the distance between branchlines in bays #1 and #6, as shown in Figure 9.1. The distance between the branchlines and the column centerlines in bays #1 and #6 is calculated as follows:

$$\tfrac{1}{2}(L) = \tfrac{1}{2}(12 \text{ ft-6 in.})$$

$$= 6 \text{ ft-3 in., as shown in Figure 9.1}$$

Note that the dimensions in bay #1 and bay #6 add up to 25 ft-0 in. (6 ft-3 in. + 12 ft-6 in. + 6 ft-3 in.).

Example 9.4 Branchline Spacing (L) in Bays #2, #3, #4, and #5 in Figure 9.1

Using the result for number of branchlines per bay that was obtained in Example 9.2, the L dimension is determined as follows:

$$L = \frac{\text{Total width of bay}}{\text{Number of branchlines in the bay}}$$

$$= \frac{35 \text{ ft}}{3 \text{ branchlines}}$$

$$= 11 \text{ ft 8 in., as shown in Figure 9.1}$$

Note that L in this example is less than L_{max} (15 ft), so the spacing is acceptable.

11 ft-8 in. represents the L dimension, the distance between branchlines in bays #2, #3, #4, and #5, as shown in Figure 9.1. The distance between the branchlines and the column centerlines in bays #2, #3, #4, and #5 is $\tfrac{1}{2}(L)$ or $\tfrac{1}{2}(11 \text{ ft-8 in.})$:

$$\tfrac{1}{2}(L) = 5 \text{ ft-10 in., as shown in Figure 9.1}$$

Note that the dimensions in bays #2, #3, #4, and #5 add up to 35 ft-0 in. (5 ft-10 in. + 11 ft-8 in. + 11 ft-8 in. + 5 ft-10 in.).

Maximum Permissible Distance Between Sprinklers (S_{max})

NFPA 13 permits S_{max} to be a value with a maximum permissible value of 15 ft for light hazard, 15 ft for ordinary hazard, and 12 ft for extra hazard, but the value of S is also regulated by Equation 9.1. Equation 9.1 is modified by solving for S, and the formula becomes Equation 9.4:

$$S_{max} = \frac{A_{max}}{L} \qquad (9.4)$$

EXAMPLE 9.5 Sprinkler Spacing Along Lines in Bays #1 and #6 in Figure 9.1

As shown in Examples 9.1 and 9.3, where a building of ordinary hazard was being evaluated, it has been established that for ordinary hazard, $A = 130$ ft^2 maximum area limitation per sprinkler, and as calculated in Examples 9.1 and 9.3, $L = 12$ ft-6 in. in bays #1 and #6.

Knowing that A_{max} is 130 ft^2, and having calculated L as 12 ft-6 in. For this example, the maximum distance between sprinklers on branchlines, S_{max}, in bays #1 and #6 can now be determined by use of Equation 9.4. Substituting the value of L obtained in Examples 9.1 and 9.3 into Equation 9.4:

$$
\begin{aligned}
S_{max} &= \frac{A_{max}}{L} \\
&= \frac{130 \text{ ft}^2}{12 \text{ ft-6 in.}} \\
&= 10.4 \text{ ft} \\
&= 10 \text{ ft-}4^{13}/_{16} \text{ in.}
\end{aligned}
$$

S_{max}, the maximum distance between sprinklers on a branchline for this example, is 10 ft-4^{13}/16 in., as just computed. In practice, foot-and-inch dimensions using sixteenths of an inch are not used and are usually rounded to the nearest ½ in., or sometimes to the nearest ¼ in., by some design firms.

EXAMPLE 9.6 Sprinkler Spacing Along Lines in Bays #2, #3, #4, and #5 in Figure 9.1

As shown in Examples 9.2 and 9.4, where a building of ordinary hazard was being evaluated, it has been established that for ordinary hazard, $A = 130$ ft^2 maximum area limitation per sprinkler, and $L = 11$ ft-8 in. Knowing that A is 130 ft^2, and having calculated L as 11 ft-8 in., the maximum distance between sprinklers on branchlines, S_{max}, in bays #2, #3, #4, and #5 can now be determined by using Equation 9.4. Using the value for L obtained in Example 9.2 and 9.4 and substituting into Equation 9.4:

$$
\begin{aligned}
S_{max} &= \frac{130 \text{ ft}^2}{11 \text{ ft-8 in.}} \\
&= 11.4 \text{ ft} \\
&= 11 \text{ ft-}1^{11}/_{16} \text{ in.}
\end{aligned}
$$

S_{max}, the maximum distance between sprinklers on a branchline for this example, is 11 ft-1^{11}/16 in., as just computed. In practice, foot-and-inch dimensions using sixteenths of an inch are not used and are usually rounded to the nearest ½ in., or sometimes to the nearest ¼ in.

Minimum Permissible Number of Sprinklers on Each Branchline

The minimum permissible number of automatic sprinklers on each branchline in a bay is determined by dividing the length of the bay by the maximum distance permitted between sprinklers on the branchlines (S_{max}), as shown in the following equation. For these examples, the value for S_{max} (max. distance between sprinklers) that was calculated previously in Examples 9.6 and 9.7 is determined by use of Equation 9.5.

$$\text{Minimum number of sprinklers on each branchline} = \frac{\text{Total length of bay}}{S_{max}} \qquad (9.5)$$

EXAMPLE 9.7 Minimum Number of Sprinklers on Each Branchline in Bays #1 and #6 in Figure 9.1

Using the value of S_{max}, previously obtained in Examples 9.1, 9.3, and 9.5, for a building 120 ft in length and substituting into Equation 9.5:

$$\text{Minimum number of sprinklers on each branchline} = \frac{\text{Total length of bay}}{S_{max}}$$

$$\frac{120 \text{ ft-0 in.}}{10.4 \text{ ft}} = 11.53 \text{ sprinklers}$$

Of course, it is impossible for a branchline to have a fractional sprinkler; therefore, the value obtained, 11.53 sprinklers, is rounded up to the nearest whole sprinkler. For Example 9.7, round the value of 11.53 sprinklers up to 12 sprinklers proposed to be installed on each branchline in bays #1 and #6, as shown in Figure 9.1.

EXAMPLE 9.8 Minimum Number of Sprinklers on Each Branchline in Bays #2, #3, #4, and #5 in Figure 9.1

Using the value of S_{max} previously obtained in Examples 9.2, 9.4, and 9.6, for a building 120 ft in length, and substituting into Equation 9.5:

$$\text{Minimum number of sprinklers on each branchline} = \frac{\text{Total length of bay}}{S_{max}}$$

$$\frac{120 \text{ ft-0 in.}}{11.4 \text{ ft}} = 10.77 \text{ sprinklers}$$

As discussed in Example 9.7, it is impossible to have a fractional sprinkler, so the value obtained is rounded up to the nearest whole sprinkler. For this example, round the value of 10.77 up to 11 sprinklers on each branchline in bays #2, #3, #4, and #5, as shown in Figure 9.1.

Actual Distance Between Each Sprinkler on Branchline (S_{actual})

It was previously shown how to determine the maximum permissible distance between sprinklers on a branchline. Assuming an equal distance between all sprinklers on the branchline, the actual spacing of sprinklers on each branchline, S_{actual}, is calculated by dividing the length of the bay by the number of sprinklers on a branchline in that bay, as shown next. The number of sprinklers on each branchline was obtained in Example 9.7 for bays #1 and #6, and Example 9.8 for bays #2, #3, #4 and #5, as shown in Equation 9.6.

$$S_{\text{actual}} = \frac{\text{Total length of bay}}{\text{Number of sprinklers on line}} \qquad (9.6)$$

EXAMPLE 9.9 Calculate (S_{actual}) in Bays #1 and #6 in Figure 9.1

Using result for number of sprinklers on each branchline that we obtained in Example 9.7, calculate (S_{actual}) as follows:

$$S_{\text{actual}} = \frac{120 \text{ ft-0 in.}}{12} \text{ (see Figure 9.1)}$$
$$= 10 \text{ ft-0 in.}$$

Recall from previous examples that S_{actual} is less than S_{max}, ($S_{\text{max}} = 10.4$ ft, or 10 ft-4^{13}⁄$_{16}$ in. as computed in Example 9.5), which means that this example will install sprinklers at less than the maximum mandated area limitation of 130 ft^2, a proposal that is permitted by NFPA 13.

The spacing in bays #1 and #6 works out quite nicely, with 12 sprinklers spaced uniformly at a distance of 10 ft-0 in. between sprinklers, as shown in Figure 9.1. To calculate the distance from the centerline of the sprinkler closest to the wall to the face of the wall:

$$\tfrac{1}{2}(S_{\text{actual}}) = \tfrac{1}{2}(10 \text{ ft-0 in.})$$
$$= 5 \text{ ft-0 in. (see Figure 9.1)}$$

Note that the dimensions between all sprinklers in bays #1 and #6 add up to 120 ft-0 in. Therefore, for this example, the number of sprinklers proposed to be installed in bays #1 and #6 is shown in Figure 9.1.

(12 sprinklers per line) × (2 lines per bay)
= 24 sprinklers each in bays #1 and #6.

EXAMPLE 9.10 Calculate (S_{actual}) in Bays #2, #3, #4, and #5 in Figure 9.1

Using the result for the number of sprinklers on each branchline that was obtained in Example 9.8, which determined a requirement of 11 sprinklers in bays #2, #3, #4, and #5, calculate (S_{actual}) as shown in Figure 9.1:

$$S_{\text{actual}} = \frac{\text{Total length of bay}}{\text{Number of sprinklers on line}}$$
$$= \frac{120 \text{ ft-0 in.}}{11}$$
$$= 10 \text{ ft-10}^{15}/_{16} \text{ in. (see Figure 9.1)}$$

Note that the value for S_{actual} calculated for this example, 10 ft-10^{15}⁄$_{16}$ in., is less than S_{max}, ($S_{\text{max}} = 11.14$ ft, or 11 ft-1^{11}⁄$_{16}$ in., as computed in Example 9.6), which means that this example is proposing to install sprinklers at less than the maximum mandated area limitation of 130 ft^2, which is permitted by NFPA 13.

The value of S_{actual} calculated in this example, 10 ft-10^{15}⁄$_{16}$ in., is very unwieldy, and fractional values in the sixteenths of an inch are generally not used in fire protection design. Since the S_{actual} dimension 10 ft-11 in. is also less than S_{max} and since, in terms of plan detailing and determining cutting lengths for sprinkler pipe, the value of 10 ft-11 in. is considerably easier to work with than 10 ft-10$^{15}\!/_{16}$ in., the sprinklers can be positioned at a uniform spacing of 10 ft-11 in. and still be spaced at less than the mandated 130 ft^2 sprinkler coverage area. The spacing example continues:

- S_{actual} is 10 ft-11 in., as demonstrated above.
- There are 11 sprinklers on each branchline.
- There are 10 spaces between 10 sprinklers at 10 ft-11 in., which equals 109 ft-2 in.
- The bay length of 120 ft-0 in., minus 109 ft-2 in., is 10 ft-10 in.
- Half of this distance is the distance from the end sprinklers to the end wall, or 5 ft-5 in., as shown here:

$$\tfrac{1}{2}(S_{actual}) = \tfrac{1}{2}(10 \text{ ft-10 in.})$$
$$= 5 \text{ ft-5 in., as shown in Figure 9.1}$$

Note that for this example, since the figure was rounded up to avoid unwieldy fractional values for S_{actual}, the distance between the end sprinkler and the wall is less than $\frac{1}{2}$ the distance between sprinklers, which is permissible. Note also that the dimensions between all sprinklers in bays #2, #3, #4, and #5 add up to 120 ft-0 in. The total number of sprinklers in each bay for bays #2, #3, #4, and #5 is

(11 sprinklers per line) $\times$ (3 lines per bay) = 33 sprinklers per bay

Actual Area of Sprinkler Coverage

After having completed the spacing of sprinklers in each of the six bays in Figure 9.1 by way of the previous examples, it is important to verify that the square footage of the sprinklers that have been spaced is equal to or less than the maximum square footage permissible by NFPA 13.

EXAMPLE 9.11 Calculate the Actual Area of Sprinkler Coverage in Bays #1 and #6 in Figure 9.1

Using the results for S_{actual} and L_{actual} that were obtained in Examples 9.1, 9.3, 9.5, 9.7, and 9.9, the equation becomes

$$S_{actual} = 10 \text{ ft-0 in. from Example 9.9}$$
$$L_{actual} = 12 \text{ ft-6 in., from Example 9.3}$$
$$A = S \times L$$
$$= (10 \text{ ft-0 in.}) \times (12 \text{ ft-6 in.})$$
$$= 125 \text{ ft}^2$$

Comparing A_{max}, which is equal to 130 ft^2 maximum for ordinary hazard occupancy (as was discussed earlier in this chapter), to the values of S_{actual} and L_{actual} calculated in the previous examples, where A is 125 ft^2, it is clear that the sprinklers within bays #1 and #6 were properly spaced at less than A_{max} and the proposed layout of sprinklers in these bays, as shown in Figure 9.1, is therefore in conformance with NFPA 13.

In the previous examples, it was assumed that there were no obstructions to sprinkler discharge within a bay. It is important to emphasize that the maximum square footage of sprinklers in a bay is a function of the greatest spacing encountered within the bay. Any shifting of sprinklers within the bay required to avoid obstructions to sprinkler discharge requires a reanalysis of spacing of the shifted sprinklers, to ensure that actual spacing is less than A_{max} in all portions of the bay, as discussed later in this chapter.

EXAMPLE 9.12 Calculate Actual Square Foot Coverage in Bays #2, #3, #4, and #5 in Figure 9.1

Using the result for S_{actual} and L_{actual} obtained for bays #2, #3, #4, and #5 in Examples 9.1, 9.3, 9.5, 9.7, and 9.9, the equation becomes

$$S_{actual} = 11 \text{ ft-0 in., from Example 9.10}$$
$$L_{actual} = 11 \text{ ft-8 in., from Example 9.4}$$
$$A = S \times L$$
$$= (10 \text{ ft-11 in.}) \times (11 \text{ ft-8 in.})$$
$$= 127.4 \text{ ft}^2$$

Comparing A_{max}, which is equal to 130 ft^2 maximum for ordinary hazard occupancy as was demonstrated earlier in this chapter, to the value calculated in this example, where A is 127.4 ft^2, it is clear that the sprinklers within bays #2, #3, #4, and #5 have been properly spaced at less than A_{max} and the proposed layout of sprinklers in these bays, as shown in Figure 9.1, is therefore in conformance with NFPA 13.

Effect of Obstructions on Sprinkler Spacing

In the previous examples, it was assumed that there were no obstructions to sprinkler discharge within a bay. Any shifting of sprinklers within the bay required to avoid obstructions to sprinkler discharge requires a reanalysis to ensure that actual spacing is less than A_{max} in all portions of the bay. A demonstration of the effect of shifting sprinklers to avoid obstructions within bay #1 and #6 is shown in Example 9.13.

EXAMPLE 9.13 Shifting Sprinklers to Avoid Obstructions in Bays #1 and #6 in Figure 9.1

In the real world, sprinklers are not always uniformly spaced as shown in the previous examples, since obstructions may require sprinklers to be shifted to avoid blockage of water spray from discharging sprinklers.

For this example, assume that a ceiling-mounted obstruction, such as a beam, creates an obstruction to the sprinkler closest to one of the end walls in bay #1 or #6 in Figure 9.1. Further assume that the sprinkler is required to shift 6 in. further from the end wall than was previously calculated, increasing the distance from the centerline of the sprinkler to the face of the wall from 5 ft-0 in. to 5 ft-6 in., with all other dimensions in the bay remaining the same as previously calculated. As can be seen from inspection, the shifted sprinkler has a larger protection area than the other sprinklers on the branchline.

Recall that in Example 9.9 the distance was calculated from the end sprinkler to the wall; as $\frac{1}{2}(S_{actual}) = \frac{1}{2}(10 \text{ ft-0 in.}) = 5 \text{ ft-0 in.}$ In this case, that sprinkler has been shifted further from the wall to a dimension of 5 ft-6 in. To determine the actual protection area of the shifted sprinkler, the shifted value for $\frac{1}{2}(S_{actual})$ is doubled to become 11 ft. Remembering that L is 12 ft-6 in. in bays #1 and #6, as shown in Figure 9.1, the actual protection area for the shifted sprinkler in those bays becomes

$$S \times L = A$$
$$11 \text{ ft-0 in.} \times 12 \text{ ft-6 in.} = 137.5 \text{ ft}^2 \tag{9.7}$$

In light of the fact that 137.5 ft² exceeds 130 ft², the maximum permissible protection area for ordinary hazard, the proposed shift of sprinklers in Example 9.13 violates the sprinkler spacing provisions of NFPA 13.

In all cases where the presence of an obstruction requires shifting of sprinklers to avoid the obstruction, a reevaluation is required to ensure that the shifted sprinklers do not cover a protection area in excess of the NFPA 13 mandated maximum. In the case of Example 9.13, an additional sprinkler must be added to each branchline in bays #1 and #6 where the obstruction is present, with all sprinklers spaced within their maximum protection areas and with no sprinkler spaced closer than the mandated minimum of 6 ft-0 in. for standard coverage sprinklers. The minimum spacing requirements are present in the standard to ensure that a sprinkler does not spray water on, or "cold solder," adjacent sprinklers within a bay, retarding activation of the adjacent sprinklers.

Show the Crossmains and Feedmains on Drawings

The examples in this chapter, and Figure 9.1, describe a center-fed tree sprinkler system. It is important to note that alternative piping arrangements for these examples are available options for the designer.

If the system shown in Figure 9.1 was a dry pipe system, the center-fed tree piping arrangement shown is permitted, but alternative permissible piping arrangements might include a system with looped feedmains, or if the roof is peaked, perhaps an end-fed system, with crossmains at each eave of the building and end-fed branchlines sloping up to the peak. As previously discussed, a gridded system is prohibited by NFPA 13 for dry pipe systems.

If the system shown in Figure 9.1 is a wet pipe system, the designer may determine that a gridded system is hydraulically advantageous. Other options, such as looped mains are also permissible and may provide a hydraulic advantage for this arrangement as well.

When crossmains run parallel to and below two roof beams, trapeze hangers, in accordance with NFPA 13, paragraph 9.1.1.6, and as discussed in Chapter 8, "Design and Layout of Sprinkler Systems," will permit supporting a crossmain from both beams. As an alternative, it may be possible to run a crossmain parallel and directly below a beam in such a manner as to permit direct attachment to the beam without need for a trapeze hanger.

It may not always be feasible to run a crossmain parallel and directly below a single bar joist. Bar joists are less likely (as compared to beams) to be installed perfectly straight in a bay and may not always be colinear with joists in an adjacent bay. Further, depending on the size and weight of the crossmain or feedmain being hung, and the structural capabilities of the bar joists, it may not be structurally feasible to suspend a sprinkler main from a single bar joist.

In some jurisdictions, a structural analysis may be required to ensure that a single bar joist possesses the structural capacity to support a sprinkler crossmain or feedmain. In the absence of a structural analysis, it may be a wise choice to arrange crossmain and feedmain locations such that they are suspended from no fewer than two bar joists with a trapeze hanger. It should also be stressed that bar joists are intended to support loads from the top of the joists, not from the bottom. Hangers are therefore required to be attached to the top of the bar joist.

Another factor affecting crossmain and feedmain locations not yet discussed is the presence of numerous masonry walls in a building, such as might be found in a school or a hospital. Penetrating these walls costs the installing contractor time and money for such field operations as creating an opening in the wall, installing and patching a pipe sleeve to protect the wall from settling and expansion of the sprinkler pipe, and providing a fire stopping system to seal between the sleeve and the piping to prohibit a fire on one side of the wall from spreading to the opposite side through sleeve or the wall opening. A designer should consider the time savings and cost savings associated with arranging piping to minimize the number of penetrations of masonry walls. For example, if a crossmain is in a corridor and an adjacent music room requires three branchlines, consider making one penetration of the wall, with a short crossmain in the music room to feed the three branchlines. Where numerous such situations exist on a project, the sprinkler system installation cost could be considerable.

Sprinkler Spacing for Extended Coverage Upright and Pendent Sprinklers

Extended coverage sprinklers are tested and listed to provide a larger area of water spray coverage at the floor, which is achieved through a combination of specialized deflector design and larger orifices. Most extended coverage sprinklers also have higher than typical minimum spacing requirements to prevent wetting the thermal element of an adjacent sprinkler.

Upright and Pendent Extended Coverage Sprinkler Spacing

For upright and pendent extended coverage sprinklers, it is permitted to use the $A = S \times L$ rules, as discussed in previous examples involving Equation 9.1, based on the listed even-number square (i.e., 16×16, 18×18, 20×20), as shown in Table 9.5. The maximum protection area of coverage for extended coverage sprinklers is 400 ft^2, in accordance with this table.

EXAMPLE 9.14 Extended Coverage Sprinkler Spacing

An elementary school classroom, with dimensions of 145 ft $\times$ 46 ft, is determined that an occupancy is to be classified as light hazard. An extended coverage sprinkler, listed as permissible for coverage up to 20 ft $\times$ 20 ft, is selected for this room. Assuming no obstructions, determine the number of sprinklers and the spacing of the sprinklers in the room.

Along the 46-ft dimension of the room, three sprinklers are required, since the maximum distance between sprinklers may not exceed 20 ft and the maximum

TABLE 9.5

Protection Areas and Maximum Spacing (Extended Coverage Upright and Pendent Spray Sprinklers)

Construction Type	Light Hazard Protection Area (ft²)	Spacing (ft)	Ordinary Hazard Protection Area (ft²)	Spacing (ft)	Extra Hazard Protection Area (ft²)	Spacing (ft)	High-Piled Storage Protection Area (ft²)	Spacing (ft)
Unobstructed	400	20	400	20	—	—	—	—
	324	18	324	18	—	—	—	—
	256	16	256	16	—	—	—	—
	—	—	196	14	196	14	196	14
	—	—	144	12	144	12	144	12
Obstructed noncombustible (when specifically listed for such use)	400	20	400	20	—	—	—	—
	324	18	324	18	—	—	—	—
	256	16	256	16	—	—	—	—
	—	—	196	14	196	14	196	14
	—	—	144	12	144	12	144	12
Obstructed combustible	N/A	N/A	N/A	N/A	N/A	N/A	N/A	N/A

For SI units, 1 ft = 0.3048 m; 1 ft² = 0.0929 m².

Source: NFPA 13, 2002, Table 8.8.2.1.2.

distance from any wall may not exceed 10 ft, per Table 9.5. The maximum permissible area of coverage for this sprinkler is 400 ft² (20 ft × 20 ft). Spacing along the 46-ft width is determined as follows:

$$L = \frac{\text{Total width of room}}{\text{Number of branchlines}}$$

$$= \frac{46 \text{ ft}}{3} \tag{9.8}$$

$$= 15 \text{ ft-4 in.}$$

Recall that the elementary school classroom in this example has dimensions of 145 ft × 46 ft. Spacing along branchlines oriented parallel to the 145-ft length is determined by dividing the length by the maximum distance between sprinklers:

$$\text{Sprinklers on the branchline} = \frac{\text{Total length of bay}}{\text{Max. distance between sprinklers}}$$

$$= \frac{145 \text{ ft}}{20 \text{ ft}}$$

$$= 8$$

$$\text{Spacing of sprinklers on the branchline} = \frac{\text{Total length of bay}}{\text{Number of sprinklers}}$$

$$= \frac{145 \text{ ft}}{8}$$

$$= 8 \text{ ft-}1\frac{1}{2} \text{ in.}$$

The number of extended coverage sprinklers in the room is therefore (3) × (8), or 24 sprinklers, spaced at 15 ft-4 in. × 18 ft-1½ in., or 277.9 ft^2 per sprinkler, which is less than the A_{max} of 400 ft^2 per sprinkler permissible by its listing.

The manufacturer of the sprinkler lists minimum required water flows associated with each extended coverage sprinkler (16 ft × 16 ft, 18 ft × 18 ft, and 20 ft × 20 ft), and no interpolation is permitted between flows for extended coverage sprinklers. Since one sprinkler spacing dimension in this room is 18 ft-1½ in., exceeding 18 ft, the water flow and pressure associated with 20 ft × 20 ft spacing is required when hydraulically calculating the sprinklers in this room.

Extended Coverage Sprinklers for Extra Hazard or High-Piled Storage

Extended coverage sprinklers for extra hazard or high-piled storage must be listed for that purpose. It is permitted to use the $A = S \times L$ rules, and Table 9.5 shows that a 14-ft maximum spacing applies to these occupancies and that a 196-ft^2 maximum area per sprinkler applies. Table 9.5 also permits a 14 × 14, 144 ft^2 spacing for extended coverage sprinklers listed for extra hazard or high piled storage applications.

Sprinkler Spacing—Standard Sidewall Sprinklers

Standard sidewall sprinklers are permitted to be spaced using the $A = S \times L$ methodology discussed in previous sections of this chapter, with the proviso that the maximum coverage shall not exceed 196 ft^2, as shown in Table 9.6, the maximum distance to any wall not exceed ½ the maximum distance between sprinklers (S), and the minimum distance between sprinklers shall not be less than 6 ft.

Sidewall sprinklers must be listed for the occupancy in which they are applied, with light and ordinary hazard permitted to be protected by listed sprinklers in accordance with Table 9.6. Note from the table that the room finish plays a vital role in determining the spacing of standard coverage sidewall sprinklers.

TABLE 9.6

Protection Areas and Maximum Spacing (Standard Sidewall Spray Sprinkler)

	Light Hazard		Ordinary Hazard	
	Combustible Finish	Noncombustible or Limited-Combustible Finish	Combustible Finish	Noncombustible or Limited-Combustible Finish
Maximum distance along the wall (S)	14 ft	14 ft	10 ft	10 ft
Maximum room width (L)	12 ft	14 ft	10 ft	10 ft
Maximum protection area	120 ft^2	196 ft^2	80 ft^2	100 ft^2

For SI units, 1 ft = 0.3048 m; 1 ft^2 = 0.0929 m^2.

Source: NFPA 13, 2002, Table 8.7.2.2.1.

TABLE 9.7

Protection Area and Maximum Spacing for Extended Coverage Sidewall Sprinklers

	Light Hazard				Ordinary Hazard			
	Protection Area		Spacing		Protection Area		Spacing	
Construction Type	ft²	m²	ft	m	ft²	m²	ft	m
Unobstructed, smooth, flat	400	37.2	28	8.5	400	37.2	24	7.3

Source: NFPA 13, 2002, Table 8.9.2.2.1.

Using Table 9.6 for multiple sidewall sprinklers in a room, the maximum distances shown represent the centerline distance between sprinklers in the room. Sidewall sprinklers are permitted to be installed on opposite or adjacent walls, provided they are not located within the protection area of another sprinkler and would therefore not be wetted or sprayed upon by any adjacent sprinkler. Sidewall sprinklers are not permitted to be installed back-to-back without being separated by a continuous lintel or soffit.

Sprinkler Spacing—Extended Coverage Sidewall Sprinklers

Extended coverage sidewall sprinklers are spaced in a manner similar to standard sidewall sprinklers but are constrained by the provisions of Table 9.7. By inspection of the table, we can see that permitted listings are for light and ordinary hazard and that the maximum protection area is 400 ft².

EXAMPLE 9.15 Spacing of Extended Coverage Sidewall Sprinklers

A community activity room with a smooth, unobstructed, flat ceiling, 18 ft-6 in. wide by 110 ft long, is to be protected by an automatic sprinkler system. The activity room is classified as light hazard occupancy by NFPA 13, and extended coverage sidewall sprinklers are permitted to be used in accordance with Table 9.7. One manufacturer of extended coverage sprinklers lists spacing up to 16 ft wide × 20 ft long. Extended coverage sidewall sprinklers are supplied by this manufacturer for 16 ft × 16 ft spacing, 16 ft × 18 ft spacing, and 16 ft × 20 ft spacing, all of which cover less than the 400 ft² maximum permitted coverage area. Since the width of the room exceeds 18 ft, select the sprinkler capable of protecting 16 ft wide by 20 ft long. Since the length of the room is 110 ft, and the maximum width of protection for this sprinkler is 16 ft-0 in., seven extended coverage sidewall sprinklers, spaced at 15.714 ft (15 ft-8⁹⁄₁₆ in.) apart, are needed and will flow at the minimum flow rate associated with the 16 ft × 20 ft sidewall.

Sprinkler Spacing for ESFR Sprinklers

NFPA 13 places a number of limitations on the use of ESFR sprinklers, in recognition of their unique design and in accordance with the high challenge fires they are designed to protect. ESFR sprinklers are spaced in accordance

TABLE 9.8

Protection Areas and Maximum Spacing of ESFR Sprinklers

Construction Type	Ceiling/Roof Heights up to 30 ft (9.1 m)				Ceiling/Roof Heights over 30 ft (9.1 m)			
	Protection Area		Spacing		Protection Area		Spacing	
	ft²	m²	ft	m	ft²	m²	ft	m
Noncombustible unobstructed	100	9.3	12	3.7	100	9.3	10	3.1
Noncombustible obstructed	100	9.3	12	3.7	100	9.3	10	3.1
Combustible unobstructed	100	9.3	12	3.7	100	9.3	10	3.1
Combustible obstructed	N/A		N/A		N/A		N/A	

Source: NFPA 13, 2002, Table 8.12.2.2.1.

with Table 9.8. Buildings up to 45 ft high containing up to 40 ft of storage can be protected using ESFR protection; roof decks are allowed to be either combustible or noncombustible; and ESFR sprinklers may be installed in buildings with unobstructed or obstructed construction.

There are some buildings where ESFR sprinklers may not be installed. For example, if the depth of the solid structural members is greater than 12 in., sprinklers must be installed within each bay and minimum spacing and area of coverage limitations still apply. Further, ESFR sprinklers can only be used in wet pipe systems where the ceiling slope does not exceed a pitch of 2 in 12.

TABLE 9.9

Positioning of Sprinklers to Avoid Obstructions to Discharge (ESFR Sprinkler)

Distance from Sprinkler to Side of Obstruction (A)	Maximum Allowable Distance of Deflector above Bottom of Obstruction (in.) (B)
Less than 1 ft	0
1 ft to less than 1 ft 6 in.	1½
1 ft 6 in. to less than 2 ft	3
2 ft to less than 2 ft 6 in.	5½
2 ft 6 in. to less than 3 ft	8
3 ft to less than 3 ft 6 in.	10
3 ft 6 in. to less than 4 ft	12
4 ft to less than 4 ft 6 in.	15
4 ft 6 in. to less than 5 ft	18
5 ft to less than 5 ft 6 in.	22
5 ft 6 in. to less than 6 ft	26
6 ft	31

For SI units, 1 in. = 25.4 mm; 1 ft = 0.3048 m.
Note: For (A) and (B), refer to Figure 8.12.5.1.1 [of NFPA 13].

Source: NFPA 13, 2002, Table 8.12.5.1.1.

Table 9.8 shows ESFR sprinkler spacing that is dependent on construction type and roof height. It is assumed that all occupancies are storage occupancies per NFPA 13, Chapter 12. In all cases shown in the table, the maximum protected area is 100 ft². The standard provides the possibility for 110-ft² spacing to avoid obstructions caused by trusses and bar joists, provided that the average actual floor area protected by the moved sprinkler and the adjacent sprinklers does not exceed 100 ft² and provided that the adjacent branch-lines maintain the same pattern and that the distance between any sprinkler does not exceed 12 ft.

ESFR sprinkler design is based on 12 hydraulically remote sprinklers operating plus up to 2 sprinklers that may be positioned below obstructions. The design area consists of 4 sprinklers on three lines with a minimum design area of 960 ft². ESFR sprinklers do not use design density tables and cannot be used with pipe schedule systems. ESFR sprinkler design criteria cannot be interpolated.

ESFR sprinklers are particularly affected by the presence of obstructions. Table 9.9 outlines some of the many considerations that must be addressed when using ESFR sprinklers. If sprinklers cannot comply with the obstruction rules in Table 9.9, the designer must place sprinklers under the obstruction. There are a number of exceptions listed in NFPA 13 that modify Table 9.9.

Sprinkler Spacing for Special Sprinklers

NFPA 13 permits a wide variety of sprinklers for special applications to be used and spaced in accordance with their listing. Some of the sprinklers cited by the standard include the following:

- *Residential sprinklers*—residential pendent and sidewall sprinklers are listed for use in residential occupancies and are spaced in accordance with their listing.
- *Large drop sprinklers*—large drop sprinklers are spaced in accordance with Table 9.10, the provisions of NFPA 13, and their listing.
- *In-rack sprinklers*—NFPA 13 outlines requirements for maximum system size, permissible K-factors, and the use of water shields to prevent cold soldering of in-rack sprinklers.
- *Attic sprinklers*—attic sprinklers are spaced in accordance with their listing.

TABLE 9.10

Protection Areas and Maximum Spacing for Large Drop Sprinklers

Construction Type	Protection Area		Maximum Spacing	
	ft²	m²	ft	m
Noncombustible unobstructed	130	12.1	12	3.7
Noncombustible obstructed	130	12.1	12	3.7
Combustible unobstructed	130	12.1	12	3.7
Combustible obstructed	100	9.3	10	3.1
Rack storage applications	100	9.3	10	3.1

Source: NFPA 13, 2002, Table 8.11.2.2.1.

Obstructed Construction Sprinkler Spacing

NFPA 13 requires that sprinklers be located in a way that promotes minimization of obstructions to discharge, with additional sprinklers provided where needed to ensure adequate sprinkler coverage of the hazard. Obstruction criteria vary by type of sprinkler and by type of obstruction.

Sprinklers are required to be positioned in accordance with the minimum distances and special exceptions outlined in NFPA 13, so that they are located sufficiently away from obstructions such as truss webs and chords, pipes, columns, and fixtures.

Types of Obstructions

Examples of some of the obstruction issues addressed by NFPA 13 include the following:

- *Roof construction*—sprinklers are required to be spaced to avoid structural members and web members, depending on their dimensions, as illustrated by Example 9.13.
- *Peaked roofs*—sprinklers are to be positioned not greater than 3 ft from the roof peak.
- *Deflector orientation*—sprinklers are to be positioned with deflectors parallel to the slope of the roof, and sprinklers are spaced along the slope of the roof.
- *Double joist construction*—where two levels of joist construction exist below a building roof, with no flooring over the lower set of joists, and where there exists 6 in. or more of clearance between the two levels of joists, sprinklers are required as shown in Figure 9.5.
- *Beams and obstructions to discharge*—NFPA 13 provides distances from which sprinklers shall be separated from obstructions such as beams; see Figure 9.6, Table 9.11, and Example 9.16.

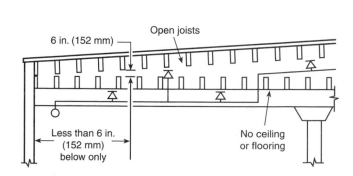

FIGURE 9.5
Arrangement of Sprinklers Under Two Sets of Open Joists—No Sheathing on Lower Joists.

(Source: NFPA 13, 2002, Figure 8.6.4.1.5.1)

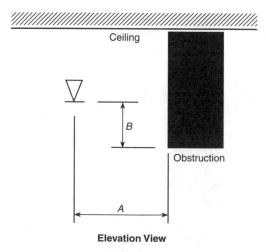

FIGURE 9.6
Positioning of Sprinklers to Avoid Obstructions to Discharge (SSU/SSP).

(Source: NFPA 13, 2002, Figure 8.6.5.1.2(a))

TABLE 9.11	
Positioning of Sprinklers to Avoid Obstructions to Discharge (SSU/SSP)	
Distance from Sprinklers to Side of Obstruction *(A)*	Maximum Allowable Distance of Deflector above Bottom of Obstruction (in.) *(B)*
Less than 1 ft	0
1 ft to less than 1 ft 6 in.	2½
1 ft 6 in. to less than 2 ft	3½
2 ft to less than 2 ft 6 in.	5½
2 ft 6 in. to less than 3 ft	7½
3 ft to less than 3 ft 6 in.	9½
3 ft 6 in. to less than 4 ft	12
4 ft to less than 4 ft 6 in.	14
4 ft 6 in. to less than 5 ft	16½
5 ft and greater	18

For SI units, 1 in. = 25.4 mm; 1 ft = 0.3048 m.
Note: For *(A)* and *(B)*, refer to Figure 8.6.5.1.2(a) [of NFPA 13].

Source: NFPA 13, 2002, Table 8.6.5.1.2.

- *Soffits and obstructions against walls*—spacing rules are provided in NFPA 13 to ensure that items under the obstruction are protected by the sprinkler; see Example 9.17 and Figure 9.7.

- *Vertical obstructions*—vertical obstructions such as floor-mounted partitions and shelf storage have spacing requirements outlined by NFPA 13.

- *Horizontal obstructions*—horizontal obstructions, such as ducts, greater than 48 in. wide require additional sprinklers, but some less than 48 in. do too, depending on the type of sprinkler.

- *Ceiling pockets*—ceiling pockets require sprinklers installed within, depending on the volume and depth of the pocket.

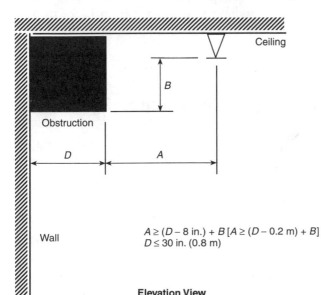

FIGURE 9.7
Obstructions Against Walls (SSU/SSP).

(Source: NFPA 13, 2002, Figure 8.6.5.1.2(b))

$A \geq (D - 8\ \text{in.}) + B\ [A \geq (D - 0.2\ \text{m}) + B]$
$D \leq 30\ \text{in.}\ (0.8\ \text{m})$

Elevation View

EXAMPLE 9.16 Beam Obstruction for a Standard Upright Sprinkler

A wet pipe system is installed in an exposed warehouse. One line of standard coverage sprinklers is 2 ft-6 in. from a beam, where the bottom of the beam is 12 in. below the deflector. Make a determination of a corrective measure, if needed.

In accordance with Table 9.11, a branchline that is 2 ft-6 in. from the side of a beam needs to have its deflector at a dimension that does not exceed 7½ in. above the bottom of the beam, to avoid water spray obstruction. The proposed arrangement is therefore noncompliant.

Assuming that the deflector elevation cannot be adjusted, the deflector will be kept at a distance of 12 in. above the bottom of the beam and the branchline must shift to 3 ft-6 in. from the edge of the beam, in accordance with Table 9.11.

As an alternative solution, assuming that the deflector elevation can be adjusted and still maintain the required deflector distance, the branchline may remain at 2 ft-6 in. from the beam, provided that the sprinkler deflector is lowered so that the distance from deflector to the bottom of the beam is 7½ in., in accordance with Table 9.11.

EXAMPLE 9.17 Resolving an Obstruction Against a Wall

A standard coverage sprinkler system is being installed in a shoe store in a shopping mall. A soffit is present at the wall, as shown in Figure 9.7, where the "D" dimension is 29 in., the "A" dimension is 36 in., and the "B" dimension is 20 in. Use the formula shown in Figure 9.7 to determine whether a sprinkler is required under the soffit:

$$A \geq (D - 8 \text{ in.}) + B$$
$$D \leq 30 \text{ in.}$$

According to Figure 9.7, if both formulae in the figure are true, then a sprinkler is not required below the soffit: D is less than 30 in., but A (36 in.) is not greater than or equal to (29 in. − 8 in.) + 20 in. = 41 in. Therefore, a sprinkler is required under the soffit, or the sprinkler must be moved further away from the wall. For example, if the A dimension is increased to 42 in., then the sprinkler is not required under the soffit, assuming that the maximum area limitation for the sprinkler is not exceeded.

Special Situations Affecting Sprinkler Spacing

Although a wide variety of spacing situations has been covered in this chapter, there remain some special situations addressed by NFPA 13 that are worthy of note.

Concealed Spaces

Combustible concealed spaces require sprinklers, with a number of exceptions outlined in NFPA 13, including the provision shown in Figure 9.5, which requires sprinklers in combustible concealed spaces where there exists 6 in. or more of clearance between the two levels of joists.

Vertical Shafts

Vertical shafts require a sprinkler at the top of the shafts, except for noncombustible or limited combustible duct, electrical, or mechanical shafts. Shafts with combustible surfaces require sprinklers at alternate floor levels. Accessible shafts require a sprinkler at the bottom of the shaft.

Stairways

Combustible stairways require sprinklers below all combustible stairs and landings. Noncombustible stairs require sprinklers at the top of the shaft and under the first landing above the bottom of the shaft.

Vertical Openings

Vertical openings, such as might be found at an escalator in a department store, require a water curtain of closely spaced sprinklers around the opening, not greater than 6 ft on center, and no closer than permitted by the listing of the sprinkler. Where sprinklers are installed closer than their listing permits, baffles are required. The purpose of this water curtain is to attempt to prevent heat from a fire on the lower level from spreading through the opening and actuating sprinklers on the level above, resulting in a diminished available water supply for the sprinklers affected by the fire.

Elevator Hoistways and Machine Rooms

Elevator pits have the potential to gather trash and perhaps grease or spilled oil, and a sidewall sprinkler is required at the bottom of the elevator shaft but may be eliminated if the shaft is noncombustible and contains no hydraulic fluids. A sprinkler is required at the top of the elevator shaft unless the hoistway is noncombustible and the car enclosure materials meet the requirements of ASME A17.1, *Safety Code for Elevators and Escalators* [1]. An ordinary- or intermediate-temperature sprinkler is required in the elevator machine room.

Spaces Under Ground Floors, Exterior Docks, and Platforms

Sprinklers are required in spaces under ground floors, exterior docks, and platforms unless they are not accessible for storage and are protected against wind-borne debris and unless the space has no equipment such as conveyors or fuel-fire heating units, the floor over the space is of tight construction, and no combustible or flammable liquids are present.

Exterior Roofs and Canopies

Exterior roofs and canopies exceeding 4 ft in width require sprinklers, unless the canopy or roof is of noncombustible. Roofs or canopies where combustibles are stored or handled, such as a loading dock, require sprinklers. Exterior exit corridors require sprinklers unless the corridor is 50 percent open and when the corridor is entirely of noncombustible construction.

Other Special Situations

NFPA 13 also includes requirements for dwelling units, library stack rooms, electrical equipment, industrial ovens and furnaces, open-grid ceilings, drop-out ceilings, old style sprinklers, and stages.

Worksheet

Box 9.1 summarizes the procedure for spacing automatic sprinklers.

BOX 9.1

Automatic Sprinkler Spacing Worksheet

- *Branchline orientation*—sprinkler branchlines are always oriented to be perpendicular to bar joists
- *Sprinkler system configuration*—configurations are the tree, the grid, and loop.
- *Occupancy classification*—occupancies are light hazard, ordinary hazard, and extra hazard, per Chapter 4, "Hazard and Commodity Classification"
- *Protection area of coverage*—light hazard occupancies are spaced in accordance with Table 9.1; ordinary hazard occupancies are spaced in accordance with Table 9.2; extra hazard occupancies are spaced in accordance with Table 9.3; high-piled storage occupancies are spaced in accordance with Table 9.4.
- *Maximum area permitted to be protected by a single sprinkler system*—light hazard and ordinary hazard sprinkler systems, 52,000 ft^2; extra hazard occupancies, 40,000 ft^2; combined occupancies shall not exceed the floor area specified for that hazard and the total area coverage shall not exceed 52,000 ft^2.
- *Area protected by each sprinkler (A):*

$$A = S \times L$$

where:

  A = area protected by the sprinkler, in square feet

  S = distance between sprinklers on the branchlines, in feet

  L = distance between branchlines, in feet

- *Number of branchlines in a bay or room:*

$$\text{Number of branchlines} = \frac{\text{Width of bay}}{L_{max}}$$

- *Actual distance between branchlines (L):*

$$L = \frac{\text{Total width of bay}}{\text{Number of branchlines in the bay}}$$

- *Maximum permissible distance between sprinklers (S_{max}):*

$$S_{max} = \frac{A_{max}}{L}$$

- *Minimum permissible number of sprinklers on each branchline:*

$$\text{Minimum number of sprinklers on each branchline} = \frac{\text{Total length of bay}}{S_{max}}$$

- Actual distance between each sprinkler on branchlines (S_{actual})

$$S_{actual} = \frac{\text{Total length of bay}}{\text{Number of sprinklers on line}}$$

Summary

The spacing of automatic sprinklers is a formulaic procedure that begins with determining branchline orientation, selecting system configuration, classifying occupancy, determining protection area of coverage, calculating the maximum area permitted to be protected by a single sprinkler system, and the area protected by each sprinkler. Layout proceeds with calculation of the number of branchlines in a bay, actual distance between branchlines, maximum permissible distance between sprinklers, minimum permissible number of sprinklers on each branchline, and actual distance between each sprinkler on branchlines. Once sprinklers have been spaced and branchlines positioned, crossmains and feedmains can then be determined.

BIBLIOGRAPHY

Reference Cited

1. ASME/ANSI A17.1, *Safety Code for Elevators and Escalators*, American Society of Mechanical Engineers, New York, 2000.

NFPA Codes, Standards and Recommended Practices

NFPA Publications National Fire Protection Association, 1 Batterymarch Park, Quincy, MA 02169-7471

The following is a list of NFPA codes, standards, and recommended practices cited in this chapter. See the latest version of the *NFPA Catalog* for availability of current editions of these documents.

NFPA 13, *Standard for the Installation of Sprinkler Systems*

10

Hydraulic Calculations

Steven Scandaliato, SET, and Morgan Hurley, PE

The purpose for hydraulically calculating sprinkler systems is to determine the pipe sizing that will be needed to ensure that adequate water and pressure are delivered to the system. Hydraulic calculations for fire sprinkler systems have only been required since the late 1970s. Previously, sprinkler pipe sizes were determined by tables of pipe schedules. These tables allowed users to assign pipe sizes to their systems based upon predetermined numbers of sprinklers per pipe diameter. Guidelines for hydraulically calculating sprinkler systems began to appear in NFPA 13, *Standard for the Installation of Sprinkler Systems*, and by the 1994 edition, all sprinkler systems were required to be calculated and pipe schedules were limited to small systems of not more than 5000 ft^2 in total area. Although pipe schedule systems are permitted to be used for new sprinkler systems on a very limited basis, the vast majority of new automatic sprinkler systems are required to be hydraulically calculated.

Determining Water Supply Requirements of a Given Sprinkler Design

Like many other building systems, calculations for fire sprinkler systems are a type of "demand" calculation. To determine the most demanding area, a portion of the system is identified as the area that the water has to work the hardest to supply. It is often referred to as the "most remote" or "most demanding" area of operation. Once this area is defined using the criteria set forth by the design professional and NFPA 13, the calculations will simulate all the sprinklers in this area discharging at the same time, creating the "worst" condition expected for that specific occupancy. This demand is created by establishing the minimum water and pressure required for this area. The factors that are involved include hazard and commodity classifications, sprinkler layouts and piping configurations, water supplies, and sprinkler type.

Hazard and Commodity Classification

Before calculations can begin, the design criteria for the system must be established. This requires determination of the hazard and commodity classification. Differing classifications may be used for different areas of a given building, or in some cases the entire building may be one predominant hazard class. The reason a hazard must be identified first is because it results in the application of a specific design density. *Density*, for the purpose of hydraulic calculations, is the amount of water that is designed to be delivered onto the floor below. It is measured in units of volume per unit time per area, such as gal/min/ft², and when SI units are used, this reduces to units of length per time, such as mm/min. An occupancy that is determined to be light hazard can be assigned a minimum density of 0.10 gpm/ft². An ordinary Group II occupancy has a minimum density of 0.20 gpm/ft². These, as well as many other minimum densities can be found in NFPA 13. The determination of occupancy and assigning densities is discussed in detail in Chapter 4, "Hazard and Commodity Classification."

Sprinkler Layout and Piping Configuration

Another factor involved in determining the demand of a sprinkler system is the sprinkler layout and piping configuration of the system. Sprinkler spacing plays a very critical role in establishing system demand, and the way in which the sprinklers are connected has a direct effect on the demand of the system as well. The journey that water must take through the piping to reach the remote area directly affects total pressure loss. System sizing can vary greatly depending on the type of piping configuration used. For instance, the pipe sizing for a large warehouse system will be much smaller if a "gridded" system is used instead of a "center" or "side-feed" tree system, whereas a smaller office building that may not be as wide or long as a warehouse would be better served with a "center" or "looped" design.

Water Supply

A third factor involved is the water supply. The majority of fire sprinkler systems are fed from public water supplies that can vary significantly in pressure and capacity and it is this tested supply that creates the challenge of fire sprinkler system design. The goal is to design a system that is most cost-effective by utilizing as much of the water possible from the given supply while at the same time providing a design that will adequately supply the system with the pressure and water needed to meet the minimum criteria.

Protected Area

The fourth and final factor that directly affects hydraulic calculations is the actual or "assigned" area protected by the sprinklers. The reason this is so important is that the very first step in the calculation process is using Equation 10.1 to establish the minimum pressure and capacity required to meet the prescribed density.

$$(d)(A) = Q_m \tag{10.1}$$

where:

d = the density required by the occupancy or commodity classification (gpm/ft²)

A = the assigned area of the most demanding sprinkler (ft²)

Q_m = the minimum gallons per minute required at the sprinkler to achieve the density required (gpm)

The most cost-effective design is one that has few branchlines and more sprinklers on each branchline, because sprinklers are cheaper than pipe. However, this logic leads to designs that are not equal in spacing. For example, a light hazard system allows standard sprinklers to be spaced up to a maximum of 225 ft²/sprinkler. An equally spaced sprinkler layout would be 12 ft-0 in. × 12 ft-0 in., as compared to a system that is skewed for fewer branchlines, like 9 ft-0 in. between sprinklers and 14 ft-0 in. between branchlines. Both designs fall under 225 ft²/sprinkler; however the latter system will have fewer branchlines and will probably cost less in material. Regardless of the spacing layout a designer chooses, assigning the correct square footage is extremely important. It is at this final factor that the process of hydraulic calculations actually begins. The system sizing and piping configuration is being established as soon as sprinklers are displayed on the drawing. A couple of examples follow.

Figure 10.1 is a partial head layout that is located in the corner of a room. The dimensions given represent the distances between sprinklers in both directions as well as the distance from the walls to the sprinklers in both directions. At first it appears that the sprinklers are laid out to 140 ft²/sprinkler. There is 10 ft-0 in. dimension between sprinklers on the branchline and 14 ft-0 in. between branchlines. However, the distance between the sprinklers on the branchline (10 ft-0 in.) is not equal when it comes to the last sprinkler when measured to the wall. This distance is 3 ft-6 in. Because of the physical characteristics of standard sprinklers, the spray pattern a sprinkler produces must be understood. Standard spray uprights and pendents are designed to spray equally in all directions (see Figure 10.2). Their circular patterned deflector also assures that they do not spray in a rectangular shape, which is how the rules for layout are applied, but rather in a circular one.

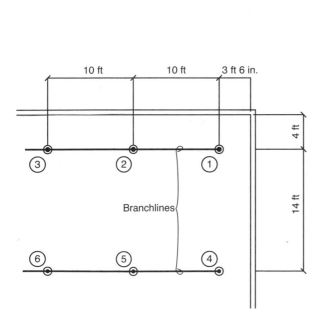

FIGURE 10.1
Partial Sprinkler Layout.

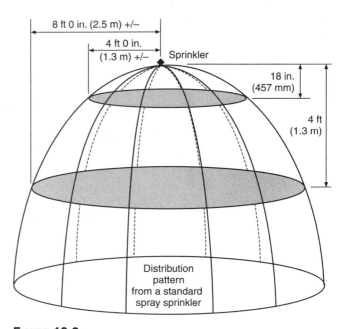

FIGURE 10.2
Typical Distribution Pattern from a Standard Spray Sprinkler.

(Source: *Fire Protection Handbook*, NFPA, 2003, Figure 10.11.18)

Since a sprinkler is designed to spray equally in all directions, the larger of the dimensions must be used when determining the length and width for the area of coverage. For example, there are 10 ft-0 in. between sprinklers and 3 ft-6 in. from the last sprinkler to the wall. Applying the rule of equality, sprinkler #1 is expected to throw a minimum of 5 ft-0 in. in one direction but only 3 ft-6 in. in the opposite direction. Since the sprinkler must throw to the greater of the two distances, 5 ft-0 in. must be used. Since the sprinkler is going to throw 5 ft-0 in. toward sprinkler #2, then it will automatically be throwing 5 ft-0 in. towards the wall. Even though it only needs 3 ft-6 in., the sprinkler will provide a minimum coverage based on the larger of the two distances. Therefore, the first dimension for determining area will be 10 ft-0 in. Taking the larger of the two distances and multiplying it by 2 gives (5 ft)(2) = 10 ft-0 in.

The second dimension is then determined by doing the same thing in the perpendicular direction. The 14 ft-0 in. dimension is the distance between the branchlines. The remaining distance from the last sprinkler to the wall is 4 ft-0 in. Since sprinkler #1 will be expected to spray at least half of the distance to sprinkler #4, it will account for 7 ft-0 in. The remaining distance to the wall is less than the distance between sprinklers #1 and #4; therefore, the larger of the two is used, which is 7 ft-0 in. This is then doubled because it is going to spray this amount in both directions, towards sprinkler #4 and the wall. So the length times width values are 10 ft × 14 ft, which leaves an "assigned" sprinkler square footage of 140 ft^2. This is quite a bit different than taking the square footage determined by dimensions to the walls.

Degree of Accuracy

Now that the factors involved have been identified, the calculation process can begin. In any calculation, engineers should be mindful of significant figures; the number of digits needed to express the number to within the uncertainty of the measurement. By looking at the variables that are involved, it should be somewhat obvious as to just what level of accuracy these calculations should produce. Two major variables are water supply and friction loss.

Accuracy of Water Supply Data

All sprinkler system hydraulic calculations are compared to the available water supply. Often, the water supply data is obtained by a hydrant or fire pump flow test where the residual pressure is measured by a cap gage and, simultaneously, a flow quantity is measured using a pitot tube.

During this flow test, the pitot tube often "bounces around" the stream of water that flows from the orifice at the test pressure. As a result, the needle on the gage is usually fluctuating between readings. The fire protection engineer should attempt to hold the pitot tube in the water stream and simultaneously make an accurate and conservative reading of the tube's pressure gage by choosing a reading at or below the midpoint of the fluctuations.

From this test, the water supply for the project is established. Because of the measurement technique, there is uncertainty in such a measurement. A prudent approach to compensate for the uncertainty is to "degrade" the test results such that they have a built-in safety factor. For additional information, see Chapter 6, "Evaluation of Water Supply."

Friction Loss

Accuracy is also influenced by the calculation of friction loss. *Friction loss* is the common term used to describe the loss of pressure due to friction resulting from the contact of flowing water with the walls of the associated pipe, tube, or hose. Generally, friction loss is determined using the Hazen-Williams formula. With computers and calculators, it is easy to calculate friction loss with many digits to the right of the decimal point. However, many of these digits are meaningless and imply a degree of precision that does not exist. Rounding to the nearest whole number is recommended.

Overview and Key Equations for Hydraulic Calculations

Background

The calculation process begins with assuming a set of design criteria. These criteria include the hazard and commodity class, the system layout, and the water supply. Prior to beginning, the design method that will be used must be decided. For standard commercial projects, there are two different methods that NFPA 13 permits. These methods involve different ways of determining the number of sprinklers that are included in the calculations. One method, called the *density/area method*, is based on providing a given minimum density over an overall floor area. For instance, the minimum criterion without permitted adjustments for light hazard occupancy is 0.10 gpm/ft^2 over a total area of 1500 ft^2. This method does not take into account walls, barriers or other obstructions to fire growth and heat transfer. It is a process of mapping out the required floor area and flowing all sprinklers that fall into that area without regard for walls. This area must be demonstrated to be the most demanding area of the system.

The second method is called the *room design method*. For the area/density method, the minimum design area for ordinary and light hazard occupancies is 1500 ft^2, with all sprinklers in that design area required to flow. In areas with numerous small rooms, the number of flowing sprinklers in that area can be considerable. The room design method (which allows for the calculation of all sprinklers in the largest room plus communicating space, where applicable) permits design areas of smaller size, based on rooms with fire-rated walls and self-closing doors or protected openings. Caution should be applied when using the room design method, since future renovations could involve removing the walls that bounded the room in question.

The density/area method of calculations starts with defining a hydraulically challenging area of sprinkler operation. From a hydraulic standpoint, the worst way a fire can grow is along or parallel to a branchline. Using the density/area method, NFPA 13 provides a simple formula for determining the size and shape of the area of operation. It should be noted that the "area of operation" has also been known as the "hydraulically most remote area" or "hydraulically most demanding area." "Area of operation" is preferred because it can be easily argued that the physically most remote area may not be the hydraulically most demanding area. So when the term *area of operation* is used, it should be assumed it is referring to the area of the system that produces the hydraulically most demanding pressure and capacity needed to meet the minimum density required for that occupancy type.

There are four major equations used in the hydraulic calculation process (shown here as Equations 10.2–10.5). Equation 10.2 is written as

$$(d)(A) = Q_m \tag{10.2}$$

where:

d = the density required by the occupancy or commodity classification (gpm/ft^2)

A = the assigned area of the most demanding sprinkler (ft^2)

Q_m = the minimum gallons per minute required at the sprinkler to achieve the density required (gpm)

If this formula is not used or is not applied correctly, the calculations from this point forward are incorrect.

Equation 10.3 is the most commonly known and used formula of sprinkler hydraulics. It is used repeatedly through the calculation process, and expressed as

$$Q_m = K\sqrt{P} \tag{10.3}$$

where:

Q_m = the minimum gallons per minute required at the sprinkler to achieve a given density (gpm).

K = the K-factor assigned to the sprinkler. This value can be found in the manufacturers' data sheets. This value is treated as a constant representing the orifice of the sprinkler.

P = the pressure required at the sprinkler (psi).

Equation 10.4 is written as

$$Q_a = (Q_l)\sqrt{\frac{P_h}{P_l}} + Q_h \tag{10.4}$$

where:

Q_a = the adjusted flow after balancing two separate flows and pressures together (gpm)

Q_l = the flow that is the lower of the two flows at a given node (gpm)

Q_h = the flow that is the higher of the two flows at a given node (gpm)

P_l = the lower of the two pressures at a given node (psi)

P_h = the higher of the two pressures at a given node (psi)

This formula is commonly referred to as the "balancing formula" and is applied whenever a node is encountered in which two demands meet. This commonly takes place with system configurations such as a gridded system, looped system, or a center-feed tree system.

Equation 10.5 is written as

$$P_f = \frac{(4.52)(Q)^{1.85}}{(C)^{1.85}(D)^{4.87}} \tag{10.5}$$

where:

P_f = the pressure lost to friction in psi/ft of pipe (psi)

4.52 = a constant value as a part of the Hazen-Williams formula (gpm)

Q = the flow rate

C = the Hazen-Williams coefficient of roughness of the pipe (dimensionless)

D = the internal pipe diameter (inches)

This formula is commonly known as the "friction loss formula" and is used to determine the loss of pressure that takes place due to friction in the pipe.

These four formulae are the basis for hydraulic calculations for sprinkler systems. Knowing how to use them and when to apply them is the next step.

Steps for Hydraulic Calculations

Figure 10.3 summarizes the steps for hydraulic calculations for fire sprinkler systems.

---

1. Determine sprinkler layout.
2. Identify area of sprinkler operation to be calculated.
3. Identify the sprinkler that will be most demanding.
4. Determine the minimum flow (Q) in gpm that is required.
5. Determine the minimum pressure (P_t) that will be required.
6. Assign a pipe size, schedule, and C factor for the piping to the next node.
7. Determine the total amount of pipe lengths, type, and number of fittings and type and number of devices that should be included to the next node.
8. Calculate the friction loss per foot.
9. Calculate the total friction loss (P_f) to the next node.
10. Calculate any pressure losses or gains due to elevation (P_e).
11. Add up the pressure column ($P_t + P_e + P_f$) to establish a new required pressure (P_t) at the next node.
12. Repeat steps 3–12 for flowing and nonflowing nodes back to the source node.

FIGURE 10.3
Summary of Steps for Hydraulic Calculations for Fire Sprinkler Systems.

Step 1: Determine Sprinkler Layout.
The piping configurations most commonly used are center-feed trees, side-feed trees, loops, and grids. Each of these has advantages and disadvantages, including their associated hydraulic performance. For buildings that have long distances between exterior walls, a gridded system may be the most cost-effective system to install, and hydraulic calculations have a large part to play in that determination.

Along with the selection of a pipe size, the type of pipe to use must be chosen as well. There are several factors that can affect this decision.

Labor, fabrication, available water supply, and material availability all must be considered.

Step 2: Identify Area of Sprinkler Operation to Be Calculated.
Before determining the sprinklers that will be calculated, it must first be determined whether the density/area method or the room design method will be used. For buildings that consist primarily of large, open spaces, the density/area method is the more likely. In buildings subdivided into spaces that are smaller in area than the minimum area requirements of the area/density curves, the room design method may be considered.

Room Design Method. To use the room design method, all of the enclosing walls must have a fire resistance rating equal to or greater than the minimum water supply duration required for the type of occupancy. Specifically, the walls in light hazard occupancies must have a minimum fire resistance rating of 30 minutes, walls in ordinary hazard occupancies must have a rating of at least 60 minutes, and walls in extra hazard occupancies must have a rating of at least 90 minutes.

Also, any openings in the walls must be suitably protected. In light hazard occupancies, nonrated doors are permitted, provided that they are automatic closing or self-closing. If openings are not protected, the hydraulic calculations must include all sprinklers in the room plus two sprinklers in the communicating space nearest each unprotected opening. If the communicating space has only one sprinkler in it, then only that sprinkler needs to be included in the design area.

In ordinary or extra hazard occupancies, automatic closing or self-closing doors with the appropriate fire rating (60 minutes for ordinary hazard or 90 minutes for extra hazard) must be provided at each opening from the room for use of the room design method to permitted.

When using the room design method, consideration should be given to possible future renovations to the space. If walls are moved or removed in a future renovation, then the basis for the hydraulic calculations can be invalidated. In such occasions, reanalysis of the hydraulic calculations would be required, which could result in the need to retrofit larger piping. Discussions with the building owner or operator can help develop an understanding of the future types of renovations that might be desired. While impressing upon the owner or operator the types of renovations that could require reanalysis of the hydraulic calculations is an option, the owner or operator could change, and agreements forgotten.

Area/Density Method. If the area/density method is used, then an area and density must be selected from the area/density curve appropriate for the occupancy. It is only necessary to meet the area and density requirements of at least one point on the appropriate area/density curve. See Chapter 4 for further information.

Once the design area is selected, the number of sprinklers that must be calculated must be determined. Assuming that the sprinkler spacing within the design area is uniform, the number of sprinklers that must be calculated is equal to the design area divided by the area protected per sprinkler. If the result is a

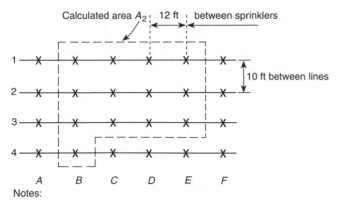

Notes:
1. For gridded systems, the extra sprinkler (or sprinklers) on branch line 4 can be placed in any adjacent location from *B* to *E* at the designer's option.
2. For tree and looped systems, the extra sprinkler on line 4 should be placed closest to the cross main.

Assume a remote area of 1500 ft² with sprinkler coverage of 120 ft²

$$\text{Total sprinklers to calculate} = \frac{\text{Design area}}{\text{Area per sprinkler}}$$

$$= \frac{1500}{120} = 12.5, \text{ calculate } 13$$

$$\text{Number of sprinklers on branch line} = \frac{1.2\sqrt{A}}{S}$$

Where:
A = design area
S = distance between sprinklers on branch line

$$\text{Number of sprinklers on branch line} = \frac{1.2\sqrt{1500}}{12} = 3.87$$

For SI units, 1 ft = 0.3048 m; 1 ft² = 0.0929 m².

FIGURE 10.4
Example of Determining the Number of Sprinklers to Be Calculated.
(Source: NFPA 13, 2002, Figure A.14.4.4)

fraction, it must be rounded up to the next whole number of sprinklers. For design areas where sprinklers are irregularly spaced, the design area must be geometrically evaluated using the actual sprinkler spacing. See Figure 10.4.

The design area must be rectangular in shape, with a dimension parallel to the branchlines equal to at least 1.2 times the square root of the area of sprinkler operation used. If the system has too few sprinklers to fulfill this requirement, then the design area must be extended until the minimum design area requirement is met.

If the number of sprinklers is such that it is not possible to construct a rectangular area, then extra sprinklers would be added to the adjacent branchline, as shown in Figure 10.5. Note that in general, it is more hydraulically demanding if the extra sprinklers are closer to the crossmain, since having them closer would result in less friction loss from the crossmain to the sprinkler, and hence a larger pressure at the sprinkler, which would result in a larger flow. This larger flow would result in greater friction from the hydraulically most demanding area to the water supply.

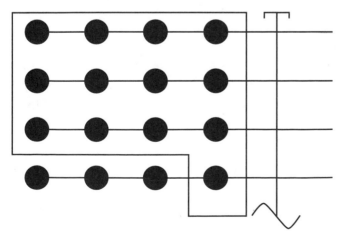

FIGURE 10.5
Nonrectangular Hydraulically Most Demanding Area.

Determination of Hydraulically Most Demanding Area. In general, for tree systems or looped systems, the hydraulically most demanding area will be on the highest floor in the physically most distant area from the water supply riser. However, this is not always the case, particularly if there are multiple occupancy classifications within the building or differences in elevation on the floor. If it is not obvious which area is the hydraulically most demanding, then several areas should be calculated, to determine the most demanding area. For gridded systems, a minimum of three sets of calculations must be prepared to ascertain which design area is hydraulically most demanding. This is commonly referred to as "peaking."

Regardless of whether the room design method or the area/density method is used, the design area and density that is used for hydraulic calculations must be selected from the area/density curve appropriate for the occupancy. If the area of the room and portions of communicating spaces required to be included in the calculation is less than the minimum area from the appropriate area/density curve, then the design density associated with the minimum design area in the appropriate area/density curve must be used. Figure 10.5 is an example that illustrates determining the hydraulically most demanding area.

Step 3: Identify the Sprinkler That Will Be Most Demanding.

Once the design area has been determined, the most demanding sprinkler within the area must be selected. A sprinkler with a larger assigned area that is closer to the main may actually be more demanding than the sprinkler that is at the end of the branchline.

Step 4: Determine the Minimum Flow (Q) in gpm That Is Required.

Using Equation 10.1, the designer must determine the minimum amount of water that must be available at this sprinkler to achieve the prescribed density, remembering to use the assigned square foot area of the sprinkler, not the actual or maximum allowed. If this value is not correct, all of the calculations done from this point forward will be incorrect.

Step 5: Determine the Minimum Pressure (P_t) That Will Be Required.

Using Equation 10.2, the pressure that will be needed in order to provide the water that is needed from step 1 must be determined, keeping in mind that these values have limits. First, NFPA 13 requires the minimum pressure at any sprinkler to be not less than 7 psi, meaning no matter what the value is after doing steps 1 and 2, the pressure must be at least 7 psi. So, for example, if the assigned area for the sprinkler is 120 ft^2 and the density required for the occupancy is 0.10 gpm/ft^2, then the minimum Q required would be 12 gpm (Equation 10.2). Taking the 12 gpm and inserting it into Equation 10.2 using a standard 5.6 K-factor sprinkler, the pressure required is 4.59 psi. This is well below the 7 psi minimum; therefore, the starting demand for this sprinkler would be defaulted to 7 psi and 14 gpm. Here, again, Equation 10.3 is used, but instead of solving for P, the equation is solved for Q, as in the following example. Some types of sprinklers have minimum pressures higher than 7 psi.

EXAMPLE 10.1

Assumptions:

Light-hazard density (0.10 gpm/ft^2)

Assigned area of sprinkler coverage equals 120 ft^2

Starting with Equation 10.2,

$$(d)(A) = Q_m \rightarrow (0.1)(120) = Q_m = 12.0 \text{ gpm}$$

Plug 12 gpm into Equation 10.3:

$$Q = K\sqrt{P} \rightarrow (12) = 5.6\sqrt{P} \rightarrow \text{Solve for } P \rightarrow P = \left(\frac{12}{5.6}\right)^2 = 4.59$$

$4.59 < 7$ psi minimum allowed by NFPA 13; therefore, $P = 7$ psi.

Using 7 psi, solve for Q:

$$Q = K\sqrt{P} \rightarrow (Q) = 5.6\sqrt{7} \rightarrow Q = 14 \text{ gpm.}$$

Once the minimum starting end sprinkler demand (psi and gpm) has been determined, the pipe sizing process begins.

Step 6: Determine the C Factor for the Piping to the Next Node.

The third value that must be identified is called the "C factor." This value is a constant derived as part of the Hazen-Williams formula. It represents the pipe roughness associated with the type of pipe chosen. NFPA 13 has assigned values for each pipe type. It also correlates the C factor for given pipe types with the system type being used. For instance, if a wet pipe system is being designed using black steel pipe, the C factor is 120. However, if a dry pipe system is being designed using the same black steel pipe, the C factor is now lowered to 100. In this case, the characteristics of a dry pipe system are being considered. The same pipe type used in a dry pipe system, where it is exposed to moist air and varying temperatures, will inevitably lead to the development of corrosion on the inside of the pipe. This increased roughness is being anticipated and therefore the C factor is adjusted to account for it.

Another example is that of copper tubing or CPVC piping. The C factors assigned to these two types of piping are much higher than that of standard black steel pipe, since they have surfaces that are much smoother and are not as likely to propagate corrosion as black steel. Therefore they are assigned a C factor of 150. When dealing with C factors, the higher the C factors, the better hydraulic performance the system will have, resulting in the use of smaller pipe.

Step 7: Determine the Total Amount of Pipe Lengths, Type and Number of Fittings, and Type and Number of Devices That Should Be Included to the Next Node.

This step includes the addition of all of the pipe lengths from the first node to the next node. Along with the actual pipe lengths, all of the

TABLE 10.1

Equivalent Schedule 40 Steel Pipe Length Chart

Fittings and Valves	Fittings and Valves Expressed in Equivalent Feet of Pipe														
	½ in.	¾ in.	1 in.	1¼ in.	1½ in.	2 in.	2½ in.	3 in.	3½ in.	4 in.	5 in.	6 in.	8 in.	10 in.	12 in.
45° elbow	—	1	1	1	2	2	3	3	3	4	5	7	9	11	13
90° standard elbow	1	2	2	3	4	5	6	7	8	10	12	14	18	22	27
90° long-turn elbow	0.5	1	2	2	2	3	4	5	5	6	8	9	13	16	18
Tee or cross (flow turned 90°)	3	4	5	6	8	10	12	15	17	20	25	30	35	50	60
Butterfly valve	—	—	—	—	—	6	7	10	—	12	9	10	12	19	21
Gate valve	—	—	—	—	—	1	1	1	1	2	2	3	4	5	6
Swing check*	—	—	5	7	9	11	14	16	19	22	27	32	45	55	65

For SI units, 1 in. = 25.4 mm; 1 ft = 0.3048 m.
Note: Information on ½-in. pipe is included in this table only because it is allowed under 8.14.19.3 and 8.14.19.4 of NFPA 13.
*Due to the variation in design of swing check valves, the pipe equivalents indicated in this table are considered average.

Source: NFPA 13, 2002, Table 14.4.3.1.1.

equivalent lengths that are associated with this piping will be added. In the hydraulic calculation process, friction loss associated with fittings and certain devices is expressed in equivalent lengths of straight pipe. For example, the equivalent length for a 1-in., 90° elbow is 2 ft. This means that the friction loss that takes place with water flowing through a 1-in., 90° ell is equal to that same amount of water flowing through 2 ft of straight 1-in. pipe. Table 10.1 provides equivalent lengths for several types of fittings and valves based on size. The manufacturer's actual losses that are recorded as part of the testing or listing process may also be used. Using the manufacturer's data will usually provide a more accurate reflection of the losses that will be experienced through the valve or fitting. In some cases, the differences are very small and do not make enough of a difference in the final numbers to justify the time it takes to look up all of these values. Except for some of the newer valves and backflow preventors that are on the market today, using the table in NFPA 13 is the most productive choice.

There are a few nuances to this step that should be mentioned. First, not all fittings have to be considered. For instance, NFPA 13 does not require that any loss be accounted for if the flow of water is straight through a tee. Even though there are losses in this situation, they are not considered to be large enough to warrant inclusion. Only if the flow of water is turned 90° in the tee should the loss be considered. Another is at the sprinkler itself. NFPA 13 does not require that the actual fitting to which the sprinkler is attached be included. For example, if the sprinkler is attached to a threaded tee that is part of the branchline, and there is no nipple in between the tee and the sprinkler, then the tee that the sprinkler is attached to does not need to be included in the overall equivalent length. One other example has to do with requirements from sources other than NFPA. For instance, some insurance agencies require that 1 ft of equivalent length be added for every grooved coupling the water passes through in the run of piping. This does not make that much difference individually, but it can make a considerable difference collectively. So, it is very important that all of the pipe, fittings, and devices such as valves and electric water flow switches be accounted for in the calculation process. Also, nominal pipe diameters do not always correspond to actual pipe diameters. Table 10.2 provides the actual pipe diameters for steel pipe based on nominal sizes.

Step 8: **Calculate the Friction Loss per Foot.**

As a result of step 5, the friction loss per foot associated with this node can be determined. Using the Q that was arrived at in step 3, Equation 10.5 is used to determine how much friction loss there is in 1 lineal ft.

EXAMPLE 10.2

$$P_f = \frac{(4.52)(Q)^{1.85}}{(C)^{1.85}(D)^{4.87}} \rightarrow P_f = \frac{(4.52)(14)^{1.85}}{(120)^{1.85}(1.049)^{4.87}} = .0673 \text{ psi/ft}$$

assuming

$Q = 14$ gpm

$C = 120$ (black steel pipe)

$D = 1.049$ (internal diameter for 1-in. Schedule 40 black steel pipe)

Step 9: **Calculate the Total Friction Loss (P_f) to the Next Node.**

Once the loss per foot has been found, it is simply multiplied by the total amount of equivalent feet determined from this beginning node to the next to determine the total P_f due to friction loss.

Step 10: **Calculate Any Pressure Losses or Gains Due to Elevation (P_e).**

There is one more category for loss that must be considered—the loss of pressure due to elevation, expressed as P_e. Elevation pressure loss is sometimes the largest hydraulic loss in the system, such as in a high rise office building. This represents the amount of pressure that a column of water exerts due to its weight. Another term that is commonly used is *head*. One foot in elevation of water will exert a pressure of 0.433 psi. So, if the system changes elevations between the two nodes, then the elevation change in feet is multiplied by 0.433 to determine the total loss in the change of elevation. If, in the direction of water flow, the water must rise in elevation, then the loss is positive. However, if the water is dropping down in elevation as it flows in the pipe, then the loss is negative. Obviously, if pressure is lost when lifting the water up, then pressure is gained when it drops down.

TABLE 10.2

Steel Pipe Dimensions

Nominal Pipe Size (in.)	Outside Diameter in.	Outside Diameter mm	Schedule 5 Inside Diameter in.	Schedule 5 Inside Diameter mm	Schedule 5 Wall Thickness in.	Schedule 5 Wall Thickness mm	Schedule 10[a] Inside Diameter in.	Schedule 10[a] Inside Diameter mm	Schedule 10[a] Wall Thickness in.	Schedule 10[a] Wall Thickness mm
½[b]	0.840	21.3	—	—	—	—	0.674	17.0	0.083	2.1
¾[b]	1.050	26.7	—	—	—	—	0.884	22.4	0.083	2.1
1	1.315	33.4	1.185	30.1	0.065	1.7	1.097	27.9	0.109	2.8
1¼	1.660	42.2	1.530	38.9	0.065	1.7	1.442	36.6	0.109	2.8
1½	1.900	48.3	1.770	45.0	0.065	1.7	1.682	42.7	0.109	2.8
2	2.375	60.3	2.245	57.0	0.065	1.7	2.157	54.8	0.109	2.8
2½	2.875	73.0	2.709	68.8	0.083	2.1	2.635	66.9	0.120	3.0
3	3.500	88.9	3.334	84.7	0.083	2.1	3.260	82.8	0.120	3.0
3½	4.000	101.6	3.834	97.4	0.083	2.1	3.760	95.5	0.120	3.0
4	4.500	114.3	4.334	110.1	0.083	2.1	4.260	108.2	0.120	3.0
5	5.563	141.3	—	—	—	—	5.295	134.5	0.134	3.4
6	6.625	168.3	6.407	162.7	0.109	2.8	6.357	161.5	0.134[c]	3.4
8	8.625	219.1	—	—	—	—	8.249	209.5	0.188[c]	4.8
10	10.750	273.1	—	—	—	—	10.370	263.4	0.188[c]	4.8
12	12.750	—	—	—	—	—	12.090	—	0.330	—

[a]Schedule 10 defined to 5-in. (127-mm) nominal pipe size by ASTM A 135, *Standard Specification for Electric-Resistance-Welded Steel Pipe.*
[b]These values applicable when used in conjunction with 8.14.19.3 and 8.14.19.4 of NFPA 13.
[c]Wall thickness specified in 6.3.2 and 6.3.3 of NFPA 13.

Source: NFPA 13, 2002, Table A.6.3.2.

Step 11: Add up the Pressure Column ($P_t + P_e + P_f$) to Establish a New Required Pressure (P_t) at the Next Node.

The final step in the first calculation is to add up the pressure loss column to determine the total pressure that will be present at the second node. This will be the pressure that is required to meet the demand of downstream sprinklers, not necessarily the pressure that will be required by the sprinkler at the second node. In many cases, applying Equation 10.2 to this sprinkler reveals that it requires less pressure than is needed to supply downstream sprinklers. The new P_t is used to begin the next line of calculations.

Step 12: Repeat the Process for Flowing and Nonflowing Nodes Back to the Source Node.

There are several other issues that play a part in this process; however, in its simplest form, this is the calculation process. The minimum water (gpm) that is needed to meet the density requirement is determined. It is then flowed through a total equivalent length of

| | Schedule 30 | | | | Schedule 40 | | | |
| | Inside Diameter | | Wall Thickness | | Inside Diameter | | Wall Thickness | |
in.	mm	in.	mm	in.	mm	in.	mm
—	—	—	—	0.622	15.8	0.109	2.8
—	—	—	—	0.824	21.0	0.113	2.9
—	—	—	—	1.049	26.6	0.133	3.4
—	—	—	—	1.380	35.1	0.140	3.6
—	—	—	—	1.610	40.9	0.145	3.7
—	—	—	—	2.067	52.5	0.154	3.9
—	—	—	—	2.469	62.7	0.203	5.2
—	—	—	—	3.068	77.9	0.216	5.5
—	—	—	—	3.548	90.1	0.226	5.7
—	—	—	—	4.026	102.3	0.237	6.0
—	—	—	—	5.047	128.2	0.258	6.6
—	—	—	—	6.065	154.1	0.280	7.1
8.071	205.0	0.277	7.0	7.981	—	0.322	—
10.140	257.6	0.307	7.8	10.020	—	0.365	—
—	—	—	—	11.938	—	0.406	—

pipe, fittings, and devices, which are sized initially by guessing and then added up for a subtotal pressure loss, as shown in subsequent examples. The loss or gain for elevation change is then added in, yielding a total pressure loss between two given points.

If there are several portions of sprinkler piping, such as identical branchlines, an equivalent K-factor can be established for each branchline to simplify calculations. This equivalent K-factor can be calculated using a rearranged version of Equation 10.3, specifically:

$$K = \frac{Q}{\sqrt{P}}$$

where:

Q = the required flow

P = the required pressure

For example, if several identical branchlines must be calculated, it would be simplest to calculate the pressure and flow requirements of the first branchline, then determine the equivalent K-factor for the branchline. This equivalent K-factor could be used for subsequent branchlines in lieu of calculating individual sprinklers.

Example Calculation

All of the major factors involved with hydraulic calculations for sprinkler systems have now been discussed. Using Figure 10.6, these steps can be applied to determine the demand for the system shown.

First the designer identifies the nodes to be considered. Figure 10.6 shows that each sprinkler being flowed has been given a node tag. These have been designated as nodes 1, 2, 3, 6, 7, and 8. Changes of pipe size have been identified with node tags as well. The calculations will begin with the assumption that sprinkler #1 is the most demanding.

Using Equation 10.2, the designer determines the minimum Q that will be required at this node. As previously discussed, the assigned area for sprinkler #1 is 180 ft^2 (15′ × 12′). The minimum Q required is 18 gpm (0.10 gpm/ft^2 × 180 ft^2). Using the minimum Q required, the minimum P that will be needed can be found. Using Equation 10.3 and an assumed sprinkler K-factor of 5.6, the minimum pressure is determined to be 10.33 psi (18.0/5.6)^2.

Having the two values that make up the minimum end sprinkler demand, the designer fills in the worksheet in the appropriate fields. See Figure 10.7. Uppercase Q represents the total water (gpm) required at a given node for that node plus all of the nodes preceding it. Lowercase q represents the flow (gpm) present at that node from the sprinkler. P_t represents the total pressure required

FIGURE 10.6
Hydraulic Calculation
Exercise.

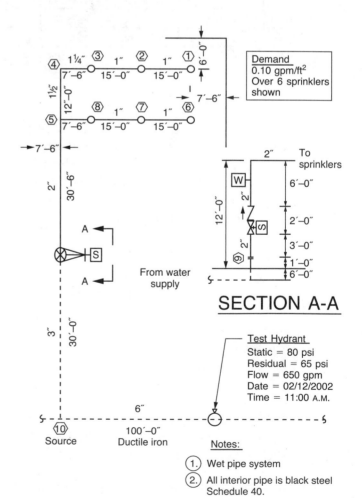

Contract Name: _____ Sheet _____ Of _____

Step No.	Nozzle Ident. And Location		Flow in gpm	Pipe Size	Pipe Fittings and Devices	Equiv. Pipe Length	Friction Loss psi Foot	Pressure Summary	Normal Pressure	Notes	Ref. Step
	1		q 18.0	1	15-0	L 15-0		Pt 10.33	Pt	(d)(sq.ft.) = Qmin	
1	to	1	———	1.049	0	F 0	0.1071	Pe 0	Pv	(.1)(180) = 18.0gpm	
	2		Q 18.0	40	0	T 15-0		Pf 1.607	Pn	$(18.0/5.6)^2 = p =\ >10.33$	
	2		q 19.35	1	15-0	L 15-0		Pt 11.937	Pt	$Q = k\sqrt{p}$ solve for Q	
2	to	2	———	1.049	0	F 0	0.4133	Pe 0	Pv	$Q = 5.6\sqrt{11.937}$	
	3		Q 37.35	40	0	T 15-0		Pf 6.2	Pn	Q = 19.35	
	3		q 23.85	1 1/4	7-6	L 7-6		Pt 18.14	Pt	$K = Q/\sqrt{p}$	
3	To		———	1.38	0	F 0	0.27	Pe	Pv	$K = 61.2/\sqrt{20.17}$	
	4		Q 61.20	40	0	T 7-6		Pf 2.03	Pn	K = 13.6	
	4		q			L		Pt 20.17	Pt		
4			———			F		Pe	Pv		
			Q			T		Pf	Pn		
	5		q			L		Pt	Pt		
5			———			F		Pe	Pv		
			Q			T		Pf	Pn		

FIGURE 10.7
Calculation of First Branchline.

at the identified node. Now that these two values have been established, the designer can begin working toward the water source.

Next is the assignment of a pipe size and type to the piping that takes place between nodes 1 and 2. In the example, the designer has guessed at 1-in. Schedule 40. The nominal pipe size, actual internal diameter, and associated schedule and C factor should all be entered into this portion of the worksheet. The C factor for Schedule 40 pipe is 120.

Moving across the worksheet in Figure 10.7 left to right, the pipe, fittings, and devices must now be calculated. Taking into consideration the fittings and pipe that the water must flow through to get from node 2 to node 1, each type is identified. In this example there are 15 ft. of pipe from node 2 to node 1. There are no fittings. Remember, NFPA 13 does not require the fitting to which the sprinkler is physically attached to be included in this count. Therefore, the total length of pipe can be carried over into the Equivalent Pipe Length column. Likewise, the total equivalent lengths for fittings and devices can be brought over as well. Since there are none, this column is then totaled vertically into the "T" field: $L + F = 15$ ft.

The next step is to calculate the friction loss per foot using Equation 10.5. Inserting 18 gpm flowing through 1-in. Schedule 40 pipe into formula 10.5 yields 0.1071 of pressure loss per foot. Taking the loss per foot times the total equivalent pipe loss, the P_f for this step is ($0.1071 \times 15 = 1.607$). The final value needed is the pressure loss due to elevation. Since there is no elevation change between nodes 1 and 2, this value is 0.

The final step in this first line of the calculations is to total up the Pressure Summary column. As yet, none of the values for the Normal Pressure column have been addressed. This is something to consider, but it is not critical to learning the basics of sprinkler calculations and is therefore ignored for the purpose of this example. Adding $P_t + P_e + P_f$, the new total pressure required at node 2 is 11.937 psi. (*Note*: For the sake of this example, the values are carried out up to two decimal places to help the user obtain the same numbers as the example without dealing with rounding issues and different settings on calculators. In real life, these values should be rounded to the nearest whole number. Please see the discussion about the degree of accuracy for hydraulic calculations.)

Now that a required pressure at node 2 has been established, the designer can calculate the pressure losses from node 2 to node 3. See Figure 10.9.

Knowing the K-factor and the pressure required, the designer can use Equation 10.3 to determine the required q that will be present at node 2. The minimum amount of water that will flow from node 2 is 19.35 gpm.

This new q is placed in the lowercase q field because it represents the water just for node 2. The node identification for this line of the calculations is filled in, and the two nodes being considered at this point.

Having the minimum water required at node 1 and the amount of water that will be at node 2 because of the demand at node 1, the total amount of water required thus far can be determined by adding the gpm column for a new Q (18.0 + 19.35 = 37.35 gpm).

The next guess of 1-in. Schedule 40 pipe along with the pipe ID and C factor is entered into the Pipe Size column. Also filled in are the pipe, fittings, and device lengths for the Individual Lengths column. See Figure 10.7.

Having determined that there is 15 ft. of pipe from node 2 to node 3 and no fittings or devices, the total equivalent pipe lengths can be carried across and added down. The total equivalent pipe length for this step is 15 ft. Again the friction loss per foot is determined applying Equation 10.5, using 37.35 gpm for Q, 120 for C, and 1.049 for D. The loss per foot equals 0.4133 psi/ft.

The total friction loss of 6.2 psi is established by multiplying the loss per foot by the total equivalent length. There are no elevation changes between these two nodes, so $P_e = 0$. Now the Pressure Summary column is added down to the next step, and the total pressure required at node 3 equals 18.14 psi. Since node 3 is also a flowing sprinkler head, Equation 10.3 is applied again. This time it is solved for q using 5.6 for K and 18.14 for P. This equals a q of 23.85 at node 3. This is the pressure that will be at node 3 in order to achieve the minimum pressure necessary at node 1.

The q field is filled in with the new flow at node 3. Step 3 and the nodes that are being considered are identified. In this step the designer is looking at what happens between node 3 and node 4. The total water required at node 2 is added to the individual water that is present at node 3; this equals 61.2. The pipe size and type information is filled in along with the pipe, fitting, and devices that are located between these two nodes. The designer increased the pipe sizing to 1¼-in. Schedule 40. This was done because the friction loss per foot was nearing values that would likely put the total demand of the system over the available supply curve. A rule of thumb is that when the friction loss per foot nears or goes over 0.30 psi/ft., the pipe size should be increased by one size. Obviously, this is dependent on the available water supply; however, based on typical pressures of approximately 60 psi to 80 psi static, the type of demand that will be needed can be anticipated. If there is a fire or booster pump involved, it will make it easier to keep the sizing smaller.

The total equivalent lengths for the pipe and fittings are carried over and then added up to determine the total equivalent length. In this case, there is 7 ft-6 in.

of pipe between nodes 3 and 4. Although there is an elbow, this will be counted when the flow between nodes 4 and 5 is considered. Therefore, the total equivalent length of pipe between nodes 3 and 4 is 7 ft-6 in. With 61.2 gpm flowing through pipe with an internal diameter of 1.38 in., the friction loss will be 0.2710 gpm/ft. The total friction loss through 7 ft-6 in. will be 2.03 psi, which results in a total pressure at node 4 of 20.17 psi. Adding this to the pressure at node 3 results in a total pressure required at node 4 of 20.17 psi. See Figure 10.7.

Branches 1-2-3-4 and 6-7-8-5 are identical. Therefore, an equivalent K-factor can be calculated for these branches to simplify the calculation. Since branch 1-2-3-4 requires 61.2 gpm at 20.17 psi, the K-factor can be calculated as follows:

$$K = \frac{Q}{\sqrt{P}} = \frac{61.2}{\sqrt{20.17}} = 13.6$$

Next, the pressure loss from node 4 to node 5 is calculated. There is 12 ft of 1½-in. pipe between these nodes and an elbow, which has an equivalent length of 4 ft. Using the Hazen-Williams formula (Equation 10.5), the friction loss with 61.2 gpm flowing through 1½-in. pipe (internal diameter = 1.61 in.) equals 0.1279 gpm/ft. Multiplying this by the 16 ft in equivalent length between nodes 4 and 5 yields a total friction loss of 2.05 psi, which means that the pressure at node 5 is 22.22 psi. See Figure 10.8.

Since branches 1-2-3-4 and 6-7-8-5 are identical, the equivalent K-factor for branch 1-2-3-4 can be used to calculate the flow from branch 6-7-8-5. With a pressure of 22.22 psi at node 5 and an equivalent K-factor of 13.6, 64.11 gpm would flow through branch 6-7-8-5. Adding this to the flow from branch 1-2-3-4 yields 125.3 gpm.

FIGURE 10.8
Calculation of Second Branchline.

Contract Name: _____ Sheet ____ Of _____

Step No.	Nozzle Ident. And Location	Flow in gpm	Pipe Size	Pipe Fittings and Devices	Equiv. Pipe Length	Friction Loss psi Foot	Pressure Summary	Normal Pressure	Notes	Ref. Step
	1	q	1	15-0	L 15-0		Pt 10.33	Pt	(d)(sq.ft.) = Q min	
1 to	1		1.049	0	F 0	0.1071	Pe 0	Pv	(.1)(180) = 18.0 gpm	
	2	Q 18.0	40	0	T 15-0		Pf 1.607	Pn	$(18.0/5.6)^2 = p = >10.33$	
	2	q 19.35	1	15-0	L 15-0		Pt 11.937	Pt	$Q = k\sqrt{p}$ solve for Q	
2 to	2		1.049	0	F 0	0.4133	Pe 0	Pv	$Q = 5.6\sqrt{11.937}$	
	3	Q 37.35	40	0	T 15-0		Pf 6.2	Pn	Q = 19.35	
	3	q 23.85	1¼	7-6	L 7-6		Pt 18.14	Pt	$K = Q/\sqrt{P}$	
3 to	3		1.38	0	F 0	0.27	Pe	Pv	$K = 61.2/\sqrt{20.17}$	
	4	Q 61.20	40	0	T 7-6		Pt 2.0	Pn	K = 13.6	
	4	q —	1½	1.2	L 12		Pt 20.17	Pt		
4 to			1.61	ELL	F 4	0.1279	Pe	Pv		
	5	Q 61.20	40	0	T 16		Pf 2.05	Pn		
	5	q 64.11	2"	48-6	L 48-6		Pt 22.22	Pt	$Q = K\sqrt{P}$	
5 to	6-7-8-5		2.067	2L,CK, G	F 22	0.1426	Pe 7.79	Pv	$Q = 13.6/\sqrt{22.22}$	
	9	Q 125.3	40	0	T 70-6		Pf 10.07	Pn	Q = 64.11	
							Pt 40.03			

Between nodes 5 and 9, there is 48 ft-6 in. of 2-in. pipe, two elbows, a gate valve, and a check valve. The combined equivalent length of these fittings is 22 ft, resulting in a total equivalent length of 70 ft-6 in. Using Equation 10.5, the friction loss of 125.3 gpm flowing through 2-in. Schedule 40 pipe (internal diameter = 2.067 in.) is 0.1426 gpm/ft, which multiplied by 70 ft 6 in. results in a total friction loss between nodes 5 and 9 of 10.07 psi. There is also a difference in elevation between nodes 5 and 9 of 18 ft, which multiplied by 0.433 psi/ft equates to a pressure loss due to change in elevation of 7.79 psi. Therefore, the total pressure loss between node 5 and node 9 is 17.86 psi, which added to the pressure required at node 5 (22.22 psi) results in a pressure requirement at node 9 of 40.08 psi. See Figure 10.8.

From node 9 to node 10 (the connection to the city water supply), there is 30 ft of Schedule 10 pipe and a tee, which has an equivalent length of 15 ft. Therefore, the total equivalent length from node 9 to node 10 is 45 ft. Using equation 10.4, the friction loss with 125.3 gpm flowing through 3-in. Schedule 10 pipe (internal diameter = 3.26 in.) is 0.02 gpm/ft, which results in a total friction loss of 0.9 psi. Added to the pressure demand at node 9, this equates to a pressure demand at node 10 of 40.98 psi. See Figure 10.9.

FIGURE 10.9

Calculation Back to Water Supply.

Contract Name: _____ Sheet ____ Of _____

Step No.	Nozzle Ident. And Location		Flow in gpm	Pipe Size	Pipe Fittings and Devices	Equiv. Pipe Length	Friction Loss psi Foot	Pressure Summary	Normal Pressure	Notes	Ref. Step
	1		q –	1	15-0	L 15-0		Pt 10.33	Pt	(d)(sq.ft.) = Q min	
1	to	1		1.049	0	F 0	0.1071	Pe 0	Pv	(.1)(180) = 18.0 gpm	
	2		Q 18.0	40	0	T 15-0		Pf 1.607	Pn	$(18.0/5.6)^2 = p = {>}10.33$	
	2		q 19.35	1	15-0	L 15-0		Pt 11.937	Pt	$Q = k\sqrt{p}$ solve for Q	
2	to	2		1.049	0	F 0	0.4133	Pe 0	Pv	$Q = 5.6\sqrt{11.937}$	
	3		Q 37.35	40	0	T 15-0		Pf 6.2	Pn	Q = 19.35	
	3		q 23.85	1¼	7-6	L 7-6		Pt 18.14	Pt	$K = Q/\sqrt{P}$	
3	to	3		1.38	0	F 0	0.27	Pe	Pv	$K = 61.2/\sqrt{20.17}$	
	4		Q 61.20	40	0	T 7-6		Pt 2.03	Pn	K = 13.6	
	4		q –	1½	12	L 12		Pt 20.17	Pt		
4	to			1.61	ELL	F 4	0.1279	Pe	Pv		
	5		Q 61.20	40	0	T 16		Pf 2.05	Pn		
	5		q 64.11	2"	48-6	L48-6		Pt 22.22	Pt	$Q = K/\sqrt{P}$	
5	to	6-7-8-5		2.067	2L,CK, G	F 22	0.1426	Pe 7.79	Pv	$Q = 13.6/\sqrt{22.22}$	
	9		Q 125.3	40	0	T 70-6		Pf 10.07	Pn	Q = 64.11	
	9		q	3"	30'	30'		Pt 40.08	Pt		
6	to			3.26"	T	15'	0.02	Pe	Pv		
	10		Q 125.3	10	0	45'		Pr 0.9	Pn		
	10		q	6"	100'	100'		Pt 40.98	Pt		
7	to					–	0.0011	Pe	Pv		
	11		Q 125.3	DI		100'		Pr 0.4	Pn		

Total = 41.09 psi

At this point, the water supply has been calculated to the point of connection to the city water supply. However, the test data is for a hydrant that is 100 ft away. If the flow of water in the underground main was such that the test hydrant was downstream from the connection to the building, the calculated demand of 125.3 gpm at 40.98 psi could conservatively be used as the sprinkler system demand at the point of connection to the water supply. However, if the flow direction is not known, or if the flow direction is in the direction from the hydrant to the connection to the building, then the friction loss in the underground piping must be included.

A flow of 125.3 gpm through 6-in.-diameter pipe with a C factor of 100 results in a total friction loss of 0.11 psi, resulting in a total demand at the test location of 125.3 gpm at 41.09 psi. See Figure 10.9. This demand for the sprinkler system would be compared to the water supply in accordance with Chapter 6.

Summary

This chapter introduced four key equations and the 12-step process of hydraulically calculating automatic sprinkler systems. An example hydraulic calculation illustrated the process of determining system demand.

BIBLIOGRAPHY

NFPA Codes, Standards, and Recommended Practices

NFPA Publications National Fire Protection Association, 1 Batterymarch Park, Quincy, MA 02169-7471

The following is a list of NFPA codes, standards, and recommended practices cited in this chapter. See the latest version of the *NFPA Catalog* for availability of current editions of these documents.

NFPA 13, *Standard for the Installation of Sprinkler Systems*

11

Backflow Protection for Fire Protection Cross Connections

Jack Poole, PE, and Grant Cherkas

Over the course of many years of sprinkler system service, stagnant water in system piping can develop an undesirable concentration of metallic particles or other undrinkable or undesirable contaminants. For the purpose of this chapter, *cross connection* is the term used to describe a connection between a potable water system and an automatic sprinkler system whereby contaminants in sprinkler water can enter the potable drinking water system. Unwanted substances include any substance that can change the color or taste or add odor to the water. A *backflow* is an undesirable reversal of flow of sprinkler water to a potable drinking water distribution system.

The federal government of the United States passed the Safe Drinking Water Act that establishes jurisdiction over the public health aspects of the drinking water supply. State governments have jurisdiction over health-related aspects of the water supply, but state regulations cannot supersede the federal regulations, although they may be more stringent if the state so desires. The water purveyor, which is the public or private owner or operator of the potable water system that supplies approved water to the public, has the full responsibility to ensure safe drinking water to the consumer. "Approved" water has to meet the criteria of the Safe Drinking Water Act. The water purveyor has the overall responsibility for preventing water from unapproved sources from entering either the potable water system within the water consumer's premise or the public water supply directly. Most water purveyors have developed a Cross-Connection Control Program that provides criteria for backflow prevention to reduce the risk of cross connections.

Cross-connection and backflow prevention for fire sprinkler systems has been a hotly debated topic within the fire sprinkler and water purveyor industries for well over a decade [1,2]. Water purveyors have documented backflow incidents that were attributed to sprinkler systems and consider any potential cross connection as justification to impose backflow prevention requirements. The Environmental Protection Agency's (EPA's) Safe Drinking Water Act is often referenced to support this position. On the other hand, the fire sprinkler industry usually views backflow prevention as an expensive requirement that reduces the available water supply pressure with little net benefit in risk reduction, pointing as justification to the limited number of documented cross-connection incidents involving fire sprinkler systems. This chapter will not resolve this debate but will provide guidance on what types of backflow prevention assemblies are typically installed in fire sprinkler systems and the issues that need to be reviewed by engineers involved in fire sprinkler installations.

Potable Water Supplies and Fire Sprinkler Systems

Typically, fire sprinkler systems take their water supply from municipal potable water sources. Historically this arrangement developed out of convenience and simple economics in an effort to provide a reliable water supply to fire sprinkler systems. Since potable water supplies were available to most buildings in developed areas and were considered reliable (loss of domestic water usually prompts an investigation and quick repairs), this source proved to be an economical water supply for the fire sprinkler system. A by-product of interconnecting a potable water source with any nonpotable water system (i.e., fire sprinkler system) is the potential to contaminate the potable source. This potential for contamination was recognized fairly early in the development of fire sprinkler systems, and early standards cautioned installers not to connect nonpotable water sources, such as fire pumps drafting from a river, to sprinkler systems supplied by city water mains due to the danger of polluting the city's water supply, unless the approved "schemes" of the Authority Having Jurisdiction (AHJ) were installed to protect the potable water supply.

As fire sprinkler standards developed, various clauses appeared attempting to mitigate the risks associated with cross connections. The 2002 edition of NFPA 13, *Standard for the Installation of Sprinkler Systems*, and the 2003 edition of NFPA 20, *Standard for the Installation of Stationary Pumps for Fire Protection*, address the issue of cross-connection control by acknowledging that the water supply AHJs may require the installation of backflow prevention assemblies to mitigate the potential for a cross connection and providing detailed design guidance on how and where to install backflow prevention assemblies. The composition of antifreeze solutions used in antifreeze sprinkler systems is now restricted to less toxic chemicals when the sprinkler system is connected to a potable water source. Even though the antifreeze utilized in fire sprinkler systems is less toxic, it will affect the color, taste, and odor of the potable water, so typically the highest level of backflow prevention is required.

How and Why Backflow Incidents Occur with Fire Sprinkler Systems

Backflow incidents occur due to two separate backflow principles: backsiphonage and backpressure. *Backsiphonage* is defined as a reduction in system pressure that creates a subatmospheric pressure to exist at a site in the water system. Conditions or arrangements that may cause backsiphonage conditions include high-demand flow conditions (fire flow, hydrant testing, large peak demands during heat-wave emergencies), inadequate public water source or storage capacity, or a water main break. *Backpressure* occurs when the pressure on the downstream side of the piping system is greater than the pressure on the upstream side, which may cause a reversal of the normal direction of flow. Backpressure may be caused by a pump, elevation of piping, thermal expansion, or pressurized containers or systems.

Backflow occurs when a pressure differential drives the movement of fluid from areas of high pressure to lower pressures. This movement can occur in any pressurized system. During the loss of supply pressure, the elevation head can become the driving force to create a backflow condition. In fire sprinkler systems, backflow can occur due to the presence of fire pumps, during fire department operations when fire department connections are charged with water, during fire hydrant flow tests, or when city water mains lose pressure as a result of repairs or breaks. As a result, the contaminated water that is typically found in fire sprinkler systems can then be drawn back into the potable water system of the building and as far back as the city water mains, potentially affecting a large number of persons.

Cross-Connection Control Requirements and the Water Purveyor

Presently there is wide variability in cross-connection control requirements for fire sprinkler systems across North America. Although water purveyors increasingly require a backflow prevention assembly of some kind to be installed on a fire sprinkler system, the exact requirements depend on the water purveyor. The fire protection industry has considered a single, listed, rubber-faced check valve or a UL-listed alarm check valve to serve as a backflow prevention device. However, the water industry does not classify these devices as backflow prevention assemblies because they are not testable and have not been designed to meet the standards of a backflow prevention assembly. The water industry typically requires cross connections to be eliminated or provided with an approved backflow prevention assembly. The term *approved backflow prevention assembly* means an assembly that has been investigated or approved by an administrative AHJ.

The water industry typically relies on two resources to provide recommendations and guidance to water purveyors on information, procedures, and practices for developing, implementing, and enforcing a Cross-Connection Control Program that will meet the federal, state and local regulations governing cross-connection control. These two resources are the *Recommended Practice of Backflow Prevention and Cross-Connection Control* (M14) developed by the American Water Works Association [3] and the *Manual of Cross-Connection Control* developed by the Foundation for Cross-Connection Control and Hydraulic Research of the University of Southern California [4]. Many water purveyors or jurisdictions have additional requirements that are over and above the recommendations of M14 or the USC *Manual*, so it is important to understand the individual requirements of the jurisdiction. As a result, the engineer should check with the water purveyor to confirm what level of cross-connection control is required and what assemblies will be accepted as a backflow preventer for the intended application. The inquiry should be made early in the project, preferably before the initial design stage. The installation of backflow prevention assemblies results in additional friction loss in the water supply, which can affect pipe sizing and sprinkler spacing, increase the cost of the fire sprinkler installation, and increase the physical space requirements.

Basic Methods of Backflow Control

There are three basic methods used by the fire protection industry to control backflow: an air gap, a double check valve assembly, and a reduced-pressure-principle backflow prevention assembly.

Air Gap. An air gap is physical separation between the free-flowing discharge end of a pipe and an open end of a pipe or nonpressurized vessel. An air gap in a fire protection system would be employed when the potable water system is being utilized to maintain the water level of a fire protection storage tank. Since a physical separation is required, an air gap is not the method of choice for a sprinkler system. The physical separation distance of an approved air gap must be at least twice the diameter of the effective opening size of the water outlet, but never less than 1 in.

Double Check Valve Assembly (DC). The DC consists of two internally loaded, independently operating, approved check valves; two resilient-seated shutoff valves attached at each end of the combined check valves; and four properly located test cocks. See Figure 11.1. The DC is designed to protect against low hazard cross-connection conditions.

Reduced-Pressure-Principle Backflow Prevention Assembly. The third method of backflow control is a reduced-pressure-principle backflow prevention assembly (RP), as shown in Figure 11.2 and Figure 11.3. The RP consists of two internally loaded, independently operating, approved check valves together with a hydraulically operating, mechanically independent, pressure differential relief valve located between the check valves and below the first check valve; two resilient-seated shutoff valves attached at each end of the combined check

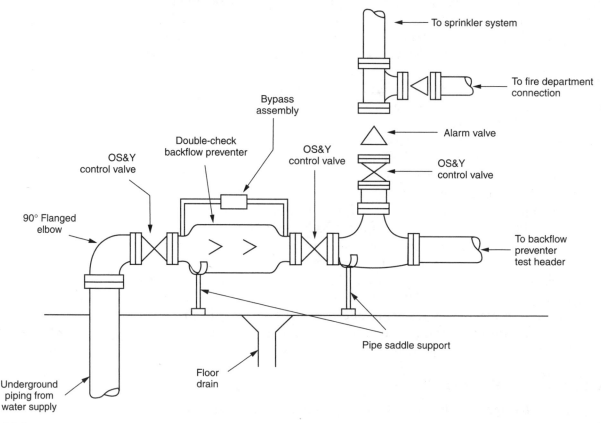

FIGURE 11.1
Schematic of a Double Check Valve Backflow Preventer for a Wet Pipe Automatic Sprinkler System.

FIGURE 11.2
Reduced-Pressure-Principle Backflow Preventer.
(Source: IRI and Cla-Val)

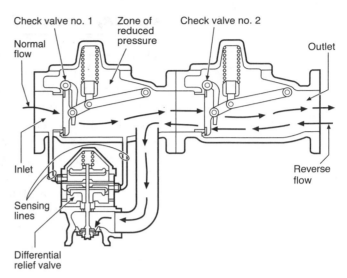

FIGURE 11.3
Reduced-Pressure-Principle Backflow Preventer Diagram.
(Source: Cla-Val)

valves; and four properly located test cocks. The RP is designed to protect against high hazard cross-connection conditions.

Sister Assemblies. Both the DC and the RP have a sister assembly, the DC detector assembly (DCDA) and the RP detector assembly (RPDA), respectively, which contain a bypass meter and a separate DC or RP on the bypass line. The method of protection is the same; however, a low-flow (typically less than 3 gpm) water meter is included to monitor for water usage. Many water purveyors may require a DCDA or an RPDA on fire protection systems to monitor for low-flow water usage.

Cross-Connection Risks, Consequences, and Professional Obligations

In general, the overall risk of contamination from a typical wet pipe sprinkler system is low due to the low probability of a cross connection between a fire sprinkler system and the potable water system and the relatively minor consequences to public health if a cross connection does occur [5,6]. For systems other than standard wet pipe sprinkler systems, the consequences of a cross connection are higher, especially if chemicals such as antifreeze or corrosion inhibitors are present in the system.

Documented cross-connection incidents between potable water systems and fire sprinkler systems have occurred and can potentially occur in the future unless the backflow prevention assembly is commensurate with the hazards present [7,8]. As a result, ensuring that the potable water supply does not become contaminated is both a legal and moral obligation for both the professional engineers designing the system and the sprinkler contractors installing the system.

Codes, Standards, and Other Resources

There are a number of additional resources on fire sprinkler systems and cross connections available. NFPA standards such as NFPA 13, *Standard for the Installation of Sprinkler Systems*; NFPA 20, *Standard for the Installation of Stationary Pumps for Fire Protection*; NFPA 22, *Standard for Water Tanks for Private Fire Protection*; NFPA 24, *Standard for the Installation of Private Fire Service Mains and Their Appurtenances*; NFPA 25, *Standard for the Inspection, Testing, and Maintenance of Water-Based Fire Protection Systems*; and NFPA 750, *Standard on Water Mist Fire Protection Systems*, provide detailed design guidance and installation requirements for backflow preventers in applicable systems when required by local or national codes. *Backflow Protection for Fire Sprinkler Systems* [8], printed by the National Fire Sprinkler Association, provides a detailed and well-documented discussion of the types and installation of backflow prevention devices on fire sprinkler systems. Typically, the NFPA codes and standard will not specify when backflow prevention assemblies are required, but will provide installation or testing guidance when they are required by the water purveyor. As stated earlier in this chapter, the quality of the potable water to be delivered to the consumer is the sole responsibility of the water purveyor under the Safe Drinking Water Act.

Retrofitting Backflow Prevention Assemblies onto Existing Fire Sprinkler Systems

The AWWA *M14 Manual* [3] recommends that an air gap, reduced-pressure-principle backflow prevention assembly, or reduced-pressure principle detector backflow prevention assembly be provided where there is a high hazard, but not for systems presenting a low hazard only and having a modern UL-listed alarm check valve that contains no lead, provided the check valve is maintained in accordance with NFPA 25. When an existing sprinkler system with an alarm check valve is significantly expanded or modified, requiring a comprehensive hydraulic analysis, a double check valve assembly should be installed. For existing systems that present a low hazard and that have an alarm check valve containing lead, it is recommended that a double check valve assembly be installed.

If the water purveyor requires the retrofit of a backflow prevention assembly on an existing sprinkler system, then a detailed engineering hydraulic analysis is required prior to installing a backflow prevention device. The 2002 edition of NFPA 13 (paragraph 8.16.4.6.2, Retroactive Installation) explicitly requires a thorough hydraulic analysis, including revised hydraulic calculations, new fire flow data, and all necessary system modifications to accommodate the additional friction loss. The existing sprinkler system would have been designed based on either a pipe schedule system or by a hydraulic analysis method. In either case, the system needs to be assessed to confirm that with the added friction loss from the backflow prevention assembly in the system, there is adequate pressure available to the sprinkler system to deliver the required water volume at the required pressure. In some cases, additional piping modifications may be required. The engineer should keep in mind that the new backflow prevention assembly can serve as the system control valve and the existing check valve or alarm check valve can be removed.

Design Guidance

Backflow prevention assemblies must be listed for fire protection service by a recognized listing agency, as required by NFPA 13. The testing performed by Underwriters Laboratories (UL) or Factory Mutual Global (FM) is limited, generally consisting of body strength and friction loss only, and does not test to confirm that the assemblies comply with the specific design standard to prevent backflow. The testing of backflow prevention assemblies performed by the Foundation for Cross-Connection Control and Hydraulic Research of the University of Southern California or the American Society of Sanitary Engineers (ASSE) is in accordance with the specific manufacturing and performance standard developed by AWWA or ASSE.

Backflow prevention devices are listed for installation in vertical, horizontal, or sometimes both orientations. The manufacturer's literature needs to be consulted to ensure the installation arrangement will be acceptable and the device will function properly.

Guidance from AWWA *M14 Manual* and *Manual of Cross-Connection Control*

The second edition of the AWWA *M14 Manual* and the ninth edition of the USC *Manual of Cross-Connection Control* use six broad categories to define fire protection systems as they relate to the water source and arrangement of the water supply. The recently released third edition of the AWWA *M14 Manual* includes many changes, which are discussed later in this section. Table 11.1 summarizes the six classes of fire protection systems as defined by the second edition of the AWWA *M14 Manual* and the current ninth edition of the USC *Manual of Cross-Connection Control*.

Generally, Class 1 and 2 fire protection systems are those systems that generally and ordinarily do not require an approved backflow protection assembly. However, the water industry has recognized that special conditions may exist on the site of a Class 1 or 2 fire sprinkler system that may warrant the installation of an approved backflow prevention assembly.

The water utility will generally require minimum protection (double check valve assembly) on a Class 3 system to prevent stagnant waters from backflowing into the public potable water system.

A Class 4 system will normally require backflow protection at the service connection. The type (air gap, RP, or DC) will depend on the quality of the auxiliary supply.

Class 5 systems normally need maximum protection (air gap or RP) to protect the public potable water system.

Class 6 system protection depends on the requirements of both industry and fire protection, and could only be determined by a survey of the premises.

The *M14 Manual* and the USC *Manual of Cross-Connection Control* also indicate that where the fire sprinkler system piping is not an acceptable potable water system material, a backflow prevention assembly should be installed to isolate the fire sprinkler system from the potable water system. Generally, the water purveyor will require the maximum protection (RP) if there are chemicals such as foam, antifreeze, or other biological or chemical substances added to the fire sprinkler system.

TABLE 11.1

Summary of Six Classes of Fire Protection Systems

Class	Characteristics
1	■ Direct connections from public water mains only ■ No pumps, tanks, or reservoirs ■ No physical connection from other water supplies ■ No antifreeze or other additives of any kind ■ All sprinkler drains discharging to atmosphere, dry wells, or other safe outlets
2	Same as Class 1, except: ■ Booster pumps may be installed in the connections from the street mains. Note that booster pumps do not affect the potability of the system; however, it is necessary to avoid drawing so much water that pressure in the water main is reduced below 10 psi.
3	■ Direct connection from public water supply main, plus one or more of the following: — Elevated storage tanks — Fire pumps taking suction from aboveground reservoirs or tanks — Pressure tanks ■ All storage facilities are filled by or connected to public water only, the water in the tanks to be maintained in a potable condition. ■ Otherwise, Class 3 systems are the same as Class 1.
4	■ Directly supplied from public mains similar to Class 1 and 2, with an auxiliary water supply dedicated to fire department use or available to the premises; or ■ An auxiliary supply located within 1,700 feet of the fire department connection.
5	■ Directly supplied from public mains, and interconnected with auxiliary supplies (such as pumps taking suction from reservoirs exposed to contamination, or rivers and ponds) ■ Driven wells ■ Mills or other industrial water systems ■ Where antifreeze or other additives are used
6	■ Combined industrial and fire protection systems supplied from the public water mains only, with or without gravity storage or pump suction tanks.

Guidance from AWWA *M14 Manual*

The third edition of the AWWA *M14 Manual* (see Figure 11.4) indicates that fire hydrants are installed primarily to provide a water supply for fire-fighting purposes. The *Manual* further acknowledges that fire hydrants are used for other purposes (e.g., for construction water, dust control, water hauling, jumper connections for super chlorination of mains, pressure testing, and temporary service) that provide access for contaminants to enter the water distribution system, and the *Manual* recommends that fire hydrants be monitored regularly and maintained by the water purveyor and the fire authority. The water purveyor must also consider fire-fighting equipment and the use of chemicals with that equipment to ensure the potable water system is not contaminated. The water industry typically does not require cross-connection control when fire hydrants are being utilized for fire suppression purposes, but does require the use of a backflow prevention assembly if a fire hydrant is being used to fill a tanker used for construction, dust control or other reasons.

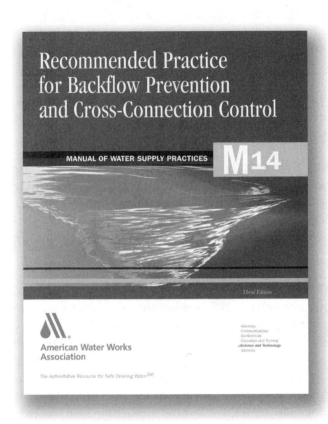

FIGURE 11.4
Recommended Practice for Backflow Prevention and Cross-Connection Control.

(Reprinted with permission. Copyright American Water Works Association)

The third edition of the *Manual* acknowledges that the cross-connection risks posed by new and existing fire suppression systems are equal, and thus, there is no justification for different levels of public health protection. This change from the water industry may result in the water purveyor requiring more existing systems to be retrofitted with backflow prevention assemblies. However, the third edition of the *Manual* goes on to say that for existing wet pipe fire sprinkler systems determined to pose only a low hazard threat, the water purveyor may consider an alternate proposal to the installation of an approved, modern, UL-listed alarm check valve, which has a rubber disc.

The third edition of the *Manual* recommends that an RP be utilized where there is a high hazard (e.g., risk of chemical addition) and a DC be used on all other systems. However, it further indicates that a second approach to identifying the required protection may be taken if the water purveyor performs a detailed assessment of each type of fire suppression system. Guidance is provided in the third edition of the *M14 Manual* for the types of systems listed next. It is understood that some of the terminology utilized in the third edition of the *M14 Manual* may not be consistent with that of the fire protection industry:

- New wet pipe fire sprinkler systems
- Existing wet pipe fire sprinkler systems
- Dry pipe nonpressurized fire suppression systems
- Dry pipe pressurized fire suppression systems

New Wet Pipe Fire Sprinkler Systems. Water contained in closed or non-flow-through fire systems may be stagnant or contaminated beyond acceptable drinking water standards. Some of the contaminants found in fire sprinkler systems are antifreeze, chemicals used for corrosion control or as wetting agents, oil, lead, cadmium, and iron. Therefore, for new wet pipe sprinkler systems a reduced-pressure-principle backflow prevention assembly, reduced-pressure-principle detector backflow prevention assembly, or air gap is recommended where there is a high hazard, and a double check valve assembly, double check detector backflow prevention assembly, or air gap is recommended for all other closed or non-flow-through systems.

Existing Wet Pipe Fire Sprinkler Systems. For existing wet pipe fire sprinkler systems determined to pose only a low hazard threat, the water purveyor may consider an alternate proposal to the installation of an approved, modern, UL-listed alarm check valve, which has a rubber disc—or require a DC. For existing wet pipe fire sprinkler systems considered to pose a high hazard threat, an RP or an air gap is recommended.

Dry Pipe, Nonpressurized Fire Suppression Systems (Deluge). Generally, deluge systems that are directly connected to a public potable water main do not present a health hazard to the public water system unless chemicals are added to the water as it enters the system. Therefore, no backflow protection is required if there is not an addition of chemicals. Where chemicals are added or are likely to be added, an RP or an air gap is recommended. As with existing wet pipe systems, the backflow protection recommended for deluge systems applies to new systems and does not generally apply to existing systems that have some acceptable form of directional flow control in place until the systems are substantially altered.

Dry Pipe, Pressurized, and Preaction Fire Suppression Systems. Dry pipe and preaction systems have been classified in the third edition of the *M14 Manual* as presenting a low hazard threat to the public water supply. Since they are considered a low hazard, a DC is recommended. As with existing wet pipe systems, the backflow protection recommended for dry pipe and preaction systems applies to new systems and does not generally apply to existing systems that have some acceptable form of directional flow control in place until the systems are substantially altered.

The *Manual* further indicates that a dry pipe sprinkler system may inappropriately be operated as a wet system during most of the year, then be charged with air for the months of freezing weather, or a maintenance contractor could inappropriately add antifreeze to a wet system that was designed to operate without added chemicals. Even though this is not normally the intent of the fire protection industry for dry pipe systems, this is what the water industry perceives may be taking place; therefore, it is extremely important that the systems be used and maintained as originally designed.

In summary, the third edition of the *Manual* indicates that the backflow prevention requirements for new sprinkler systems do not apply to existing systems that have some acceptable form of directional flow control in place until the systems are substantially altered. It further indicates that all new systems that present a high hazard threat should be protected with an RP and that all systems that present a low hazard threat be protected with a DC. Deluge systems have been classified as presenting no health threat to the potable water system, and no backflow protection is required. It might be argued that a dry pipe or preaction system is similar to a deluge system because under normal conditions the system does not contain any significant quantities of liquid to backflow into the potable water system.

Backflow Prevention Device Layout Considerations

Inspection, Testing, and Maintenance

A backflow prevention device installed in a fire sprinkler system requires some forethought during the design phase in order to accommodate the acceptance testing and future inspection, testing, and maintenance activities that will be required. There are five basic tests required for backflow prevention assemblies when they are installed in a fire sprinkler system: forward flow testing capacity, check valve integrity testing, flow switch and valve tamper switch testing, and isolation valve inspection.

Forward Flow Testing Capacity. NFPA 13 and NFPA 25 require a forward flow test to be performed during acceptance testing and at regular intervals to ensure the system is capable of supplying the required volume and pressure of water. A concern with any device located directly in the water supply piping is that it will fail in a manner that impedes the flow of water to the sprinkler system. In order to achieve forward flow testing, a test connection or test header is required. Alternatively, a dedicated test header can be designed and installed. It is recommended that the test header be sized to match the size of the backflow prevention assembly. Even though the theoretical flow through a 2-in. main drain may be large at higher pressure, the actual flow through the main drain is much less than the theoretical flow due to the angle valve and the numerous fittings. One of the primary reasons for forward flow testing is to exercise the internal components on the backflow prevention assembly to ensure they are capable of providing the minimum fire flow demand.

Check Valve Integrity Tests. Plumbing codes require that the backflow prevention assembly be tested, usually annually, to ensure that it will perform its intended function and close tight under backflow conditions. This test involves accessing the backflow preventer with a specialized test gauge kit and measuring the resistance to an induced backflow with the device isolated from the system. If repairs are required, sections of the backflow prevention assembly need to be disassembled and cleaned and possibly have specific components replaced. As a result, reasonable access needs to be provided around the backflow prevention device, preferably, at waist level directly over a floor drain with a minimum of 18 in. of clear space around the device for access. Some models require additional clearance to remove the internal check valve assemblies, so the manufacturer's literature must be carefully reviewed. An RP is provided with a relief valve to discharge water when backflow conditions are present and the second check valve is not holding tightly. The quantity of water that may discharge may be several hundreds of gallons a minute. The design engineer must ensure that adequate floor drains or draining provisions are provided in the areas were RPs are installed. Most floor drains in a mechanical space will not handle the maximum water discharge from a relief valve on a 6-in. RP.

Flow Switch and Valve Tamper Switch Testing. Valve tamper switches and flow switches, if the system is so equipped, need to be inspected and tested regularly in accordance with NFPA 25 or local codes. Access for this testing is therefore required. The isolation valves on the backflow prevention assembly shall be of the indicating type and should be supervised.

Isolation Valve Inspection and Exercise. The isolation valves on the backflow prevention assembly need to be inspected and exercised regularly in accordance with NFPA 25 or local codes. Access is therefore required.

Support of the Backflow Prevention Assembly

Backflow prevention assemblies are usually much heavier than an equivalent length of water-filled pipe. As a result, their weight requires adequate support of the device and of the connected piping. Typically, due to the weight of the device the supports need to be engineered as per NFPA 13 (paragraph 9.1.1.2) and an adequate means of transferring this load to the building structure needs to be devised as per NFPA 13 (paragraph 9.2.1.3.1).

Backflow prevention assemblies up to 2½ in. in size are usually supported by the piping system, with a pipe support located immediately before and after the inlet and outlet isolation valves. Assemblies 3 in. and larger are typically supported off the floor on dedicated pipe stands, with independent supports for the piping that connects to the backflow preventer. Larger-size backflow prevention assemblies are usually installed using an overhead crane or hoist; therefore, the installation is simplified if the backflow preventer can be positioned and supported independently of the piping.

An important consideration in supporting backflow preventers is ensuring they can be installed and serviced in the chosen location. For large assemblies, a location near the floor on dedicated pipe stands allows quick and safe positioning during construction and eases replacement if required in the future. A general rule for the installation of fire protection backflow preventers is to locate them approximately 2 ft above the finished floor. This allows good access from above and below the device and a convenient elevation for testing.

Drainage

Currently there is no single, standard method of providing forward flow testing capabilities for backflow preventers. It is therefore advisable to check with the project AHJ to ensure that the arrangement envisioned will be acceptable. If the fire department connection bypass valve method is utilized, however, some caution must be exercised. Most listed fire department connections on the market utilize internal swing clappers that act like a check valve on each inlet. The purpose of the clappers is to allow a two-inlet fire department connection to be charged with one fire hose and not have the water spray out the unconnected inlet. If this type of fire department connection is used with the pipe and valve arrangement, then only one 2½-in. outlet would be available for forward flow testing the backflow prevention device. Depending on the size of the backflow preventer, the available water pressure, and the water demand, this may not be adequate. Some listed flush mount fire department connections utilize individual clappers on each inlet, which would effectively prevent any water from exiting the fire department connection.

For backflow preventers installed in close proximity to an outside door, a number of jurisdictions have accepted the following arrangements. The hose valves are connected to 2½-in. fire hoses which are then connected to portable test devices or portable meters. The number of 2½-in. hose valves required to provide adequate water flow will depend on the forward flow demand, the available water supply, the length of hose run, and the test device. The number of hose valves required can be calculated using the standard hydraulic calculation methods of NFPA 13 or 14. The hose path and location of the test device require planning at the initial design stage.

A more expensive option is to install a dedicated backflow preventer test header similar to a fire pump test header. NFPA 13 specifies antifreeze chemicals that are permitted to be mixed with water to achieve an antifreeze solution. The type of chemical depends on whether there is a connection to a potable water source. Subsection 7.5.2 of NFPA 13 should be consulted before the antifreeze additive is chosen.

Antifreeze Loops, Expansion Tanks, and Backflow Prevention Devices

Antifreeze loops utilizing backflow prevention devices require a means of addressing the thermal expansion of the antifreeze solution to ensure pressure boundary integrity. A relief valve is not the preferred choice since this option typically would cause the expensive antifreeze solution to drain and eventually reduce the antifreeze solution strength, thereby raising the freezing point of the system.

NFPA 13 requires a listed expansion chamber downstream of the backflow preventer on glycol loops (see Figure 11.5). NFPA 13, paragraph 7.5.3.3, requires the expansion chamber to be listed; some jurisdictions require listed expansion tanks to be installed. For devices such as drain valves that are not required for the successful operation of the fire sprinkler system, there is no listing requirement, and some follow a similar line of reasoning for expansion tanks. They reason that the failure of an expansion tank is a maintenance issue and will not prevent the system from operating. AHJs enforce the NFPA 13 requirement for the listing of expansion tanks in recognition of the fact that if an expansion tank fails, the sprinkler system fails. Failures that have the potential to prevent successful operation include freezing due to leakage of solution and

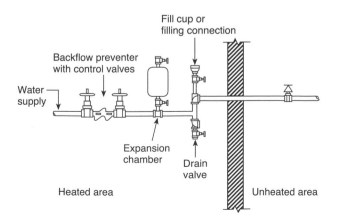

FIGURE 11.5

Arrangement of Supply Piping with Backflow Device.

(Source: NFPA 13, 2002, Figure 7.5.3.2)

overpressurization. The failure of an expansion tank is an event that can prevent proper operation of the sprinkler system until expansion tank repairs are completed.

There is currently only one manufacturer producing listed expansion tanks for fire protection systems in North America [9]. Expansion tanks employed in a fire sprinkler system must have a working pressure of 175 psi, or higher if the system is designed for pressures in excess of 175 psi, even if there is no fire pump installed. Fire departments usually conduct their operations in sprinklered buildings in accordance with NFPA 13E, *Recommended Practice for Fire Department Operations in Properties Protected by Sprinkler and Standpipe Systems*, which directs the pumper operator to charge the fire department connection to 150 psi upon arrival.

Expansion tanks need to be properly sized and adequately precharged with air to accept the volume of expanding antifreeze solution through the anticipated temperature extremes to which the system will be exposed. Expansion tank manufacturers usually publish charts indicating the acceptance volume of their tanks and the amount of precharge required. Alternatively, this can be calculated using standard expansion tank sizing methods.

Summary

Cross-connection control is a requirement in most jurisdictions, but the means of achieving an acceptable level of backflow prevention varies. The installation of backflow prevention devices requires planning at the initial design stage to ensure that the code requirements are met and there will be adequate water pressure available for the design of the system downstream of the backflow preventer.

BIBLIOGRAPHY

References Cited

1. *Proceedings of Fire & Water '97*, NEWWA, NFPA & AWWA, August 7–8, 1997 (available at http://cee.wpi.edu/FW_Proceedings).
2. Poole, J., "Is there a Cross-Connection Between Fire Protection Systems and Backflow Prevention?" *PM Engineer*, October 2002, pp. 34–38.
3. *Recommended Practice for Backflow Prevention and Cross-Connection Control, Manual of Water Supply Practices M14*, American Water Works Association, www.awwa.org, 800-926-7337.
4. *Manual of Cross-Connection Control*, Foundation for Cross-Connection Control and Hydraulic Research of the University of Southern California.
5. Duranceau, S. J., Poole, J., Foster, J. V., "Wet-Pipe Fire Sprinklers and Water Quality," *Journal AWWA*, Vol. 91, No. 7, July 1999, pp. 78–90.
6. *Impact of Wet-Pipe Fire Sprinkler Systems on Drinking Water*, American Water Works Association, ISBN 0-89867-943-5, Catalog Number 90752, 1998.
7. Fleming, R., "Expansion Chambers on Antifreeze Systems," *Sprinkler Quarterly*, National Fire Sprinkler Association, Patterson, NY, Spring 1992, pp. 26–28.
8. *Backflow Protection for Fire Sprinkler Systems*, National Fire Sprinkler Association, Patterson, NY.
9. Young Engineering Manufacturing Inc., 1-800-337-8743, http://www.youngeng.com/bet.htm.

NFPA Codes, Standards, and Recommended Practices

NFPA Publications National Fire Protection Association, 1 Batterymarch Park, Quincy, MA 02169-7471

The following is a list of NFPA codes, standards, and recommended practices cited in this chapter. See the latest version of the *NFPA Catalog* for availability of current editions of these documents.

NFPA 13, *Standard for the Installation of Sprinkler Systems*

NFPA 13E, *Recommended Practice for Fire Department Operations in Properties Protected by Sprinkler and Standpipe Systems*

NFPA 20, *Standard for the Installation of Stationary Pumps for Fire Protection*

NFPA 24, *Standard for the Installation of Private Fire Service Mains and Their Appurtenances*

NFPA 25, *Standard for the Inspection, Testing, and Maintenance of Water-Based Fire Protection Systems*

NFPA 750, *Standard on Water Mist Fire Protection Systems*

Additional Reading

FM Global Property Loss Prevention Data Sheet 3-3, "Cross Connections," FM Global, Norwood, MA, September 2000.

12

The Role of Fire Pumps in Automatic Sprinkler Protection

Michael A. Crowley, PE

Water supply is one of the most important parts of an automatic sprinkler system design. In many cases, the municipal water supply is adequate to meet the demand of an automatic sprinkler system. Many sprinkler systems protect light hazard, one- to three-story buildings without automatic standpipe systems. For automatic sprinkler systems with pressure or water flow demands that exceed the municipal water supply system, fire pumps are required to boost the water supply system pressure. Water storage tanks are required for water flow demand of a sprinkler system that exceeds the volume of water flow available.

This chapter will cover the basics in fire pump selection and installation arrangement. For additional information, see the Additional Readings section at the end of this chapter, which includes resources such as *Pumps for Fire Protection Systems*, prepared by the National Fire Sprinkler Association and the National Fire Protection Association and published by the NFPA. Also see relevant chapters in the *Fire Protection Handbook*.

History of Fire Pumps

In the late 1800s, the early fire sprinkler systems operated on the water pressure available from public water supplies. Many early sprinkler systems were manually activated, and some of the early automatically activated sprinkler systems were less successful than others, primarily because the available water supply was either inadequate or was not adequately distributed over the fire. Early sprinkler systems in mill buildings of New England were piping systems with open orifices along the pipe. A valve was manually opened to charge the pipe with water, and all the orifices flowed water to control a fire. In buildings close to the water source (and power source for the mill), a water supply was readily available, but as manufacturing moved away from rivers and used other power sources for the mill machines, adequate water pressure and water flow were not always accessible to the mills. It became apparent that a reliable means of providing water at the needed pressures for fire sprinkler systems was needed.

Fire protection systems were mostly installed in industrial settings during the late 1800s. Industrial plants of that era predominately had their own method of power generation, which varied from steam to electrical power, produced by the most economical method available to the plant. Early fire pumps adapted the available method of power generation and applied it to water pumps for fire protection use. These pump types varied from positive displacement to centrifugal-type pumps. The pumps at each individual plant generally followed the technology available at that plant or in that general area.

The first NFPA standard for automatic sprinklers was published in 1896, and contained paragraphs on steam and rotary fire pumps, but the reliability of the fire pumps became an issue. The Committee on Fire Pumps was organized in 1899, and NFPA 20, *Standard for the Installation of Stationary Pumps for Fire Protection*, was subsequently developed. Early fire pumps were only secondary supplies for sprinklers, standpipes, and hydrants, and were started manually. The modern fire pump has evolved into two major categories, centrifugal pumps and vertical turbine pumps. The physics of both pump styles are very similar, but the arrangement of the pumps is very different.

The power for fire pumps has also been standardized. Modern fire pumps are equipped with either an electric drive or a diesel engine. Internal combustion engines can be powered by diesel fuel gasoline, or natural gas. Prior to 1974, gasoline-fueled internal combustion engines were allowed, but currently NFPA 20 does not permit gasoline or natural gas engines. NFPA 20 does not require the removal of these gasoline drivers, but the storage of the fuel must comply with NFPA 30, *Flammable and Combustible Liquids Code*, and local fire codes.

Selecting the Fire Pump Size

Chapter 6, "Evaluation of Water Supply," addressed how to determine water demand for an automatic sprinkler system. Once the fire sprinkler system demand has been determined and the water supply available has been evaluated, the need for the fire pump may become evident. The fire pump is designed to supplement the pressure available from the water supply in order to meet the demand of the fire sprinkler system.

Fire pumps have three major variables—rated flow, rated net pressure, and rated speed—that need to be selected in order to correctly specify the fire pump, as shown in Table 12.1. Fire pump capacities may be selected from Table 12.1.

Rated Flow

The first variable is the rated flow in gallons per minute (gpm). The rated flow is a value supplied by the manufacturer that represents the flow in gallons per minute at a rated net pressure. Rated fire pump flows vary from 25 gpm to up to 5000 gpm.

Rated Net Pressure

The rated net pressure boost increases the suction pressure at the rated flow. If the pump's net rated pressure is 50 psi and the pump suction pressure from the water supply is 30 psi, then the discharge pressure would be 80 psi. This is a net pressure boost of 50 psi at rated flow. The rated pressures on fire pumps range from 40 psi to 200 psi net pressure boost. Figure 12.1 shows the relationship between net pressure boost, the discharge, and suction pressures.

TABLE 12.1			
Centrifugal Fire Pump Capacities			
L/min	gpm	L/min	gpm
95	25	3,785	1,000
189	50	4,731	1,250
379	100	5,677	1,500
568	150	7,570	2,000
757	200	9,462	2,500
946	250	11,355	3,000
1,136	300	13,247	3,500
1,514	400	15,140	4,000
1,703	450	17,032	4,500
1,892	500	18,925	5,000
2,839	750		

Source: NFPA 20, 2003, Table 5.8.2.

Rated Speed

The last major item in pump selection is the rated speed. Rated speed is the revolutions per minute of the pump driver at the rated flow and pressure point of output. The rated speed varies between 1770 revolutions per minute and 3550 revolutions per minute, depending on manufacturer and type of driver.

The Fire Pump Curve

A fire pump curve is comprised of three points: the rated point, the churn point, and the maximum load point. Fire pumps are allowed to perform in a range along the fire pump curve, from the churn point, to the rated point, to the maximum load point, as shown in Figure 12.2. This range is identified in NFPA 20.

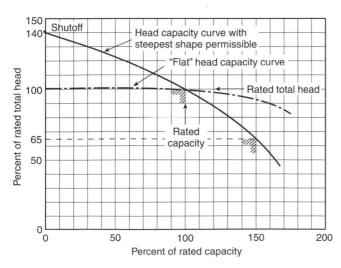

FIGURE 12.1
Pump Characteristics Curves.
(Source: NFPA 20, 2003, Figure A.6.2)

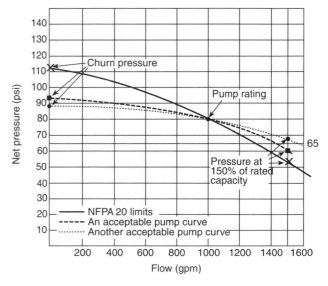

FIGURE 12.2
Acceptable Flow Curves for a 1000-gpm Pump.
(Source: *Pumps for Fire Protection Systems*, NFPA, 2002, Figure 4-1)

Rated Point. The rated point is the 100 percent mark in a pump performance evaluation and is composed of the rated flow at the rated pressure of the fire pump, as identified by the "pump rating" point in Figure 12.2.

Churn Point. The churn point, as identified by the "churn pressure" point in Figure 12.2, represents the no-flow condition. NFPA 20 allows a maximum of a 140 percent rating of the net pressure at churn. The churn point is the pressure boost measured with the fire pump running and a no-water-flow condition.

Maximum Load Point. The maximum load point represents the point of greatest flow (150 percent of the rated flow) permitted by NFPA 20, as shown in Figure 12.2. NFPA 20 requires a maximum load point determined by a minimum of 65 percent net rated pressure when the pump is producing 150 percent of its rated flow capacity (maximum load flow).

Figure 12.3 shows the effect a fire pump has on a city water supply. Curve 1 is a city water supply curve from a flow test (as described in Chapter 6). Curve 2 is a typical fire pump curve. Curve 3 is the water supply available downstream of the fire pump. Curve 3 is the combined supply curve representing the effect the fire pump has on the performance of the water supply and is obtained by adding the city and pump pressures at the rated point, the churn point, and the maximum load point. The fire pump will increase the pressure

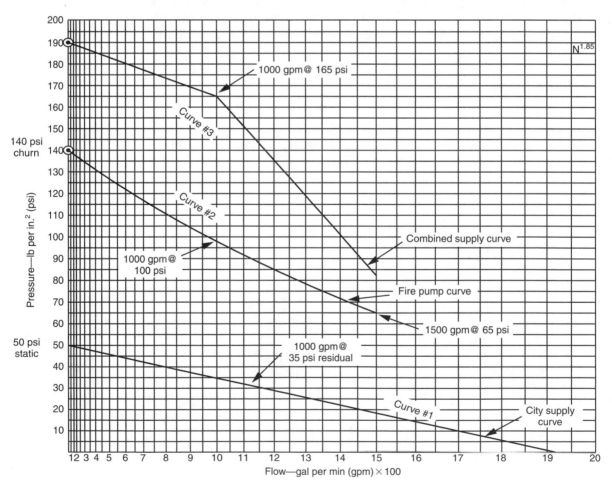

FIGURE 12.3
Effect of Fire Pump on City Water Supply.

available at a given city flow. The fire pump cannot create water flow in excess of what is available from the city supply. Note that neither curve 2 nor curve 3 extends beyond 1500 gpm, the maximum pump capacity in this example.

The Manufacturer's Pump Curve

With the performance criteria of net pressure at churn (no flow), rated net pressure at the rated flow (100 percent flow), and the net pressure at the maximum load (150 percent flow), a water supply curve for a rated fire pump can be drawn. In Figure 12.2, the solid line shows the NFPA 20 limits for a rated fire pump performance curve. Figure 12.2 also shows two acceptable manufacturer's pump curves that meet or exceed the performance criteria of NFPA 20. All three of the curves shown in Figure 12.2 represent a fire pump rated at 1000 gpm and 80 psi. NFPA 20 requires that the pump not generate more than 140 percent of 80 psi, or 112 psi at churn (no flow). NFPA also requires the pump to generate a minimum of 65 percent of 80 psi, or 52 psi at 150 percent of rated flow, or 1500 gpm. The solid line joining these three points in Figure 12.2 represents the limits of a fire pump performance curve as described in NFPA 20.

Every fire pump rated for 1000 gpm and 80 psi does not follow this exact curve. Both the dashed lines and dotted lines in Figure 12.2 represent actual manufacturer's fire pumps rated at 1000 gpm and 80 psi. Both of these pumps are acceptable because they produce less than 140 percent of the rated pressure at churn and more than 65 percent of the rated pressure at maximum flow point. As shown in Figure 12.1, there are major differences in the actual fire pump performance and the limitations dictated by NFPA 20.

Fire pumps produce differing supply curves due to the shape and configuration of the fire pump impeller. The width, curve angle, number of vanes, and diameter of the impeller shape the performance. Figure 12.4 shows the relationships between the impeller variables and the fire pump's performance. Some commercially available hydraulic programs use the NFPA 20 curve in the application of pumps. This could cause the program to assume there is more pressure available at flow points below the 100 percent rated point. The

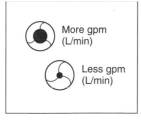

Diameter of eye

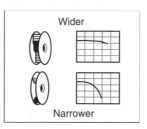

Width of impeller

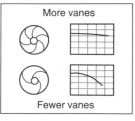

Number of vanes

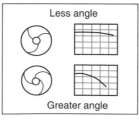

Angle of vanes

FIGURE 12.4
Effect of Impeller Design on Head/Discharge Curves for Fire Pumps.

(Source: *Fire Pump Handbook*, NFPA, 1998, Exhibit S2-7)

use of actual manufacturer's pump curves is strongly recommended in the final evaluation of a water supply system. Each manufacturer's pumps have unique characteristics and produce unique flow and pressure curves.

Technique for Selecting Fire Pump Size

A six-step technique for selecting a fire pump size follows:

1. Calculate the system demand to the fire pump discharge flange.
2. Calculate the water supply from the point of water flow test to the pump suction flange.
3. Select a fire pump such that the system flow demand is less than 150 percent of the rated flow of the fire pump and such that the sprinkler system demand pressure does not exceed the fire pump maximum load point.
4. Using the manufacturer's fire pump curve, select the fire pump's net pressure at the system demand flow.
5. Add the suction pressure at the demand flow to the net pressure at the demand flow to get the fire pump discharge pressure at the demand flow.
6. If the discharge pressure is greater than the most demanding calculation for sprinkler system flow and pressure, then this pump is sized to meet or exceed the automatic sprinkler system demand.

Step 1. Calculate the system demand to the fire pump discharge flange.
This will require a detailed hydraulic calculation of the proposed fire sprinkler system, as discussed in Chapter 10. The theoretical or estimated demand method should not be used in sizing a fire pump because it may cause the fire pump to be oversized or undersized in either pressure or flow capability. A method to size the fire pump in the very early stages of the design process is discussed later in this chapter (see "Fire Pump Sizing Example"). Even though fire pumps are typically specified early in the building construction process, the engineer is advised to perform a layout of the most remote and most demanding design areas and perform a computerized calculation to best ensure proper pump size. This procedure is significantly more accurate than estimating flow and pressure.

Step 2. Calculate the water supply from the point of water flow test to the pump suction flange.
This will require taking the water supply data and determining the actual flow point of the water supply test. The designer will then calculate the friction loss from the flow point to the pump suction flange, through the piping fittings and valves at the automatic sprinkler system maximum flow. Water system authorities, such as public works, city water departments, and municipal water districts, sometimes adjust the minimum water supply pressure to accommodate maximum domestic usage and anticipated low-water levels. The maximum water flow available at the minimum water main pressure permitted by the water system authority must exceed the automatic sprinkler system flow demand. If the water system authority flow exceeds the designed system demand, only a fire pump is needed to produce the required design pressure. If the water system authority flow is less than demand, a water storage tank may be required in addition to the fire pump.

Step 3. **Select a fire pump such that the sprinkler system flow demand is less than 150 percent of the rated flow of the fire pump and such that the system demand pressure does not exceed the fire pump maximum load point.**

It is recommended to use a maximum of 140 percent of the rated flow to take into account miscellaneous losses in the system. NFPA 20 recommends using between 90 and 140 percent of the rated flow when selecting a fire pump. Also, the local authority having jurisdiction (AHJ) should be consulted to determine if it would permit fire pump sizing below the 100 percent rated point. Engineers who specify fire pumps where the greatest system flow is less than the 100 percent rated point should reevaluate the pump selection. A small flow rate pump with adequate pressure may cost less.

Step 4. **Using the manufacturer's fire pump curves, select the fire pump's net pressure at the system demand flow.**

Step 5. **Add the suction pressure at the demand flow to the net pressure at the demand flow to get the fire pump discharge pressure at the demand flow.**

Step 6. **If the discharge pressure is greater than the demand calculation for system pressure, then this pump is sized to meet or exceed the automatic sprinkler system demand.**

If the discharge pressure is less than the system demand pressure, a different pump must be selected or a more efficient sprinkler design should be considered. If discharge pressures are significantly greater than the demand, the static pressure at churn should be checked to confirm the system pressure at churn is below 175 psi. Pressures higher than 175 psi are not permitted in automatic sprinkler systems unless special high-pressure valves, sprinklers, and fittings are installed.

Fire Pump Sizing Example with Supply Pressures Exceeding 175 psi

An automatic sprinkler system for a warehouse has been determined by hydraulic calculations to have a demand of 1010 gpm at 125 psi calculated to the discharge flange of the fire pump. Water supply for the building consists of a 100,000-gal gravity tank with an 80-ft vertical elevation from the fire pump suction flange to the bottom of the gravity tank. The gravity tank is sized to exceed the fire sprinkler system water demand for the duration of 90 minutes (1010 gpm × 90 min), as required by NFPA 13, *Standard for the Installation of Sprinkler Systems*. Although the tank static pressure will be higher when the tank is full, the calculation is required to be based on the endpoint in the duration, as the tank is empty at the 90-min duration point. The full tank static pressure must be checked to confirm that system pressures do not exceed 175 psi. If the system design pressure exceeds 175 psi, the designer has the option of selecting another fire pump with a different performance curve or providing high-pressure-rated pipe and fittings plus pressure regulating valves for the sprinkler systems. Figure 12.5 shows a schematic depiction of this example.

The water demand for the system is 1010 gpm at 125 psi at the fire pump discharge flange. As previously shown, the 100,000-gal volume exceeds the system demand and required water supply duration. The pressure available at the suction side of the pump is 80 ft × 0.433 psi per foot, or 34.6 psi. Friction loss from the

FIGURE 12.5
Pump Sizing for a Warehouse.

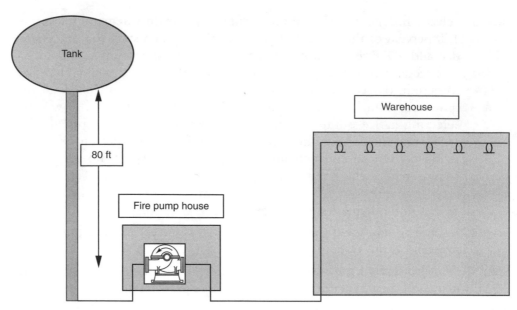

Demand: 1010 gpm at 125 psi at
fire pump discharge flange

tank to the suction flange is calculated to be 1.5 psi. There are numerous fire pump flow capacities that would meet the 1010 gpm demand. The smallest-size pump that would meet the demand would be a 750-gpm pump. The maximum flow out of the rated 750-gpm pump is 150 percent, or 1125 gpm. A 1000-gpm pump would also meet the system demand. Figures 12.6 and 12.7 show manufacturer's performance curves for a 750-gpm fire pump and a 1000-gpm fire pump. A larger-capacity fire pump may be considered in cases where future expansion or change of building occupancy is anticipated.

Figure 12.6 is the performance curve of a 750-gpm, 115-psi fire pump. This pump will produce a maximum of 149 psi at churn and 81 psi at a maximum

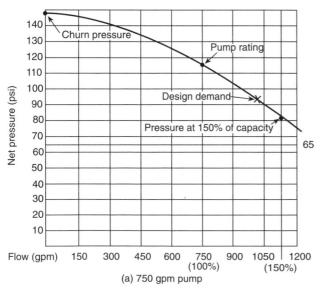

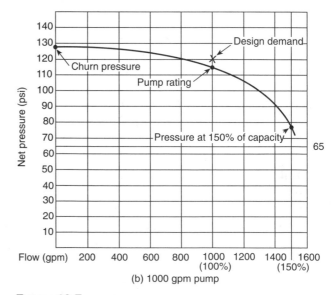

FIGURE 12.6
Fire Pump Curve for Pump Rated at 750 gpm and 115 psi.

(Source: *Pumps for Fire Protection Systems*, NFPA, 2002, Figure 4-3)

FIGURE 12.7
Fire Pump Curve for Pump Rated at 1000 gpm and 115 psi.

(Source: *Pumps for Fire Protection Systems*, NFPA, 2002, Figure 4-3)

flow rating of 1125 gpm. This curve will be used to determine the net pressure available at the system demand point. Moving across the x axis of Figure 12.7 to the 1010 gallon point, this curve shows that the pump produces a net pressure of 92 psi. This net pressure of 92 psi added to the suction pressure of 34.6, minus the friction loss of 1.5 psi, produces a discharge pressure of 125.1 psi at 1010 gpm flowing (see Figure 12.2). The churn pressure for this pump arrangement is 149 psi plus 34.6 psi (no friction loss due to no flow), or 183.6 psi. This exceeds the maximum system pressure of 175 psi.

When pressures in automatic sprinkler systems exceed 175-psi, high-pressure fittings, valves, sprinklers, and components are needed for the system design. The cost of this high-pressure equipment can be avoided with a well-engineered fire pump design. The following example demonstrates the proper pump size.

Fire Pump Sizing Example with Supply Pressures Less Than 175 psi

Figure 12.7 is another fire pump performance curve of a 1000-gpm pump rated at 115 psi. At churn pressure this pump produces a pressure of 129 psi, and at a maximum flow point of 150 gpm, this pump creates 78 psi. At the sprinkler system demand point of 1010 gpm, the pump produces 112 psi. This added to 33.1 psi for the suction side produces a discharge pressure at 1010 gpm of 145.1 psi. This exceeds the minimum 125 psi needed in the system demand calculations. This pump will meet the requirements for system demand. In addition to meeting system demand requirements, the pump will not over-pressurize the sprinkler system. At churn pressure, the pump produces only 129 psi; 129 psi plus the maximum pressure from the elevated tank of 34.6 psi provides a pressure on the system of 163.6 psi, which is below the 175-psi limitation of the normal system components. This system must also be checked for the static no-flow condition with the gravity tank at its maximum fill point. A tall water storage tank will have a significantly higher static pressure when full. Static pressure at the full point should be checked to verify that it does not exceed 175 psi. Also, a supply calculation for the automatic sprinkler system should be done to evaluate whether the water supply tank is sized for the required flow duration. The full-tank condition on tall tanks may cause excess flow due to the higher static pressure. The increased flow may cause the tank to empty before the NFPA 13 required duration. The fire pump in Figure 12.7 would be the better selection of the two pump curves presented.

Alternative Sizing Methodology—Conceptual Design

Engineers are sometimes asked to specify fire pump sizes early in a project, before detailed building plans are available and before final hydraulic calculations can be performed. An alternative method of specifying the pump size can be used in the early concept design and specifying stages of a project. This alternative makes several assumptions that must be verified during the detailed design of the fire sprinkler system. The first assumption is that the fire pump does not produce any pressure less than the rated pressure at any flow less than the rated flow. The specific manufacturer's pump curves should always exceed this pressure assumption. For flows greater than the rated flow, the engineer should assume that the pump will achieve 65 percent of its rated pressure at maximum

FIGURE 12.8
Maximum Pressure Produced
by Fire Pump.

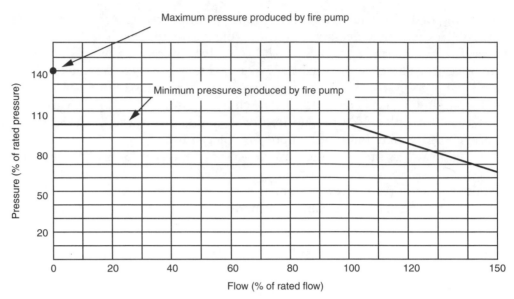

FIGURE 12.8
Maximum Pressure Produced
by Fire Pump.

flow (150 percent of rated flow), and use a linear interpolation between this maximum flow point and the rated flow point. Once again, this is a worst-case assumption because the pump will actually produce slightly more pressure than 65 percent of its rated pressure at maximum flow.

The last consideration is the maximum total pressure that the pump will put into the fire protection system. The worst-case scenario is to consider the maximum pressure from the pump as 140 percent of the rated net pressure. This maximum pressure would occur at churn condition (no flow in the fire protection system). The total pressure into the fire protection system would be the churn pressure added to the maximum static pressure of the system; for sprinkler systems using standard components, this value must not exceed 175 psi.

This pump sizing alternative method will produce a fire pump for pricing and schematic design purposes. A more detailed analysis must be done when a detailed hydraulic calculation of the sprinkler system can be performed and a specific fire pump model can be selected for the project. This analysis should be done to verify that the selected fire pump will not overpressurize the system. Figure 12.8 shows the maximum and minimum pressure points produced by the fire pump to be evaluated in this alternative method for sizing a fire pump. Even in cases where detailed building plans are not available and a final sprinkler calculation cannot be performed, it is advisable to develop a "worst-case scenario" sprinkler layout and to perform a detailed hydraulic calculation of that layout to more accurately size the fire pump. This calculation can be revised or redone when detailed building plans become available.

Effects of the Fire Pump Size on the Suction Pressure

The water supply system must be capable of providing adequate flow to a fire pump, at a positive gauge pressure, at the suction flange to the fire pump. Positive pressure at the suction flange is critical for proper fire pump operation, as required by NFPA 20. Fire pumps cannot create water flow and can only boost the pressure of the water provided to the pump suction side at a positive pressure. If the municipal water supply can only provide 500 gpm and a 750-gpm fire pump is installed, this fire pump will not operate correctly and will probably cause damage to both the pump and possibly the city mains. Evaluation of the city water supply should be done at the 150 percent rated point of the fire pump. Many municipalities require evaluations to be conducted for fire pump designs to limit the minimum pressure in the street main to a minimum of 20 psi. The local authority having jurisdiction should be consulted to determine the minimum main pressures allowed in the jurisdiction. These minimum pressures are designed to protect the water main system from low pressure and may affect other entities using the public water system. Many municipalities require some means of controlling the fire pumping system to limit or cut off flow when low suction pressure is sensed. The use of low suction cutoff valves is not permitted by NFPA 20 (NFPA 20, 2003 edition, subsection 5.14.9, permits the use of throttling valves in the suction piping), but may be required by some local municipalities to protect city mains. The low-pressure cutoff valve is arranged to sense a low pressure in the city main and automatically adjust a valve to reduce the fire pump flow. Depending upon the location of this valve, this may cause damage to the fire pump; it may compromise the fire protection system by limiting the water to the system; and it might possibly endanger fire fighters if they are using the fire protection system hose demand to fight the fire.

Other options to address with respect to city water authorities include a low-suction throttling valve, which reduces the flow at the discharge side of the fire pump, thus protecting the fire pump. Although this low-suction throttling valve will help maintain minimum 20 psi pressure within the street mains, it must also be taken into consideration when designing the fire pump to meet flow and pressure demands of the fire protection system. If using this throttling-type valve, the automatic sprinkler system should be designed to a point such that the low-suction throttling valve is never engaged at the maximum demand point of the automatic sprinkler system.

Water authorities are concerned about public health in maintaining a minimum pressure of 20 psi in their mains. It should be remembered that the 20-psi minimum is at the water main and not at the suction point of the pump. Equation 12.1 can be used to determine the suction pressure at the suction flange of the pump.

$$P_S = P_R - P_L \qquad (12.1)$$

where:

P_S = suction pressure

P_R = the residual pressure of the water supply system at the maximum flow of the fire pump (150 percent of the rated flow capacity)

P_L = the pressure loss (combination of friction loss and elevation) between the water supply and the fire pump suction flange at the maximum flow of the fire pump (150 percent of rated flow)

EXAMPLE 12.1. Calculating Suction Pressure

A 1500-gpm fire pump is taking suction from a street main. The suction pipe is 100 ft of 8 in. PVC (underground) with two 45 degree elbows, one 90 degree elbow, and one tee plus 10 ft of steel pipe running aboveground with

Figure 12.9
Schematic of Suction
Pressure for a Fire Pump
Arrangement.

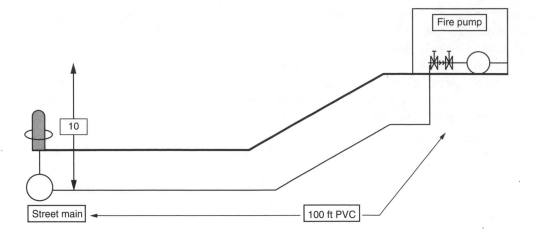

one elbow and one backflow preventer (8 psi friction loss at 2250 gpm). The fire pump is elevated 10 ft over the point where the pressure reading was taken for the street main. The street main has static pressure of 90 psi and a residual pressure of 35 psi at 2000 gpm, and at 2250 gpm, 23 psi is available. The water authority in this situation does not want the pressure in the mains to drop below 20 psi. Can this fire pump be installed in this location with this suction pipe arrangement? Figure 12.9 is a schematic of the example problem.

Equation 12.1 ($P_S = P_R - P_L$) is used to evaluate this problem. P_S must be greater than or equal to zero. In this problem,

P_R = 23 psi

P_L = the sum of all the pressure losses due to elevation and friction

   = 4 psi (due to the change in elevation) + 5 psi in the friction loss of the underground piping + 1 psi in the friction loss due to the aboveground piping + 8 psi loss due to the backflow preventer

   = 18 psi

$$P_S = 23 - 18$$
$$= 5 \text{ psi}$$

5 psi is greater than zero.

Yes, this fire pump can be installed in this location with this suction arrangement. If P_S were less than zero, design changes would have been required to get the pressure at the flange greater than zero. This might have included changing pipe size, relocating the pump to a lower elevation, or resizing the pump to have a lower flow rate at 150 percent of the rated capacity.

Another unique phenomenon that occurs due to low pressure at the suction side of the pump is pump cavitation. *Pump cavitation* is spinning of the pump impeller in an air or vapor pocket, which is normally caused by low suction pressure or inadequate water supply to the pump. Pump cavitation can cause major damage to the pump impeller and driver due to the excess speeds of the impeller spinning in air. Pump cavitations cause vibration and loud noises as the impeller spins within the pump casing.

Effects of the Pump Size on Discharge Pressure

Some building configurations make it impossible to avoid pressures in excess of 175 psi. A very tall building, for example, where the fire pump must be positioned at a low elevation, could require a pump discharge pressure in excess of 175 psi in order to ensure adequate pressure at the highest sprinkler. The key to the appropriate selection of a fire pump also evaluates the maximum pressures produced by that pump and the effect of these maximum pressures on the fire sprinkler system components. The fire pump should not produce pressures in the system greater than the system components can handle. Most systems are rated for a maximum operating pressure of less than 175 psi, and when operating pressures exceed 175 psi, special piping and fittings are required. Pressure regulating valves are required for NFPA 13 compliance for sprinkler systems being served by high-pressure bulk water supply mains. (In storage occupancies the pressure cannot exceed 175 psi regardless of the ratings of the components, as required by NFPA 13.) The evaluation of maximum pressure needs to be done at the no-flow state in the system. The maximum system static pressure will occur at the water supply static pressure plus the fire pump's churn pressure (pressure at zero fire pump flow). NFPA 20 requires a maximum churn pressure not to exceed 140 percent of the rated pressure. Many manufacturers offer fire pumps with relatively flat pump curves (churn pressure considerably less than the NFPA 20 allowed 140 percent maximum of rated pressure). To achieve a flat pump curve, larger-horsepower drivers or different impeller configurations are provided for the pump.

Pressure relief valves have been used to address overpressurization in the system due to diesel fire pump overspeed and, in some cases, incorrect design of the fire pump. NFPA 20 requirements for installation of pressure relief valves have changed with every edition of NFPA 20 since 1993. Pressure relief valves will no longer be common on systems, even those with diesel drivers. NFPA 20 specifically prohibits the use of pressure relief valves to address overpressurization. Instead, systems with diesel drivers will have an overspeed cutoff device to shut down the diesel engine. The fire sprinkler system should not need pressure relief valves because NFPA 20 requires that fire pumps be selected so that the maximum pressure produced by the pump does not exceed the ratings of the components. The only time a pressure relief valve will be allowed is for safety reasons when a diesel engine is used as a driver and the pressure produced by the fire pump when turning at 121 percent of the rated fire pump speed is greater than the pressure rating of the components.

Pressure relief valves discharge water if excess pressure is built up in the system. The relief valve setting is usually less than the maximum pressure allowed on the fire sprinkler system. This relief valve must discharge to a drainage system capable of handling the full flow of the valve. In systems fed by a water tank, it may be feasible to discharge relief valve flow back into the tank. This relief valve must be serviced and maintained in accordance with NFPA 25, *Standard for the Inspection, Testing, and Maintenance of Water-Based Fire Protection Systems*. An improper functioning of the relief valve can affect the system pressure and flow available to the sprinkler system. Prior to the 1996 edition of NFPA 20, the pressure relief valve could discharge back to the suction side of the fire pump. This would allow the recirculation of overpressured water into the system. In some configurations, the temperature of the water would increase, causing other problems in the system. The 1996 edition of NFPA 20 eliminated the allowance of discharging the pressure relief valve to the suction side of the fire pump. However, the 1999 edition was revised to permit a pressure relief valve in very specific configurations to discharge into the suction side of the fire pump provided a thermal relief to an independent drain system is provided.

Types of Fire Pumps

Rated fire pumps come in varying sizes, from 25 gpm to 5000 gpm, and pressure ratings, from 40 psi to 390 psi net boost. This range of fire pump flows and pressure ratings comes from the commercially available UL listed and Factory Mutual approved fire pumps. There are different styles of pumps to address different configurations and arrangements. The typical pump arrangement in an industrial commercial application is a horizontal split-case pump, as shown in Figure 12.10. Figure 12.10 shows the pump, but not the engine that drives it, which can require a significant amount of floor space. A variation of this

FIGURE 12.10
Horizontal Split-Case Pump.

(Source: *Pumps for Fire Protection Systems*, NFPA, 2002, Figure 3-12; photograph courtesy of ITT A-C Pump)

FIGURE 12.11
Vertical In-line Pump with an Electric Motor Driver.

(Source: *Pumps for Fire Protection Systems*, NFPA, 2002, Figure 3-13; photograph courtesy of ITT A-C Pump)

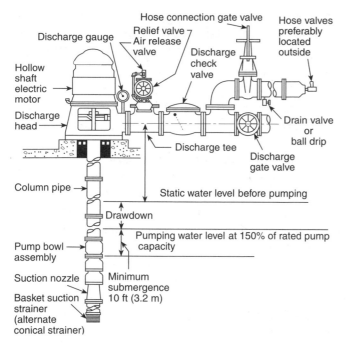

FIGURE 12.12
Vertical Shaft Turbine Pump.

(Source: *Pumps for Fire Protection Systems*, NFPA, 2002, Figure 1-3)

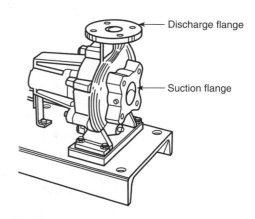

FIGURE 12.13
End Suction Pump.

(Source: *Pumps for Fire Protection Systems*, 2002, Figure 3-11)

horizontal split-case pump is the vertical in-line pump shown in Figure 12.11, which shows the engine mounted above the pump. This pump is used where space constraints dictate a minimal footprint for the fire pump arrangement.

The vertical shaft turbine fire pump is a very specialized pump that ranges in size from 500 gpm to 5000 gpm with pressure ratings up to 300 psi net pressure. This range of fire pump flows and pressure ratings comes from the commercially available UL listed and Factory Mutual approved fire pumps (see Figure 12.12.). The vertical turbine pump is constructed with multiple stages to achieve the desired pressure boost. Vertical turbine pumps are generally used to lift water out of reservoirs, drafting basins, cisterns, or even wells.

End suction fire pumps as shown in Figure 12.13 are specialty pumps, used when the area for pump layout is limited. End suction fire pumps come in sizes from 50 gpm to 750 gpm and pressure boosts from 40 psi to 150 psi net pressure boost. This range of fire pump flows and pressure ratings comes from the commercially available UL listed and Factory Mutual approved fire pumps.

Vertical turbine and split-case horizontal fire pumps are the most common fire pumps used. Each type of fire pump has slightly different general layout arrangements.

Fire Pump Driver Selection

Drivers are the motors or prime movers that power the fire pump. Fire pump driver selection is a balance among cost, maintenance, and available reliable power source. For facilities that have reliable electrical power or a secondary electrical power source such as a generator, an electric driver would generally be the first choice. Diesel drivers are typically used when a secondary electrical power is not available or the secondary electrical power must be transmitted long distances. Diesel drivers are considered a reliable power source for a fire pump.

Diesel Drivers

The use of diesel drivers for the fire pump introduces a need for fuel storage, exhaust, methods for cooling the driver, and a battery charging system for the starting batteries for the prime mover. Diesel fuel storage must be in accordance with NFPA 30, *Flammable and Combustible Liquids Code*, or the local fire code. Fuel storage requires secondary containment, ventilation, and a method for refilling the storage tanks. Designers must also provide adequate cooling air for the diesel driver to the fire pump room and should carefully locate the exhaust such that the fumes will not be introduced to other parts of the building. The diesel drive fire pump should be located in an area separated from the remainder of the building by a minimum of 2-hour fire-resistive separation from nonsprinklered areas or 1-hour fire-resistive separation from sprinklered areas. *Separate* fire pump buildings that are not attached to any other structure should be separated by 50 ft from other structures (NFPA 20, Table 5.12.1.1, which appears here as Table 12.2).

Electric Drivers

Electric drivers for fire pumps have special arrangements for the connection to the power source, shown schematically in Figure 12.14. Electric pump drivers should be located in an area separated from the remainder of the building in the manner stated above for diesel driven fire pumps. The arrangement shown in Figure 12.14 complies with NFPA 70, *National Electrical Code®*. There are no disconnects or fuses between the main power source and the fire pump controller, the device that starts and stops a fire pump. For electric drivers, the fire pump controller also contains the overcurrent protection and means of power disconnects. The electrical connection without disconnects is unique to electric driver fire pumps. The special overcurrent protection in the controller is intended to increase the reliability of the fire pump.

Figure 12.15 is a typical fire pump controller and transfer switch arrangement. This is the arrangement when secondary power is supplied by an on-site

TABLE 12.2
Equipment Protection

Pump Room/House	Building(s) Exposing Pump Room/House	Required Separation
Not sprinklered Not sprinklered Fully sprinklered	Not sprinklered Fully sprinklered Not sprinklered	2 hour fire-rated or 15.3 m (50 ft)
Fully sprinklered	Fully sprinklered	1 hour fire-rated or 15.3 m (50 ft)

Source: NFPA 20, 2003, Table 5.12.1.1.

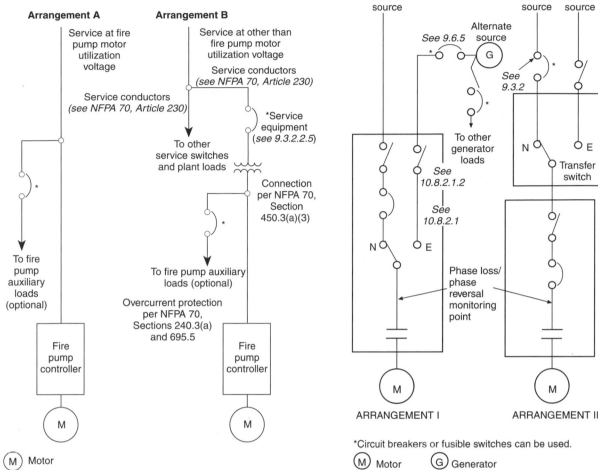

FIGURE 12.14

Typical Power Supply Arrangements from Source to Motor.

(Source: NFPA 20, 2003, Figure A.9.3.2)

FIGURE 12.15

Typical Fire Pump Controller and Transfer Switch Arrangements.

(Source: NFPA 20, 2003, Figure A.10.8)

generator. The use of power disconnects outside the controller is allowed but requires the disconnect to fulfill paragraph 9.3.2.2.3.2 of NFPA 20, which requires the proper sizing of the switch, identification of the disconnect switch, notice at the controller that a disconnect is present, and supervision of the switch. These requirements are to maintain a high degree of reliability for the source power to the fire pump.

The NFPA book *Pumps for Fire Protection Systems*, referenced in the bibliography at the end of this chapter, has extensive information to assist in the selection and layout of fire pumps and their drivers.

Supply Piping

The size of the fire pump selected must address the friction losses in the piping and components between the public water supply or water source to the fire pump suction flange. These components consist of pipe, valves, and miscellaneous

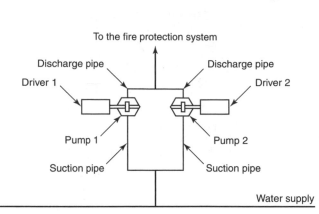

FIGURE 12.16
Pumps in Parallel.

(Source: *Pumps for Fire Protection Systems*, NFPA, 2002, Figure 5-20)

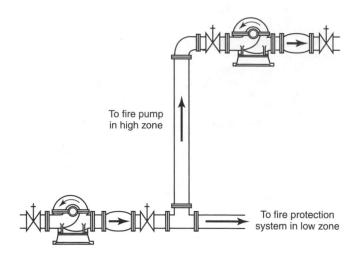

FIGURE 12.17
Pumps in Series.

(Source: *Pumps for Fire Protection Systems*, NFPA, 2002, Figure 5-21)

fittings. These losses are determined by hydraulic calculation, as discussed in Chapter 10.

Pumps may be installed in parallel (as shown in Figure 12.16) or in series (as shown in Figure 12.17). In fire pumping configurations where fire pumps are installed in series, the suction pipe of the high-pressure pump begins at the system side of a discharge valve from the low-pressure fire pump.

The size of the suction pipe and fittings are such that when the pump is run at 150 percent of rated capacity, the gauge pressure at the pump suction flange is 0 psi or higher. It must also be verified that the municipal water system, if used, is not drawn below the minimum pressure set by the local authority having jurisdiction, usually 20 psi in many jurisdictions. The pressure at the suction flange is permitted to go as low as a −3 psi gauge pressure for fire pump and tank arrangements located at the same level. The size of the portion of the suction pipe located within 10 pipe diameters upstream of the fire pump suction flange cannot be less than that specified in NFPA 20, Table 5.25(a) and 5.25(b), reproduced here as Tables 12.3 and 12.4.

The water velocity in the suction pipe must be less than 15 ft/sec within 10 pipe diameters of the fire pump suction flange. The reason for this minimum pipe size and flow speed is to minimize the turbulence within the suction pipe prior to water entering the pump. A smooth, near-laminar flow stream into the pump suction is desirable. Laminar flow is the smooth movement of a fluid through a system. The opposite of laminar flow is turbulent flow with fluid flow spinning and causing eddies as it flows through a system. Excessive turbulence may cause eddies and air pockets within the pipe and subsequent cavitation of the pump.

The type of pipe permitted for the suction side of the fire pump is outlined in NFPA 13 and NFPA 24, *Standard for the Installation of Private Fire Service Mains and Their Appurtenances*. The underground portion of this pipe can be cast iron, steel, PVC, or cement. The aboveground pipe is limited to steel. NFPA 20 stipulates that where corrosive water conditions exist, steel pipe is to be galvanized or painted prior to installation on the inside with a paint recommended for submerged surfaces.

TABLE 12.3

Summary of Centrifugal Fire Pump Data (Metric)

Pump Rating (L/min)	Minimum Pipe Sizes (Nominal)						
	Suction[1,2] (mm)	Discharge[1] (mm)	Relief Valve (mm)	Relief Valve Discharge (mm)	Meter Device (mm)	Number and Size of Hose Valves (mm)	Hose Header Supply (mm)
95	25	25	19	25	32	1–38	25
189	38	32	32	38	50	1–38	38
379	50	50	38	50	65	1–65	65
568	65	65	50	65	75	1–65	65
757	75	75	50	65	75	1–65	65
946	85	75	50	65	85	1–65	75
1,136	100	100	65	85	85	1–65	75
1,514	100	100	75	125	100	2–65	100
1,703	125	125	75	125	100	2–65	100
1,892	125	125	100	125	125	2–65	100
2,839	150	150	100	150	125	3–65	150
3,785	200	150	150	200	150	4–65	150
4,731	200	200	150	200	150	6–65	200
5,677	200	200	150	200	200	6–65	200
7,570	250	250	150	250	200	6–65	200
9,462	250	250	200	250	200	8–65	250
11,355	300	300	200	300	200	12–65	250
13,247	300	300	200	300	250	12–65	300
15,140	350	300	200	350	250	16–65	300
17,032	400	350	200	350	250	16–65	300
18,925	400	350	200	350	250	20–65	300

[1]Actual diameter of pump flange is permitted to be different from pipe diameter.
[2]Applies only to that portion of suction pipe specified in 5.14.3.4 of NFPA 20, 2003.

Source: NFPA 20, 2003, Table 5.25(a).

Installation methods for the pipe are outlined in NFPA 13 and NFPA 24. These requirements are intended to minimize the introduction of air into the pipe. The pipe should be arranged to prevent the creation of air pockets. It should also be tight to minimize air leaks. Suction pipe for the pump should be protected against freezing or exposure to temperatures below 42°F.

After the installation of the pipe, but prior to its use, hydrostatic testing of the pipe at 200 psi over 2 hours is required. Newly installed underground suction pipes to fire pump installations should be thoroughly flushed prior to connection to the pump. NFPA 24, Table 10.10.2.1.3, lists the minimum flow rates for different pipe sizes and appears here as Table 12.5.

Although fire pumps can tolerate the moving of small pebbles and sand through the pumping system, ingestion of larger rocks may damage the fire pump impeller or knock the impeller out of balance. Flushing should remove most large objects, and documentation of the flushing should be obtained from the contractor prior to initial testing of the new fire pump. On the suction side of the fire

TABLE 12.4

Summary of Centrifugal Fire Pump Data (U.S. Customary)

Pump Rating (gpm)	Minimum Pipe Sizes (Nominal)						
	Suction[1,2] (in.)	Discharge[1] (in.)	Relief Valve (in.)	Relief Valve Discharge (in.)	Meter Device (in.)	Number and Size of Hose Valves (in.)	Hose Header Supply (in.)
25	1	1	¾	1	1¼	1–1½	1
50	1½	1¼	1¼	1½	2	1–1½	1½
100	2	2	1½	2	2½	1–2½	2½
150	2½	2½	2	2½	3	1–2½	2½
200	3	3	2	2½	3	1–2½	2½
250	3½	3	2	2½	3½	1–2½	3
300	4	4	2½	3½	3½	1–2½	3
400	4	4	3	5	4	2–2½	4
450	5	5	3	5	4	2–2½	4
500	5	5	3	5	5	2–2½	4
750	6	6	4	6	5	3–2½	6
1,000	8	6	4	8	6	4–2½	6
1,250	8	8	6	8	6	6–2½	8
1,500	8	8	6	8	8	6–2½	8
2,000	10	10	6	10	8	6–2½	8
2,500	10	10	6	10	8	8–2½	10
3,000	12	12	8	12	8	12–2½	10
3,500	12	12	8	12	10	12–2½	12
4,000	14	12	8	14	10	16–2½	12
4,500	16	14	8	14	10	16–2½	12
5,000	16	14	8	14	10	20–2½	12

[1]Actual diameter of pump flange is permitted to be different from pipe diameter.
[2]Applies only to that portion of suction pipe specified in 5.14.3.4 of NFPA 20, 2003.

Source: NFPA 20, 2003, Table 5.25(b).

TABLE 12.5

Flow Required to Produce a Velocity of 10 ft/sec (3 m/sec) in Pipes

Pipe Size		Flow Rate	
in.	mm	gpm	L/min
4	102	390	1,476
6	152	880	3,331
8	203	1,560	5,905
10	254	2,440	9,235
12	305	3,520	13,323

Source: NFPA 24, 2002, Table 10.10.2.1.3.

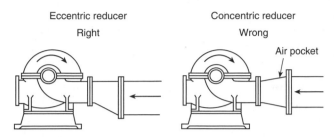

Eccentric reducer
Right

Concentric reducer
Wrong

Air pocket

FIGURE 12.18
Concentric and Eccentric Reducer.

(Source: *Pumps for Fire Protection Systems*, NFPA, 2002, Figure 5-6)

pump suction flange an eccentric reducer is required to connect the pump to the water supply piping. This eccentric reducer is used to minimize the chance of an air pocket forming near the suction portion of the fire pump. Figure 12.18 shows the correct installation of an eccentric reducer and an incorrect installation of a concentric reducer on a pump suction flange. The correctly installed eccentric reducers eliminate the possibility of an air pocket. If an air pocket is sucked into the pump, it could cause pump cavitation and damage to the pump.

Bypass Piping

In automatic sprinkler systems where the public water supply can be of material use if the fire pump is out of service, a bypass arrangement should be provided. Figure 12.19 shows two suggested fire pump arrangements with a bypass for use on a public main system. The bypass would be used as the primary source of water supply if the fire pump were temporarily out of service, or if the pump failed to start, providing some reasonable level of protection for the property with the understanding that full protection is provided only when the fire pump is operational. If the public water supply is of material use to the automatic sprinkler system and a bypass line is installed, NFPA 20 requires the control valves in the bypass be supervised in the open position.

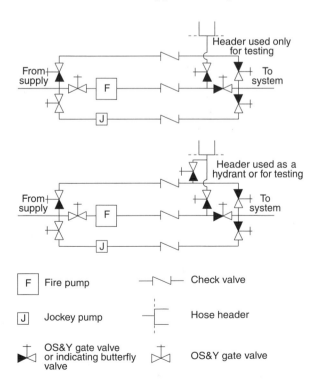

Header used only for testing

From supply

To system

Header used as a hydrant or for testing

From supply

To system

F | Fire pump

J | Jockey pump

OS&Y gate valve or indicating butterfly valve

Check valve

Hose header

OS&Y gate valve

FIGURE 12.19
Schematic Diagram of Suggested Arrangements for a Fire Pump with a Bypass, Taking Suction from Public Mains.

(Source: NFPA 20, 2003, Figure A.5.14.4)

Piping Near the Pump

The layout of the suction piping should be arranged with a straight run of pipe with a length greater than 10 times the pipe diameter away from the fire pump suction flange. Bends or elbows in the pipe in the horizontal plane leading to the suction flange are permitted, but this bend must be installed greater than 10 pipe diameters away. Vertical approaches to the pump can be from below with a bend into the suction flange, as shown in Figure 12.20.

The reason for these limitations is to minimize the turbulence of the water flow leading into the suction side of the pump, permitting the fire pump to perform in a more efficient manner. The elbows in the same plane as the pump shaft cause the loading of the impeller and pump shaft to be unbalanced. This unbalanced load can cause excessive wear to the shaft, bushings, and impellers. Also in the suction piping arrangement, at least one control valve is required. This control valve must be an OS&Y-type valve. The OS&Y valve is required because it is an indicating valve with minimum friction loss in the open position. This could be the main control valve from the water supply or an isolation valve located near the suction side of the pump. This valve is used to shut off the water supply to the fire pump for servicing of the pump. Figure 12.21 illustrates an OS&Y valve.

FIGURE 12.20

Suction Pipe Arrangements for Horizontal Split-Case Pumps.

(Source: *Pumps for Fire Protection Systems*, NFPA, 2002, Figure 5-5)

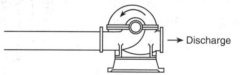

(a) Preferred arrangement

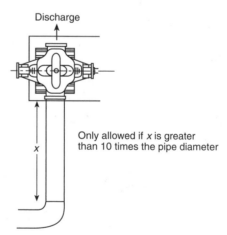

(b) Acceptable arrangement for a horizontal turn

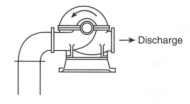

(c) Acceptable arrangement for a vertical turn

Valves and Appurtenances

The water supply authorities may require additional valves between the water supply and the suction side of the fire pump. These valves could include check valves or backflow preventers. If these valves are required, they should be installed a minimum of 10 pipe diameters away from the suction flange, or further if possible. These valves do introduce additional friction loss and the potential for additional turbulence into the water stream. Locating these valves and backflow preventers as far as possible from the suction flange of the pump will allow the water to return to a more laminar flow.

When the fire pump draws water from a storage tank, an anti-vortex plate shall be provided for the suction pipe located within the storage tank. A vortex is a rotation of the water in the tank. As the water spins around the tank, a low-pressure area is created in the water. This looks like a tornado or waterspout in the tank. This vortex can extend down into a tank and cause air to be drawn into the suction pipe and the fire pump. NFPA 22, *Standard for Water Tanks for Private Fire Protection*, provides guidance for the design of the anti-vortex plate based on the type of tank provided. The anti-vortex plate is used for atmospheric storage tanks at or near the same level of elevation as the fire pump; anti-vortex plates are not needed for gravity tanks. In general, the anti-vortex plate is a flange connected to the suction pipe and designed to minimize the creation of a vortex by the movement of water into the suction pipe leading to the fire pump. This anti-vortex plate is designed to allow the fire pump to draw all of the available fire protection water from the tank prior to ingesting air and causing the fire pump to cavitate, possibly damaging the pump. Improper design of the tank anti-vortex plate or the improper sizing of the suction

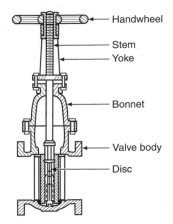

FIGURE 12.21
Typical Outside Screw and Yoke (OS&Y) Valve in the Closed Position.

(Source: *Fire Protection Handbook*, NFPA, 2003, Figure 10.3.24)

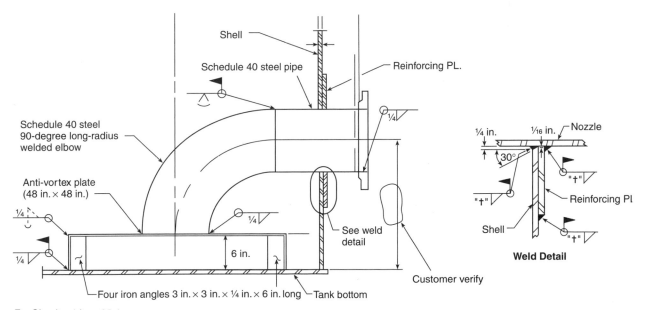

For SI units, 1 in. = 25.4 mm.

Note: Large, standard size anti-vortex plates (48 in. × 48 in.) are desirable as they are adequate for all sizes of pump suction pipes normally used. Smaller plates may be used; however, they should comply with 13.2.13.

FIGURE 12.22A
Suction Nozzle with Anti-Vortex Plate for Welded Suction Tanks.

(Source: NFPA 22, 2003, Figure B.1(o))

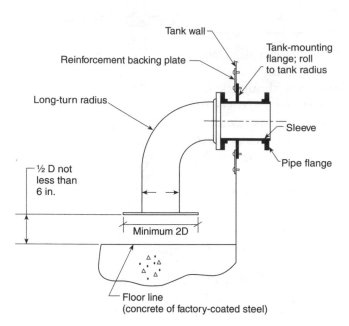

For SI units, 1 in. = 25.4 mm.

Note: Large, standard size anti-vortex plates (48 in. × 48 in.) are recommended, as they are adequate for all sizes of pump suction pipes normally used. Smaller plates may be used; however, they should comply with 13.2.13.

FIGURE 12.22B
Typical Suction Nozzle with Anti-Vortex Plate for Lap-Jointed Tanks.

(Source: NFPA 22, 2003, Figure B.1(p))

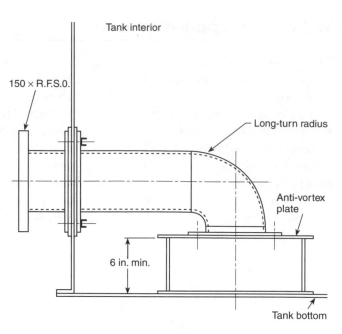

For SI units, 1 in. = 25.4 mm.

Note: Large, standard size anti-vortex plates (48 in. × 48 in.) are recommended, as they are adequate for all sizes of pump suction pipes normally used. Smaller plates may be used; however, they should comply with 13.2.13.

FIGURE 12.22C
Typical Suction Nozzle with Anti-Vortex Plate for Flange-Jointed Bolted Steel Tanks.

(Source: NFPA 22, 2003, Figure B.1(q))

line could create a vortex that would allow the ingestion of air into the water stream leading to the pump. NFPA 22 requires an anti-vortex assembly to be installed as shown in Figures 12.22a, 12.22b, and 12.22c.

NFPA 22, 2003 edition, subsection 13.2.13, requires the anti-vortex plate assembly to consist of a horizontal steel plate at least twice the diameter of the pipe outlet on an elbow fitting turning down towards the bottom of the tank. The outlet and anti-vortex plate should be at least one-half the diameter of the pipe away from the bottom of the tank, or a minimum of 6 in. from the bottom of the tank, whichever is greater.

Open Water and Raw Water Sources

The use of a raw water source such as a lake, pond, or other open body of water requires a special intake arrangement for the fire pump system. In many of these configurations where a raw source of water is used, vertical turbine pumps are installed in an intake area such as a wet pit. Figure 12.23 is Figure A.7.2.2.2 from NFPA 20, showing a typical arrangement for a vertical turbine pump drawing from a wet pit.

A means of screening out large pieces of debris is required. The screens must be installed at the entrance to the wet pit area. A minimum of two screens is required, and they should be removable for maintenance and cleaning purposes. The screens should be arranged such that they will screen the water in

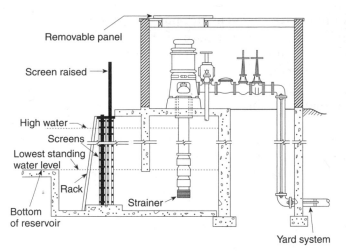

FIGURE 12.23
Vertical Shaft Turbine-Type
Pump Installation
in a Wet Pit.

(Source: NFPA 20, 2003, Figure
A.7.2.2.2.)

both the lowest and highest standing water levels. Screens should be a minimum ½-in. mesh and have an overall area of 1.6 times the net screened opening area. The screens should be designed to give an effective area equal to 1 in.^{2} for each 1 gpm of flow at the 150 percent rated point of the pump. The proper design of the wet pit and intake for the wet pit should take into consideration the ice thickness or freezing levels of the water reservoir being used. Adequate heat and protection for the pump area should be provided to ensure proper operation of the vertical turbine pump. The wet pit should be designed with proper water coverage over the intake bowls for the vertical turbine pump. The manufacturer's requirements for the pump should be consulted to determine the minimum water coverage requirements for the pump. Low-water-level monitors for the wet pit should be considered to alert the operating staff when water flow into the wet pit is being impeded. Although vertical turbine pumps are not prone to cavitation, the running of a vertical turbine pump without water can cause damage to the pump stages.

Discharge Piping

Piping on the discharge side of the pump also has many unique requirements. In general, piping is required to be steel with a maximum rated working pressure that exceeds the system static pressure plus the net churn pressure of the fire pump. The normal maximum working pressure rating of pipe is 175 psi. NFPA 20 requires a minimum of one control valve located such that it will isolate the discharge of the fire pump. This valve can be of any listed type, such as butterfly or OS&Y valves.

A check valve is required on the discharge side of the pump. The purpose of this check valve is to maintain pressure in the fire protection system after the fire pump has been shut down. Pressure relief valves shall not be used as the means to limit system pressures to below the components' pressure rating. Relief valves are allowed for diesel engine fire pumps where the total of 121 percent of the net rated shutoff (churn) pressure plus the maximum static suction pressure, adjusted for elevation, exceeds the pressure for which the system components are rated. A specific exception permits for a pressure relief valve for diesel-driven pumps when the diesel driver exceeds 121 percent of its rated speed and the system pressure is exceeded. The fire pumping system should be designed such that the maximum working pressure of the system does not exceed the pressure rating of the components. The normal pressure rating of the components is 175 psi, but in special cases, such as in some high-rise buildings, 300-psi-listed sprinklers and components are permitted to be specified on the lower levels of the building that are subject to the higher pressures. NFPA 13 specifically prohibits applications with pressures exceeding 175 psi on the system, regardless of the rating of the components. NFPA 20 permits the use of variable-speed, pressure-limiting control drivers, a speed control system used to limit the total discharge pressure by reducing the pump driver speed from rated speed, to meet the requirement of matching pump discharge pressure to the pressure ratings of components on the discharge side of the fire pump.

A circulating relief valve is a requirement for diesel-driven fire pumps with radiators and electric-driven pumps. Circulating relief valves are required to allow sufficient water to flow to prevent the fire pump from overheating in a no-flow (churn) condition. Without a circulating relief valve at no flow, the water in the fire pump casing heats up to dissipate the energy of the pump running. This heat increase can damage a pump.

There are numerous attachments and gauges required for the fire pump. Fire pumps should be equipped with a suction pressure gauge and a discharge pressure gauge. These gauges will be used to determine net pressure developed by the pump during testing. In addition, most pumps are required to have an automatic air release valve (AARV) located at the top of the horizontal split-case pump. The AARV is used to remove air accumulations at the top of a horizontal split-case, where air will naturally collect. The AARV releases the air at the start-up of the pump and closes as water reaches the AARV. This minimizes the potential for impeller cavitation.

Pump manufacturers have different methods of lubricating the pump bearings. Many of the manufacturers have water-lubricated bearings, which require a method of draining away water from the pump arrangement.

Inspection, Testing, and Maintenance

Once the fire pump is installed, it must be inspected, tested, and maintained periodically. See Table 12.6 for a summary of required inspection, testing, and maintenance procedures for fire pumps.

Testing

The method of testing the pumps should be included in the original pump design documents. Test arrangements include a test header to the exterior of the building, a flow meter to a reservoir or other area, or possibly a closed-loop metering system. A test header is not required; drains and storage tanks can be used to comply with the requirements to have a flow to open atmosphere every 3 years.

Test headers are specifically designed as a UL-listed assembly, which consists of a number of valved outlets connected to the discharge side of the pump. The piping and valves are designed hydraulically to discharge the 150 percent flow point of the fire pump. The test header should be on an exterior

TABLE 12.6

Summary of Fire Pump Inspection, Testing, and Maintenance

Item	Activity	Frequency	Reference [from NFPA 25]
Pump house, heating ventilating louvers	Inspection	Weekly	8.2.2(1)
Fire pump system	Inspection	Weekly	8.2.2(2)
Pump operation			
No-flow condition	Test	Weekly	8.3.1
Flow condition	Test	Annually	8.3.3.1
Hydraulic	Maintenance	Annually	8.5
Mechanical transmission	Maintenance	Annually	8.5
Electrical system	Maintenance	Varies	8.5
Controller, various components	Maintenance	Varies	8.5
Motor	Maintenance	Annually	8.5
Diesel engine system, various components	Maintenance	Varies	8.5

Source: NFPA 25, 2002, Table 8.1.

wall with adequate clear space to connect multiple hose lines to the test header. These hose lines will be equipped with special nozzles at the ends called play pipes. The velocity pressure of the water flowing from these play pipes will be measured using a pitot tube and calculations will be made to determine the water flow amounts.

The test header should be located so that the test header can be properly drained of water in areas that are subject to freezing. A separate shutoff valve between the pump discharge and the test header should be provided and supervised in the closed position. A ball drip or method of draining condensate from a section of pipe should be provided. Figure 12.24 shows a test header with hose setup. Figure 12.25 is a picture of an actual fire pump test, using a test header, with the water stream shown at full flow.

FIGURE 12.24
Test Header with Hose Attached.
(Source: *Pumps for Fire Protection Systems*, 2002, Figure 5-14)

FIGURE 12.25
Water Flow Streams from a Fire Pump Test.
(Courtesy of The Society of Fire Protection Engineers)

In addition to a pitot gauge to determine flow in an open flowing test, NFPA 20 permits the use of a flow meter. The flow meter must be UL listed for the application and must be designed to handle at least 175 percent of the rated flow from the fire pump. When using a flow meter, the discharge of the flow loop can be returned to a storage reservoir, open area, or discharge drain. This arrangement will test the water supply capabilities.

Although NFPA 20 permits flow meters, many authorities having jurisdiction would prefer to see a test header to verify the actual flow from the pump. NFPA 25 requires a water flow to the atmosphere once every 3 years, so allowance must be made in systems with flow meters to flow water to the atmosphere. Open drains, test headers, or storage tanks are allowed for this flow test. The local AHJ should be consulted prior to the design of the pump arrangement for water flow testing.

Controls Supervision and Pressure Maintenance

Automatic pump starting is now industry practice. In the early days of fire pumps, manual activation of the fire pump was permitted and considered normal. Today, with sophisticated fire sprinkler systems, the need for instantaneous and continual water pressure is paramount to successful operation of the system. To automatically start a fire pump, a method of sensing a drop in system pressure is needed. The drop in system pressure is an indication of water flow, thus indicating a need for the fire pump to start. Many systems have a jockey pump installed to address the normal fluctuations and pressures in the system due to expansions and contractions in heating, slow leakage through valves, and other minor system losses. Jockey pumps are generally rated at between 1 and 10 gpm and operate at pressures comparable to the fire pump. The performance objective of a jockey pump is to provide the pressure needed at the most remote sprinkler to permit accurate pressure-loss sensing when a sprinkler opens and to allow the smaller jockey pump to cycle on and off to address the minor pressure fluctuations that may occur in a system, in lieu of having the main fire pump cycle on and off for these minor pressure fluctuations. NFPA 20 specifically prohibits the use of the fire pump as a pressure maintenance pump. The cycling of the jockey pump in the nonfire condition saves wear and tear on the fire pump.

Figure 12.26 is a schematic outline of where the sensing lines should be located for the installation of a fire pump and jockey pump on a typical pumping system. The sensing lines should be a minimum ½-in. corrosion-resistant pipe. Check valves or ground face unions should be installed per paragraph 10.5.2.1.6 of NFPA 20. Another method of configuring the sensing lines is shown in Figure 12.27, which is a reproduction of NFPA 20, Figure A.10.5.2.1(a). NFPA 20 requires separate sensing lines for the fire pump and jockey pump. This allows the systems to operate at independent start pressures with the jockey pump having a specific shutoff pressure.

Many of the requirements of NFPA 20 for the installation of fire pumps address performance reliability. Valve supervision is needed to ensure the reliability of the water supply in the movement of the water through the system. Control valves such as the suction valve, discharge valve, bypass valves, and test outlet valve must be supervised. Supervision can be done by an electronic

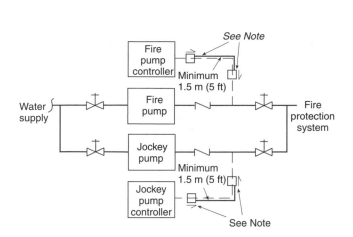

Note: Check valves or ground-face unions complying with 10.5.2.1.

FIGURE 12.26
Piping Connection for Pressure-Sensing Line.
(Source: NFPA 20, 2003, Figure A.10.5.2.1(b))

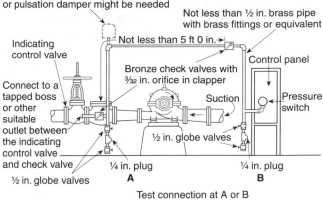

If water pulsation causes erratic operation of the pressure switch or the recorder, a supplemental air chamber or pulsation damper might be needed

Notes:
1. Solenoid drain valve used for engine-driven fire pumps can be at A, B, or inside controller enclosure.
2. If water is clean, ground-face unions with noncorrosive diaphragms drilled for ³⁄₃₂ in. orifices can be used in place of the check valves.
3. For SI units, 1 in. = 25.4 mm; 1 ft = 0.3048 m.

FIGURE 12.27
Piping Connection for Each Automatic Pressure Switch (for Fire Pump and Jockey Pumps).
(Source: NFPA 20, 2003, Figure A.10.5.2.1(a))

valve-monitoring switch that provides a local alarm or transmits an alarm to a central station service or proprietary remote-signaling service. Locking valves in the open position is another economical way to supervise these control valves, and the valves could also be sealed off behind a fenced-in enclosure or other secured area. Locking or securing the valves may require the approval of the local AHJ.

Summary

Fire pumps are critical components of a fire sprinkler system. The proper selection of a fire pump must address many variables in addition to the amount of water and pressure needed. These variables include type of drive for the pump, fire pump location, piping arrangements, flow meters, test headers, control valves, transfer switch arrangements, and power disconnects. A proper fire pump design has to address the requirements of many NFPA standards, including NFPA 13, 14, 20, 22, 24, and 70. The requirements of these standards are intended to provide a highly reliable fire pumping installation for an automatic fire sprinkler system. It is hoped that the proper sizing of a fire pump, a difficult task, has been made clearer with this chapter.

BIBLIOGRAPHY

NFPA Codes, Standards, and Recommended Practices

NFPA Publications National Fire Protection Association, 1 Batterymarch Park, Quincy, MA 02169-7471

The following is a list of NFPA codes, standards, and recommended practices cited in this chapter. See the latest version of the *NFPA Catalog* for availability of current editions of these documents.

NFPA 13, *Standard for the Installation of Sprinkler Systems*
NFPA 14, *Standard for the Installation of Standpipe and Hose Systems*
NFPA 20, *Standard for the Installation of Stationary Pumps for Fire Protection*
NFPA 22, *Standard for Water Tanks for Private Fire Protection*
NFPA 24, *Standard for the Installation of Private Fire Service Mains and Their Appurtenances*
NFPA 25, *Standard for the Inspection, Testing, and Maintenance of Water-Based Fire Protection Systems*
NFPA 30, *Flammable and Combustible Liquids Code*
NFPA 70, *National Electrical Code®*
NFPA 72®, *National Fire Alarm Code®*
NFPA 110, *Standard for Emergency and Standby Power Systems*

Additional Readings

Inspection, Testing, and Maintenance of Water-Based Fire Protection Systems: The NFPA 25 Handbook, NFPA, Quincy, MA, 2002.

Isman, K. E., and Puchovsky, M. T., *Pumps for Fire Protection Systems*, NFPA, Quincy, MA, 2002.

Jensen, J. D., "Stationary Fire Pumps," in Cote, A. E., ed., *Fire Protection Handbook*, 19th edition, NFPA, Quincy, MA, 2003, p. 10-111.

Nasby, J. S., and Puchovsky, M. T., "Power Supplies and Controllers for Motor-Driven Fire Pumps," in Cote, A. E., ed., *Fire Protection Handbook*, 19th edition, NFPA, Quincy, MA, 2003, p. 10-129.

Puchovsky, M. T., and Isman, K. E., *Fire Pump Handbook*, NFPA, Quincy, MA, 1998.

Protection of Storage

Al Moore, PE

This chapter will provide the designer an understanding of the elements of protecting storage, whether general storage or rack storage. The complexities of protecting rack storage will receive particular attention, especially the factors that are involved with choosing the appropriate protection of commodities stored on racks.

The chapter will review methods used to determine specific rack storage requirements per the 2002 edition of NFPA 13, *Standard for the Installation of Sprinkler Systems*, but will not cover every set of requirements for the wide variety of types and heights of rack storage arrangements. The designer should determine the sprinkler system design criteria for rack storage of commodities and indicate that criteria in the contract documents. This responsibility involves a variety of tasks that the designer is uniquely qualified to perform in an effort to design a system that provides the best value to the owner for the short-term and long-term use of the owner's facility.

The protection of commodities stored on racks includes ceiling sprinkler systems, in-rack sprinkler systems, and hose stations. These fire protection systems may be designed to suppress a fire or control a fire as determined by the protection requirements and considering the future use of the facility. Rack storage systems are typically found in warehouse facilities but may also be found in a wide variety of other types of buildings where some rack storage of commodities is desired. It is important to note that the requirements in this chapter are based upon NFPA 13, but that many storage facilities are insured by an insurance underwriter such as FM Global that may have additional requirements that may exceed the requirements indicated in NFPA 13.

Definitions for many of the terms used in the protection of palletized, solid-piled, bin box, or shelf storage of commodities are included in NFPA 13, Section 3.9.

General Storage Requirements

This chapter is primarily devoted to rack storage; however, it includes an overview of the requirements with regard to general storage as well. NFPA 13, Chapter 12, "Storage," is a new chapter in the 2002 edition that includes all the design criteria for protection of storage. The first section in Chapter 12, "General," includes information that applies to all storage arrangements and commodities, and subsequent sections within NFPA 13, Chapter 12, may modify requirements in the general section. NFPA 230, *Standard for the Fire Protection of Storage*, has additional information with regard to storage.

The general information includes information on roof vents and draft curtains, building height, hose connections, wet pipe systems, adjacent occupancies, dry pipe and preaction systems, ceiling slope, multiple adjustments, protection of idle pallets including wood and plastic pallets, miscellaneous storage and storage of Class I through IV commodities up to 12 ft in height, high expansion foam systems, in-rack sprinklers, and storage applications.

Idle Pallets

A pallet may be constructed of wood or plastic, and are used as a base to support products that are stored thereon. Pallets are approximately 4 in. (101 mm) tall and have openings that will allow material handling equipment, such as a forklift truck, to pick up the pallets for moving or storage. See Figure 13.1 showing a conventional wood pallet, Figure 13.2 showing a plastic pallet, and Figure 13.3 showing a fire-retardant plastic pallet.

NFPA 13, paragraph 5.6.2, has information with regard to pallet types and classification of commodity units using plastic pallets. The sprinkler design criteria for idle pallet storage are indicated in NFPA 13, Chapter 12. The classification of the commodity may need to be increased by one or two classes depending on several factors, such as the commodity type stored on the pallets, whether unreinforced or reinforced polypropylene or high-density polyethylene pallets are used, and the sprinkler protection.

The following text from NFPA 13, A.12.1.9, points out the severe challenges that idle pallet storage inside a building presents to a sprinkler system. Pallets should be stored outside and away from a building where possible.

> **A.12.1.9** Idle pallet storage introduces a severe fire condition. Stacking idle pallets in piles is the best arrangement of combustibles to promote rapid spread of fire, heat release, and complete combustion. After pallets are used

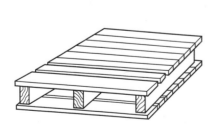

FIGURE 13.1
Wood Pallet.

(Source: *Fire Protection Handbook*, NFPA, 2003, Figure 10.12.18)

FIGURE 13.2
Plastic Pallet.

(Source: *Automatic Sprinkler Systems Handbook*, NFPA, 2002, Exhibit 5.4)

FIGURE 13.3
Classified Fire-Retardant Plastic Pallet (40 in. × 48 in.) Equivalent to Wood Pallet.
(Source: *Automatic Sprinkler Systems Handbook*, NFPA, 2002, Exhibit 5.5; courtesy of ORBIS CORPORATION/Nucon Beverage Products Group)

for a short time in warehouses, they dry out and edges become frayed and splintered. In this condition they are subject to easy ignition from a small ignition source. Again, high piling increases considerably both the challenge to sprinklers and the probability of involving a large number of pallets when fire occurs. Therefore, it is preferable to store pallets outdoors where possible.

A fire in stacks of idle plastic or wooden pallets is one of the greatest challenges to sprinklers. The undersides of the pallets create a dry area on which a fire can grow and expand to other dry or partially wet areas. This process of jumping to other dry, closely located, parallel, combustible surfaces continues until the fire bursts through the top of the stack. Once this happens, very little water is able to reach the base of the fire. The only practical method of stopping a fire in a large concentration of pallets with ceiling sprinklers is by means of prewetting. In high stacks, this cannot be done without abnormally high water supplies. The storage of empty wood pallets should not be permitted in an unsprinklered warehouse containing other storage.

Pallets may be stored outside, in a detached structure, or inside where protected in accordance with NFPA 13.

Wooden Pallets. The protection of wood pallets stored indoors is indicated in NFPA 13 Table 12.1.9.1.2(a) (reproduced here as Table 13.1) when using standard spray sprinklers, Table 12.1.9.1.2(b) (reproduced here as Table 13.2) when using control mode specific application sprinklers, and Table 12.1.9.1.2(c) (reproduced here as Table 13.3) when using ESFR sprinklers, unless the following conditions are met:

1. Pallets shall be stored no higher than 6 ft (1.8 m).
2. Each pallet pile of no more than four stacks shall be separated from other pallet piles by at least 8 ft (1.4 m) of clear space or 25 ft (7.6 m) of commodity.

TABLE 13.1

Control Mode Density-Area Protection of Indoor Storage of Idle Wood Pallets

Type of Sprinkler	Location of Storage	Nominal K-Factor	Maximum Storage Height		Sprinkler Density		Areas of Operation				Hose Stream Demand		Water Supply Duration (hours)
							High Temperature		Ordinary Temperature				
			ft	m	gpm/ft²	mm/min	ft²	m²	ft²	m²	gpm	L/min	
Control mode density/ area	On floor	K 8 or larger	Up to 6	Up to 1.8	0.2	8.2	2000	186	3000	279	500	1900	1½
		K 11.2 or larger	6 to 8	1.8 to 2.4	0.45	18.3	2500	232	4000	372	500	1900	1½
			8 to 12	2.4 to 3.7	0.6	24.5	3500	325	6000	557	500	1900	1½
			12 to 20	3.7 to 6.1	0.6	24.5	4500	418	—	—	500	1900	1½

Source: NFPA 13, 2002, Table 12.1.9.1.2(a).

TABLE 13.2

Control Mode Specific Application Protection of Indoor Storage of Idle Wood Pallets

Type of Sprinkler	Location of Storage	Nominal K-Factor	Maximum Storage Height		Maximum Ceiling/ Roof Height		Type of System	Number of Design Sprinklers by Minimum Pressure			Hose Stream Demand		Water Supply Duration (hours)
								25 psi (1.7 bar)	50 psi (3.4 bar)	75 psi (5.2 bar)			
			ft	m	ft	m					gpm	L/min	
Large drop	On floor	11.2	20	6.1	30	9.1	Wet	15	15	15	500	1900	1½
							Dry	25	25	25	500	1900	1½

Source: NFPA 13, 2002, Table 12.1.9.1.2(b).

TABLE 13.3

ESFR Protection of Indoor Storage of Idle Wood Pallets

Type of Sprinkler (Orientation)	Location of Storage	Nominal K-Factor	Maximum Storage Height		Maximum Ceiling/ Roof Height		Minimum Operating Pressure	Hose Stream Demand		Water Supply Duration (hours)
			ft	m	ft	m	psi	gpm	L/min	
ESFR (pendent)	On floor or rack without solid shelves	14.0	25	7.6	30	9.1	50	250	946	1
			25	7.6	32	9.8	60			
			35	10.7	40	12.2	75			
		16.8	25	7.6	30	9.1	35			
			25	7.6	32	9.8	42			
			35	10.7	40	12.2	52			
ESFR (upright)	On floor only	14.0	20	6.1	30	9.1	50			
			20	6.1	35	10.7	75			

Source: NFPA 13, 2002, Table 12.1.9.1.2(c).

If these conditions are met, then no additional protection is required over what is required for the occupancy being protected in the area with the pallet storage. It may be unreasonable to assume that such conditions will be strictly observed on a daily basis.

NFPA 13, paragraph 12.1.9.1.3, states that idle wood pallets shall not be stored in racks unless they are protected in accordance with the appropriate provisions of Table 12.1.9.1.2(c) (Table 13.3).

Plastic Pallets. The storage of idle plastic pallets presents a more challenging fire condition than the storage of idle wood pallets. NFPA 13 has specific criteria for indoor storage of plastic pallets that should be reviewed by the designer.

Designer Considerations. Prior to the determination of design criteria for a storage facility, the designer should ask the owner where idle pallets are to be stored, and what specific type of pallets are to be used initially and possibly in the future. The designer should encourage the owner to store idle pallets outdoors and away from the building. Fire codes, insurance underwriter guidelines, and the local authority having jurisdiction (AHJ) may have additional requirements for the storage of pallets.

Miscellaneous Storage and Storage of Class I through IV Commodities up to 12 ft (3.66 m) in Height

Storage Applications. NFPA 13 defines *miscellaneous storage* as storage that does not exceed 12 ft (3.66 m) in height and is incidental to another occupancy use group. Such storage shall not constitute more than 10 percent of the building area or 4000 ft^2 (372 m^2) of the sprinklered area, whichever is greater. Such storage shall not exceed 1000 ft^2 (93 m^2) in one pile or area, and each such pile or area shall be separated from other storage areas by at least 25 ft (7.62 m).

Miscellaneous storage applies to a building in which storage is only a part of the building's use. NFPA's *Automatic Sprinkler Systems Handbook* [1] gives examples such as the back room of a mercantile facility or a manufacturing operation that uses a portion of its facility to store small amounts of finished products and raw materials.

Discharge Criteria. Miscellaneous storage is treated virtually the same as storage under 12 ft (3.7 m). The discharge criteria for miscellaneous storage up to 12 ft (3.7 m) in height of Group A plastic, rubber tires, rolled paper, and storage of idle pallets up to 6 ft (1.4 m) in height are indicated NFPA 13, Table 12.1.10.1.1 (reproduced here as Table 13.4), and Figure 12.1.10 (referenced here as Figure 13.4). For protection of storage of Class I through IV commodities up to 12 ft (3.7 m) in height, the discharge criteria in Table 12.1.10.1.1 and Figure 12.1.10 apply. The hose stream demands and durations are also noted in Table 13.4.

Protection of Commodities That Are Stored Palletized, Solid Piled, in Bin Boxes, or on Shelves

Class I through IV Commodities. NFPA 13 has design criteria in Section 12.2 for protection of palletized, solid piled, bin box, or shelf storage of Class I through IV commodities. The criteria include criteria for using the control mode density-area method, large drop and specific application sprinklers, and ESFR sprinklers.

TABLE 13.4

Discharge Criteria for Miscellaneous Storage and Commodity Classes I through IV Storage 12 ft (3.7 m) or Less in Height[1]

Commodity		Type of Storage	Storage Height		Maximum Ceiling Height		
			ft	m	ft	m	
Class I to IV							
Class I		Palletized,	≤12	≤3.7	—	—	
Class II		bin box, shelf,	≤10	≤3.05	—	—	
Class II		and rack	>10 to ≤12	>3.05 to ≤3.7	—	—	
Class III			≤12	≤3.7	—	—	
Class IV			≤10	≤3.05	—	—	
Class IV		Palletized, bin box, and shelf	>10 to ≤12	>3.05 to ≤3.7	—	—	
Class IV		Rack	>10 to ≤12	>3.05 to ≤3.7	—	—	
Miscellaneous Group A Plastic Storage							
Cartoned	Solid and expanded	Palletized, bin box, shelf, and rack	≤5	≤1.5	—	—	
			>5 to ≤10	>1.5 to ≤3.05	15	4.6	
			>5 to ≤10	>1.5 to ≤3.05	20	6.1	
			>10 to ≤12	>3.05 to ≤3.7	17	5.2	
			>10 to ≤12	>3.05 to ≤3.7	17	5.2	
		Palletized, bin box, and shelf	>10 to ≤12	>3.05 to ≤3.7	27	8.2	
		Rack	>10 to ≤12	>3.05 to ≤3.7	—	—	
Exposed	Solid and expanded	Palletized, bin box, shelf, and rack	≤5	≤1.5	—	—	
		Palletized, bin box, and shelf	>5 to ≤8	>1.5 to ≤2.4	—	—	
		Palletized, bin box, shelf, and rack	>5 to ≤10	>1.5 to ≤3.05	15	4.6	
Miscellaneous Group A Plastic Storage							
Exposed	Solid	Palletized, bin box, shelf, and rack	>5 to ≤10	>1.5 to ≤3.05	20	6.1	
	Expanded	Rack	>5 to ≤10	>1.5 to ≤3.05	20	6.1	
	Solid and expanded	Palletized, bin box, and shelf	>10 to ≤12	>3.05 to ≤3.7	17	5.2	
			>10 to ≤12	>3.05 to ≤3.7	17	5.2	
		Rack	>10 to ≤12	>3.05 to ≤3.7	17	5.2	
			>10 to ≤12	>3.05 to ≤3.7	—	—	
Miscellaneous Tire Storage							
Tires		On floor, on side	>5 to ≤12	>1.5 to ≤3.7	—	—	
		On floor, on tread or on side	≤5	≤1.5	—	—	
		Single-, double-, or multiple-row racks on tread or on side	≤5	≤1.5	—	—	
		Single-row rack, portable, on tread or on side	>5 to ≤12	>1.5 to ≤3.7	—	—	

Design Curve Figure 12.1.10	Note	Inside Hose (gpm)	Total Combined Inside and Outside Hose (gpm)	Duration (minutes)
Curve 2		0, 50, or 100	250	90
Curve 2		0, 50, or 100	250	90
Curve 3		0, 50, or 100	250	90
Curve 3		0, 50, or 100	250	90
Curve 3		0, 50, or 100	250	90
Curve 3		0, 50, or 100	500	90
Curve 4		0, 50, or 100	500	90
Curve 3		0, 50, or 100	500	90
Curve 4		0, 50, or 100	500	120
Curve 5		0, 50, or 100	500	120
Curve 5		0, 50, or 100	500	120
Curve 3	+1 level of in-rack	0, 50, or 100	500	120
Curve 5		0, 50, or 100	500	120
Curve 3	+1 level of in-rack	0, 50, or 100	500	120
Curve 3		0, 50, or 100	500	90
Curve 5		0, 50, or 100	500	120
Curve 5		0, 50, or 100	500	120
Curve 5		0, 50, or 100	500	120
Curve 3	+1 level of in-rack	0, 50, or 100	500	120
Curve 5		0, 50, or 100	500	120
Curve 5		0, 50, or 100	500	120
Curve 3	+1 level of in-rack	0, 50, or 100	500	120
Curve 3	+1 level of in-rack	0, 50, or 100	500	120
Curve 4		0, 50, or 100	750	180
Curve 3		0, 50, or 100	750	180
Curve 3		0, 50, or 100	750	180
Curve 4		0, 50, or 100	750	180

continues

TABLE 13.4

(Continued)

Commodity	Type of Storage	Storage Height		Maximum Ceiling Height		
		ft	m	ft	m	
	Single-row rack, fixed, on tread	>5 to ≤12	>1.5 to ≤3.7	—	—	
		>5 to ≤12	>1.5 to ≤3.7	—	—	
Miscellaneous Rolled Paper Storage						
Heavy and medium weight	On end	≤10	≤3.05	—	—	
Tissue and light weight	On end	≤10	≤3.05	—	—	
Idle Pallet Storage						
Wooden pallets	Single-row rack, fixed	≤6	≤1.8	—	—	
Plastic pallets	Single-row rack, fixed	≤4	≤1.2	—	—	

Source: NFPA 13, 2002, Table 12.1.10.1.1.

For control mode density-area protection, this section includes design curves for standard and high-temperature sprinklers. For large drop, specific application sprinklers, and ESFR sprinklers, the section includes information with regard to minimum system discharge requirements, sprinkler K-factors, storage and ceiling heights, number of design sprinklers with associated minimum pressures, hose stream demand, and water supply durations.

There is an example for control mode density-area protection in NFPA 13, paragraph 12.2.2.1, to aid the designer in understanding the application of the design criteria. NFPA's *Automatic Sprinkler Systems Handbook* also has helpful commentary and information.

Plastic and Rubber Commodities. NFPA 13 has further design criteria for protection of palletized, solid piled, bin box, or shelf storage of plastic and

FIGURE 13.4
Miscellaneous Storage and Commodity Classes I through IV Storage 12 ft (3.7 m) or Less in Height—Design Curves.

(Source: NFPA 13, 2002, Figure 12.1.10)

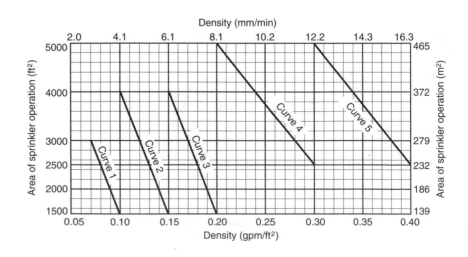

Design Curve Figure 12.1.10	Note	Inside Hose (gpm)	Total Combined Inside and Outside Hose (gpm)	Duration (minutes)
Curve 4		0, 50, or 100	750	180
Curve 3	+1 level of in-rack	0, 50, or 100	750	180
Curve 3		0, 50, or 100	500	120
Curve 4		0, 50, or 100	500	120
Curve 3		0, 50, or 100	500	90
Curve 3		0, 50, or 100	500	90

rubber commodities also using the control mode density-area method, for using large drop and specific application sprinklers, and for using ESFR sprinklers. NFPA 13 includes a decision tree for the storage of plastics up to 25 ft (7.62 m) in height protected with spray sprinklers. This section also has several examples to aid the designer in understanding the application of the design criteria. See Figure 13.5.

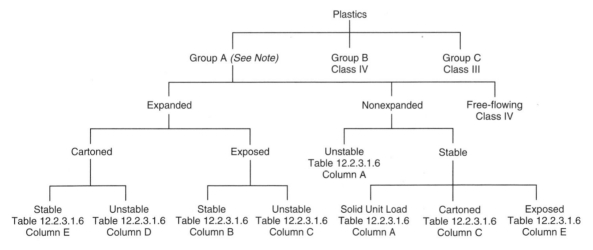

Note: Cartons that contain Group A plastic material shall be permitted to be treated as Class IV commodities under the following conditions:

(a) There shall be multiple layers of corrugation or equivalent outer material that would significantly delay fire involvement of the Group A plastic

(b) The amount and arrangement of Group A plastic material within an ordinary carton would not be expected to significantly increase the fire hazard

FIGURE 13.5
Decision Tree.
(Source: NFPA 13, 2002, Figure 12.2.3.1.1)

Brief History of Rack Storage Protection

Early fire tests to establish the first requirements for palletized storage of commodities were performed by Factory Mutual (FM) and Underwriters Laboratories Inc. (UL) in large test facilities built in 1947 and 1954, respectively. These fire tests were paid for by insurance companies and cooperating industries and companies interested in fire protection of stored commodities, including commodities stored on racks. The first NFPA standard with requirements for general storage was the 1965 edition of NFPA 231, *Standard for General Storage* (now incorporated in NFPA 13). The requirements for palletized storage were found in the appendix of NFPA 231. Prior to NFPA 231, the National Board of Fire Underwriters, the Factory Insurance Association, and Factory Mutual published several standards that addressed general storage and rack storage of various commodities [2]. To date, FM Global, UL, and other insurance companies and testing agencies continue to perform fire tests and maintain their own standards and guidelines for protection of commodities on racks. UL performs testing but does not maintain protection requirements, other than those in sprinkler listings.

In 1971, NFPA 231C, *Standard for Rack Storage of Materials*, became the first NFPA sprinkler standard based entirely on full-scale fire test data [2]. With the advent of early suppression fast-response (ESFR) sprinklers, which were included in the 1986 edition of NFPA 231C, the use of in-rack sprinklers decreased. In 1999, the NFPA 231 series of documents including NFPA 231, *Standard for General Storage*, and 231C, were incorporated into NFPA 13.

Rack Design Considerations

To determine the appropriate fire protection design criteria, the designer must understand the various elements and configurations of rack storage, such as single row, double row, and multiple row. There are also various types of ceiling and in-rack sprinkler systems to consider and different ways to install in-rack sprinkler piping. It is important to design a system that will minimize potential damage to an in-rack sprinkler system from the loading and unloading of products from rack storage units.

The *Automatic Sprinkler System Handbook* has a large amount of useful information, including definitions and figures, to help the designer understand each pertinent aspect of rack storage configurations. There are over 40 figures and 20 tables to consider when reviewing the rack storage requirements in NFPA 13. The information is carefully organized by height of storage, type of commodity, and type of sprinkler system. The purpose of this chapter is to make it easier for designers to understand and use the tables and figures for high-piled rack storage design.

The tables in NFPA 13 reference "in-rack" sprinklers and "ceiling" sprinklers. In-rack sprinklers, as shown in Figure 13.6, are sprinklers installed on sprinkler system piping that is located within the rack structure. Ceiling sprinklers (see Figure 13.7) are sprinklers that are located at the ceiling or roof of the building.

FIGURE 13.6
In-Rack Sprinkler Equipped with a Listed Water Shield and Guard.

(Source: *Automatic Sprinkler Systems Handbook*, NFPA, 2002, Exhibit 8.24)

FIGURE 13.7
Upright Sprinklers.

(Source: *Automatic Sprinkler Systems Handbook*, NFPA, 2002, Exhibit 3.32)

Maximum Storage Height

NFPA 13 defines the available height for storage as the maximum height at which commodities can be stored above the floor and still maintain adequate clearance from structural members and the required clearance below sprinklers. ESFR and large drop sprinklers require at least 36 in. (914 mm) between the sprinkler deflector and the top of storage. Standard and extended-coverage upright and pendent sprinklers require at least 18 in. (457 mm) between the sprinkler deflector and the top of storage. Where rubber tires are stored, the minimum clearance must be at least 36 in. (914 mm). Where NFPA refers to top of storage, it is not referring to the top of the highest rack load beam or rack structural member, but to the top of the item stored on that member.

FIGURE 13.8
Rack Storage in Food
Warehouse.

(Courtesy of Gordon Food Service)

Rack Types

There are many possible rack storage configurations and just as many different heights for rack storage of commodities. Some are over a hundred feet tall and the rack structures are a part of the structural support of the building, as indicated in Figure 13.8. A few of the more typical types of rack configurations are described next.

Single-Row Rack. The simplest type of rack configuration is called a single-row rack. NFPA 13 defines single-row racks as racks that have no longitudinal flue space and that have a width up to 6 ft (1.8 m) with aisles at least 3.5 ft (1.1 m) from other storage. These types of racks allow loading of the rack from either side.

Double-Row Rack. NFPA 13 defines double-row racks as two single-row racks placed back-to-back, having a combined width up to 12 ft (3.7 m) with aisles at least 3.5 ft (1.1 m) on each side. Note the vertical "flue space" between the rack units. See Figure 13.9, which shows a double-row rack.

Multiple-Row Rack. NFPA defines multiple-row racks as racks greater than 12 ft (3.7 m) wide or single- or double-row racks separated by aisles less than 3.5 ft (1.1 m) wide having an overall width greater than 12 ft (3.7 m). See Figure 13.10, which shows a multiple-row rack.

Portable Rack. NFPA 13 defines portable racks as racks that are not fixed in place. They can be arranged in any number of configurations.

Miscellaneous Racks. There are several other rack arrangements, including movable racks, automatic storage-type racks, flow racks, and pick module racks. Pick module racks have tiers that workers can use to access storage commodities on each tier of the storage rack. Figure 13.11 illustrates a movable rack, and Figure 13.12 illustrates flow-through racks and portable racks.

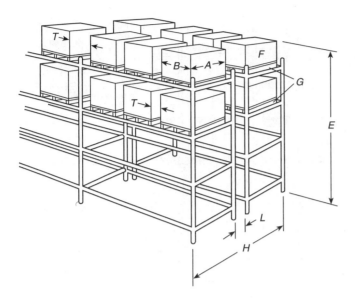

A Load depth
B Load width
E Storage height
F Commodity
G Pallet
H Rack depth
L Longitudinal flue space
T Transverse flue space

FIGURE 13.9
Double-Row Racks Without Solid or Slatted Shelves.

(Source: NFPA 13, 2002, Figure A.3.10.8(b))

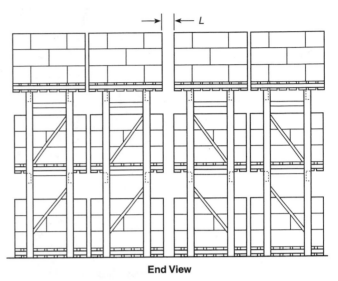

End View

L Longitudinal flue space

FIGURE 13.10
Multiple-Row Rack to Be Served by a Reach Truck.

(Source: NFPA 13, 2002, Figure A.3.10.8(f))

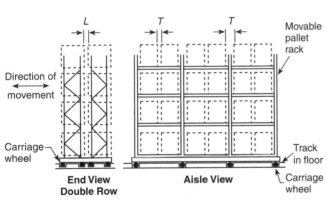

T Transverse flue space
L Longitudinal flue space

FIGURE 13.11
Movable Rack.

(Source: NFPA 13, 2002, Figure A.3.10.8(k))

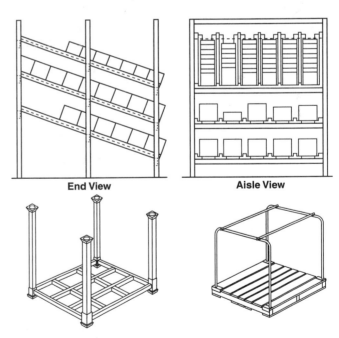

End View   **Aisle View**

FIGURE 13.12
Flow-Through Racks (Top) and Portable Racks (Bottom).

(Source: NFPA 13, 2002, Figure A.3.10.8(i))

FIGURE 13.13
Double-Row Rack
Arrangement with
Longitudinal and Transverse
Flue Spaces.

(Source: NFPA 13, 2002, Figure
A.3.10.7)

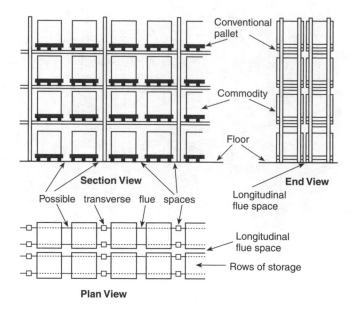

Flue Spaces

There are vertical spaces between rack sections and between pallet loads in a rack storage unit; these vertical spaces are called *flue spaces*. Flue spaces are important in rack configurations because they allow heat from a fire to rise vertically through a rack, which will then activate sprinklers at the ceiling. They also allow a path for water to get down into a rack unit to reach a fire. Where a flue space is not provided, a fire will travel horizontally down the length of the rack and up the rack face. This severely delays the activation of the ceiling sprinklers. There are two types of flue spaces, longitudinal and transverse.

Longitudinal Flue Space. NFPA 13 defines a longitudinal flue space as the space between rows of storage perpendicular to the direction of loading. Longitudinal flue spaces run the length of the racks from end to end between the racks. A longitudinal flue space is not required for multiple-row racks or racks with storage below 25 ft (7.6 m). Figure 13.13 shows rack configurations with longitudinal and transverse flue spaces.

Transverse Flue Space. NFPA 13 defines a transverse flue space as the space between rows of storage parallel to the direction of loading. Transverse flue spaces are required for all single-, double-, and multiple-row racks at each rack upright and also between pallet loads through the width of racks. See Figure 13.13, which shows rack configurations with longitudinal and transverse flue spaces.

Obstructions to Flue Spaces

Solid shelves, slatted shelves, or rack storage product can obstruct flue spaces. When product is allowed to be pushed too far into rack units such that the longitudinal flue space is obstructed, water cannot penetrate down into the racks to reach a fire. Maintaining the longitudinal flue space may be difficult unless the rack storage units have a "stop" or some means of keeping pallet loads from being pushed into the flue space. The designer should review the rack manufacturer's drawings to determine if adequate flue spaces are being maintained. See Figures 13.14 and 13.15, which show rack configurations with solid and slatted shelves.

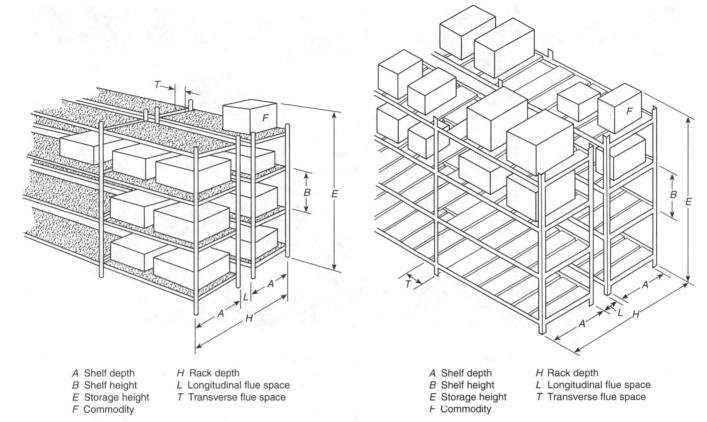

A	Shelf depth	*H*	Rack depth
B	Shelf height	*L*	Longitudinal flue space
E	Storage height	*T*	Transverse flue space
F	Commodity		

FIGURE 13.14
Double-Row Racks with Solid Shelves.
(Source: NFPA 13, 2002, Figure A.3.10.8(c))

A	Shelf depth	*H*	Rack depth
B	Shelf height	*L*	Longitudinal flue space
E	Storage height	*T*	Transverse flue space
F	Commodity		

FIGURE 13.15
Double-Row Racks with Slatted Shelves.
(Source: NFPA 13, 2002, Figure A.3.10.8(d))

Solid Shelves

Per NFPA 13, *solid shelving* is fixed in place, slatted, wire mesh, or other type of shelves located within racks. The area of a solid shelf is defined by the perimeter aisle or flue space on all four sides. Solid shelves having an area equal to or less than 20 ft² (6.1 m²) are defined as open racks. Shelves or wire mesh, slates, or other materials more than 50 percent open and where the flue spaces are maintained are defined as open racks.

Solid shelves in single-, double-, and multiple-row racks that exceed 20 ft² (1.9 m²) but do not exceed 64 ft² (5.9 m²) in area do not require sprinklers below every shelf, but do require sprinklers to be installed at the ceiling and below shelves at intermediate levels not more than 6 ft (2 m) apart vertically. Where solid shelves exceed 64 ft² (5.9 m²) or exceed 6 ft (2 m) apart, sprinklers are to be installed at the ceiling and below each level of shelving. Transverse flue spaces between and at rack uprights shall be maintained in single-row, double-row, and multiple-row racks.

The 2002 edition of the *Automatic Sprinkler Systems Handbook* has several paragraphs of explanatory information that helps to determine how much solid shelving is allowed before it becomes a significant obstruction. It states that the limit of 20 ft² (1.9 m²) is established as representing the amount of blockage typically presented by a standard pallet load, which is accepted based on traditional testing. Above this area limit, some in-rack protection is required. Where close-spaced shelves or tiers are used, the need for sprinklers under each tier can

be mitigated provided flues are spaced at reasonable intervals to prevent uninterrupted spread of fire within the tier. The 64 ft² (5.9 m²) area is considered a reasonable maximum area for this purpose. Figure 13.16 shows double-row storage racks that consist of solid shelves and inadequate flue spaces.

Horizontal Barriers. Per NFPA 13, horizontal barriers used in conjunction with in-rack sprinklers to impede vertical fire development must be constructed of a solid barrier in the horizontal position covering the full length and width of the rack. Per NFPA 13, paragraph A.12.3.1.12, in some situations a horizontal barrier may be required in racks when the ceiling is more than 10 ft (3 m) above the maximum height of storage. They can be constructed of sheet metal, wood, or similar materials. They extend the full length and width of the rack and are fitted within 2 in. (50 mm) horizontally around the rack uprights. Sprinklers are to be installed under horizontal barriers.

Ceiling and In-Rack Sprinkler System Types

The type of sprinkler system used at the ceiling and in the racks for the protection of commodities in rack storage can be wet pipe, dry pipe, or preaction. The area protected by a single in-rack sprinkler zone is limited to 40,000 ft² (3716 m²).

Wet Pipe Systems

Wet pipe systems provide the best protection since the water is present in the piping and at the sprinkler, and there is not a delay prior to water application when a sprinkler opens.

Dry Pipe and Preaction Systems

Dry pipe systems are filled with pressurized air, which must be evacuated in advance of water delivery. The delay in water delivery to an actuated sprinkler on a dry pipe or preaction system depends primarily on the volumetric capacity of the system. The delay in water delivery results in a larger fire in advance of water application, and for this reason dry pipe systems are discouraged for rack applications. In recognition of the delay in water delivery and the resulting larger fire size, the hydraulic design area of operation of the ceiling sprinkler system must be increased by 30 percent. For rack storage dry pipe and preaction systems, this is an important consideration when determining the size of sprinkler zones. The size of a dry pipe sprinkler zone will be greatly affected by the water pressure available, the system piping configuration, the air pressure carried in the system piping, and the model of dry pipe valve being utilized. It is not unusual for dry and preaction sprinkler systems to be significantly smaller than wet sprinkler zones because they are required to deliver water to an inspector's test connection within a short period of time. Dry pipe and preaction zones may need to be considerably smaller than the maximum zone size allowed in NFPA 13. Chapter 8, "Overview of Sprinkler System Layout," has a detailed explanation of dry pipe and preaction system zone sizes.

Dry pipe and preaction systems are restricted for use in areas subject to freezing, such as cold storage warehouses. If the sprinkler system piping were to fill with water either by accident or due to a fire condition, there may be frozen water in the pipes in a short period of time. This may require the removal and replacement of piping, cracked fittings, and damaged sprinklers. The owner may be willing to raise the temperature in a cooler or freezer where perishable or frozen products are stored during the time that repairs are to be made. Alternatively, the replacement may have to be made in low temperatures, which is more difficult. A single or double interlocked preaction system that is controlled by a detection system is designed to prevent the piping from filling with water in the event that a sprinkler pipe or head is damaged by product being loaded into the rack units. If a dry pipe in-rack sprinkler system is to be installed, consideration should be given to damage that might occur if an in-rack sprinkler were accidentally broken, therefore filling the piping with water.

Factors Involved in Rack Storage Design

General Storage, Rack Storage, Rubber Tire Storage, Roll Paper Storage, and Baled Cotton Storage

Sprinklers located at the ceiling have specific requirements in NFPA 13 with regard to K-factor, response, and type, depending on the density required at the ceiling. Sprinklers with K-factors of 5.6 do not produce as many large water droplets as sprinklers with larger K-factors, especially at higher water pressures. Where standard-response upright or pendent spray sprinklers are to be installed for design densities between 0.20 gpm/ft^2 (8.1 mm/min) and 0.34 gpm/ft^2 (13.9 mm/min), the K-factor of the sprinklers must be at least 8.0. For design densities over 0.34 gpm/ft^2 (13.9 mm/min), standard-response upright or pendent spray sprinklers with K-factors of at least 11.2 that are listed for storage applications are required. See NFPA 13, paragraphs 12.1.13.1 through 12.1.13.3.

Roof Vents and Draft Curtains

Draft curtains can alter the operating pattern and actuation time of sprinklers, and in some cases this can have a detrimental effect on the level of protection. The sprinkler system criteria in NFPA 13 are based on the assumption that roof vents and draft curtains are not being used, because most of the full-scale fire tests used as a basis for NFPA 13, Chapter 12, have been performed without these features. If roof vents or draft curtains are used, experienced judgment and consideration of the effect on the sprinkler performance should be exercised in determining the final design criteria.

Riser Connections

A riser connection is a vertical pipe served by a control valve that supplies a system of ceiling sprinklers or a system of in-rack sprinklers. Ceiling and in-rack sprinklers are required to be supplied by separate risers when more than 20 in-rack sprinklers are to be installed. This makes a lot of sense because it is not uncommon for a fork truck operator to accidentally break an in-rack sprinkler pipe or head when loading product into racks, which would then necessitate the shutdown of an entire sprinkler system for repairs if the ceiling and large in-rack sprinkler systems were combined. Figure 13.17 shows a separate riser set up with a ceiling and in-rack sprinkler system. Figure 13.18 shows a combined ceiling and in-rack sprinkler system.

Adjacent Occupancies

The design criteria used for a ceiling sprinkler system is to be extended at least 15 ft (4.6 m) beyond the rack storage area. An exception is made for an area where a partition is installed that is capable of preventing heat from a fire in the storage area from actuating sprinklers in the nonrack storage area. See NFPA 13, subsection 12.1.5.

Where high temperature sprinklers are used at the ceiling of the rack storage area, the use of the high temperature sprinklers should also extend beyond the racks. See Table 13.5, which is taken from the annex of NFPA 13. The table displays the distance beyond the perimeter of the rack-storage occupancy for high temperature–rated sprinklers.

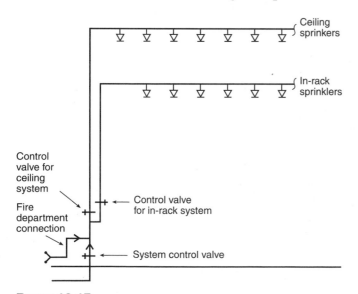

FIGURE 13.17
Separately Controlled Ceiling and In-Rack Sprinklers Where More Than 20 In-Rack Sprinklers Are Installed.

(Source: *Automatic Sprinkler Systems Handbook*, NFPA, 2002, Exhibit 8.34)

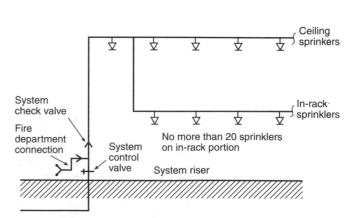

FIGURE 13.18
In-Rack Sprinklers Supplied Directly from Ceiling System Without a Separate Control Valve.

(Source: *Automatic Sprinkler Systems Handbook*, NFPA, 2002, Exhibit 8.35)

TABLE 13.5			
Extension of Installation of High-Temperature Sprinklers Over Storage			
Design Area for High-Temperature–Rated Sprinklers		**Distance Beyond Perimeter of High-Hazard Occupancy for High-Temperature–Rated Sprinklers**	
ft²	m²	ft	m
2000	185.8	30	9.14
3000	278.7	40	12.2
4000	371.6	45	13.72
5000	464.5	50	15.24
6000	557.4	55	16.76

Source: NFPA 13, 2002, Table A.12.1.5.

In-Rack Sprinklers

In-rack sprinklers are standard or quick-response sprinklers that are physically located within the rack structure. In-rack sprinklers are limited to upright or pendent-type sprinklers with K-factors of 5.6 or 8.0. Obstruction rules for ceiling sprinklers do not apply to in-rack sprinklers. In-rack sprinklers are "intermediate-level" sprinklers or sprinklers with a water shield or horizontal barrier installed directly above them where more than one level of in-rack sprinklers is installed. For more information, see the section "In-Rack Sprinkler Location" later in this chapter. Intermediate-level sprinklers or rack-storage sprinklers are sprinklers equipped with integral shields approximately 3 in. (75 mm) in diameter located directly above the sprinkler to protect their operating elements from the discharge of sprinklers installed at higher elevations. The shield keeps water from ceiling or upper-level in-rack sprinklers from wetting sprinklers installed below, and thus retarding their response time. See Figure 13.6 earlier in this chapter.

Flue Space Sprinklers

Where in-rack sprinklers are located in the longitudinal or transverse flue space between the rack units they are called *flue space* sprinklers.

Face Sprinklers. Face sprinklers are standard sprinklers that are located in transverse flue spaces along the aisle or in the rack, are within 18 in. (0.46 m) of the aisle face of storage, and are used to oppose vertical development of fire on the external face of storage. Face sprinklers are to be located a minimum of 3 in. (75 mm) from rack uprights.

Protection of In-Rack Sprinklers

A common problem with in-rack sprinklers and in-rack sprinkler piping is that forklift trucks accidently hit them periodically when they are loading and unloading commodities in the rack units. NFPA 13 requirements for placement of in-rack sprinklers are primarily concerned with fire spread within the rack units and proper water spray distribution within a rack. However, a designer's responsibility includes positioning of the sprinkler piping to enhance protection of the sprinkler piping and heads. A sprinkler head guard will not stop a fork-lift truck moving a pallet of commodities from breaking a sprinkler, bending piping, or breaking fittings on the sprinkler pipe. It is difficult for a lift operator to see sprinkler piping and sprinklers located within a rack when the operator is placing items on upper levels of the rack. It is important to carefully locate in-rack sprinkler piping and sprinklers to avoid damage. In-rack piping may be installed horizontally in the racks by attaching it to the rack upright and beam members. The piping could be located in the longitudinal flue space behind rack horizontal beams to minimize damage.

When rack storage heights exceed 25 ft (7.6 m) and protection is provided with in-rack sprinklers, face sprinklers may be required. An alternative to using face sprinklers can be an ESFR sprinkler system at the ceiling. However, face sprinklers may be required in rack storage configurations when storage heights exceed the height limits for ESFR sprinklers (up to 40 ft—12.2 m—depending on the listing requirements of each individual ESFR sprinkler). Where face sprinklers are required in double- or multiple-row racks, it is difficult to extend the pipe out to the front of a rack from the flue space without placing the pipe where it will be hit by the loading and unloading of commodities. An alternative method of in-rack piping is to install the piping vertically down to each face sprinkler in the racks. A horizontal main must be installed at the ceiling over each rack unit to supply the vertical in-rack sprinkler piping. The vertical branchline piping is installed such that it extends down into the rack structure at the appropriate rack upright. This can be a more expensive option because a vertical pipe must extend down at the location of each face sprinkler, which will normally require more pipe than a horizontally piped installation. When face sprinkler piping is installed in this manner, it hides the piping behind the rack upright near the face of the rack. The pros and cons of each installation method and associated costs should be discussed with the owner, with the agreed pipe arrangement indicated on the contract documents.

Commodity Classification

One of the more important parts of determining the protection for commodities stored in racks is the classification of the commodities. Per NFPA 13, commodities are categorized into seven classifications: Class I, Class II, Class III, Class IV, Group A plastics, Group B plastics, and Group C plastics. The designer must have a thorough understanding of items that are to be stored in order to determine the correct classification for the commodities. The designer should also consider future commodities that the owner may store. NFPA 13 has a wide-ranging list of commodities and their classifications. This chapter does not provide guidance on the classification of various commodities, but the classifications are reviewed in Chapter 4, "Hazard and Commodity Classification."

Hose Stations

Hose stations for use with rack storage of commodities consist of water supply piping to 1½ in. (38.1 mm) fire department hose valves with a 1½-in. hose connected to the valve. The valves are generally supplied by the ceiling wet sprinkler systems or by a separate wet pipe system and riser assembly. Hose stations are used primarily for fire extinguishment (mop-up) operations by the fire department but are also used for first-aid fire fighting where a facility has a trained fire brigade. The installation of rack storage hose stations is not intended to serve as standpipes and is not required to follow the installation of Class II standpipes per NFPA 14. Hose stations are not required for protection of Class I, II, III, and IV commodities up to 12 ft (3.65 mm) in height. See NFPA 13, subsection 8.16.5.

Location

NFPA 13 states that hose stations are to be located such that all portions of the storage area can be reached. It does not indicate a particular length of hose and spray, as indicated for Class II standpipe systems Fire hose for hose stations can be obtained in 50-, 75-, and 100-ft (15.2 m, 22.9 m, and 30.5 m, respectively) increments. Most hose stations can reliably deliver hose streams up to 25 ft (7.6 m) under pressure conditions in the 60 to 100 psi (41 to 6.9 bar) range. If hose stations are to be located at the end of rack storage units, it may be necessary to provide protection from damage by forklift trucks moving products. (See Figure 13.19.) A bollard, or concrete-filled pipe mounted to the floor adjacent to a hose station, is a widely-used protection scheme.

Although the installation of hose stations is an NFPA requirement, AHJs are permitted to make decisions that differ from NFPA requirements. There have been some instances where the local AHJ has determined that hose is not to be provided at the hose stations because they want the occupants of the facility to leave the building immediately during a fire condition rather than to use hose stations. They may not have assurance that the occupants have the necessary training to properly fight a fire that could put their lives at risk. Alternatively, hose stations may be clearly marked that they are for fire service use only. Prior to issuing the design documents, the designer should review proposed locations for the hose stations and evaluate personnel staffing, training, and competence with the AHJ.

Water Supply for Hose Stations

Hose stations can be supplied from one of the following: outside hydrants (not common in areas with freezing temperatures), a separate piping system for small hose stations, valved connections on sprinkler risers where connected upstream

FIGURE 13.19
Typical Hose Racks and Valve Connections.

(Source: *Fire Protection Handbook*, NFPA, 2003, Figure 10.18.6; courtesy of Potter Roemer, Inc.)

of the sprinkler control valve, an adjacent ceiling or in-rack sprinkler system, or from the ceiling sprinkler system in the same area as the hose station where in-rack sprinklers are provided in the same area and are separately controlled.

The connection of a hose system to a wet ceiling sprinkler system does not require a shutoff valve. Where hose stations are connected to an adjacent sprinkler system main, the minimum main size that the hose station piping may be connected to is 2½ in. (63.5 mm). If the system is a hydraulically calculated gridded sprinkler system, then the minimum main size may be 2 in. (50 mm).

As to the water flow rate, an allowance of 50 gpm (189.2 L/min) is added to the ceiling sprinkler demand for a single hose station, and 100 gpm (378.5 L/min) for multiple hose stations. The first 50 gpm (189.2 L/min) hose demand is added at the remote fire hose station, and where multiple hose stations are installed, the second 50 gpm is added where the next most remote hose station piping is connected to the hose station supply piping. The *Automatic Sprinkler Systems Handbook* states that for all practical purposes, the inside hose connections to sprinkler systems should be viewed as large sprinklers with no pressure requirements of their own but that are to flow the required amount of water at the pressure available at the hose outlet. The required flow at the hose outlets is to be carried through the calculations back to the system riser.

Hose Station Pipe Sizing

Pipe sizing for pipes supplying hose stations may be hydraulically calculated except when the piping is connected to the ceiling sprinkler system's mains. When hose station piping is to be connected to a ceiling sprinkler system main, the hose station pipe sizing is specified in NFPA 13. The designer can easily indicate hose station supply piping since it is scheduled by "length of run" in NFPA 13. The pipe size increases as the length of the run of pipe gets longer, but the pipe size does not exceed 1½ in. (38 mm) for any length of pipe for multiple hose stations. The vertical piping to each individual hose station may be as small as 1 in. in size. This looks a little strange when a 1½-in. (38-mm) valve is connected to a 1-in. pipe (25.4 mm). See Table 13.6.

Pressure Restriction

Where static or residual pressures exceed 100 psig (6.9 bar) at a hose station, the recoil associated with operating a fire hose is significant and the pressure at the outlet must be reduced to 100 psig (6.9 bar) by a listed device for the

TABLE 13.6

Hose Station Pipe Sizing		
	Number of Hose Stations	
Length of Pipe Run	One	Multiple
Horizontal runs up to 20 ft (6.1 m)	1 in. (25.4 mm)	1.5 in. (38 mm)
Entire runs from 20 ft to 80 ft (6.1 m to 24.4 m)	1.25 in. (33 mm)	1.5 in. (38 mm)
Entire run over 80 ft (24.4 m)	1.5 in. (38 mm)	1.5 in. (38 mm)

safety of the operator. This can be accomplished by using a pressure regulating hose valve. These valves typically include a diaphragm that maintains static and residual valve outlet pressures at a desired setting. They can be purchased in an adjustable pressure model or a fixed pressure model. The adjustable model can be adjusted on-site to provide the desired outlet pressure. The fixed pressure model requires the designer to determine the pressures available at the valve inlet, and the appropriate valve model that will reduce the inlet pressure to the desired outlet pressure. If a pressure regulating valve is adjusted to an incorrect setting or an incorrect model is installed, it will cause either too much or too little pressure at the valve outlet, which may not provide adequate water to operate the hose station. Pressure regulating valves are required to be tested at a full flow rate every 5 years, according to NFPA 25, *Standard for the Inspection, Testing, and Maintenance of Water-Based Fire Protection Systems*.

Freezing Conditions

Keep in mind that hose stations may not be connected to dry pipe or preaction sprinkler systems, and in unheated warehouses and freezer and cooler storage areas wet hose station piping cannot be installed. One option is to provide an electrically actuated deluge valve with pull station releases at each hose station. The piping remains dry until a pull station activates a solenoid valve on the deluge valve and floods the piping with water. A releasing control panel is used to supervise wiring to the pull stations. Information with regard to the operation of the release station and the operation of the deluge valve should be provided at the release stations. Another option is to connect the hose station piping to a wet riser located in a heated area with a normally closed shutoff valve, but operators may not be aware of the operation procedure. Signage could be installed at each hose station indicating where the control valve is located to turn on the hose station water supply, but the delay in operation could increase risk to the operator. These applications should be reviewed by the local AHJ, as it is not specifically addressed in NFPA 13.

Steel Column Protection

When structural steel temperatures approach a critical melting temperature, the structural integrity of the building is at risk. For that reason, either spray-on fire proofing or water spray protection of structural steel columns is required within storage racks of Class I through Class IV and plastic commodities. See NFPA 13, paragraph 12.3.1.7. Per NFPA 13, paragraph 12.3.1.7.1, where sprinkler protection of building columns within the rack structure or vertical rack members supporting the building are required in lieu of fireproofing, sprinkler protection in accordance with one of the following shall be provided:

- Placing a sidewall sprinkler at the 15-ft (4.6 m) elevation, pointed toward one side of the steel column. See Figure 13.20. The flow from a column sprinkler is not required to be included in the hydraulic calculations.
- Provision of ceiling sprinkler density for a minimum 2000 ft² (186 m²) with ordinary 165°F (74°C) or high-temperature 286°F (141°C) rated

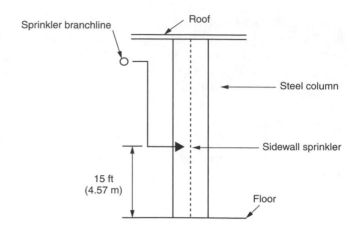

FIGURE 13.20
Sidewall Sprinkler Protecting
a Column.

sprinklers for storage heights above 15 ft (4.6 m), up to 20 ft (6.1 m). See
Table 13.7.

- Installing large drop, specific application control mode, or ESFR (suppression mode) sprinklers at the ceiling.

In certain situations where large drop and ESFR sprinklers are installed, it
may negate the requirement for in-rack sprinklers as well, which can be a cost-efficient solution to a complex storage scenario.

Protection for Rack Storage of Class I through IV Commodities up to and Including 25 ft (7.6 m) in Height

There are many possible rack storage configurations and a large number of different rack storage heights that are used with each rack storage configuration.
NFPA 13 indicates several different methods of providing minimum required
protection for rack storage. The designer must decide which method is appropriate and most cost-effective for the facility being considered. Therefore, the designer needs to be familiar with the level of protection provided by rack storage

TABLE 13.7

Ceiling Sprinkler Densities for Protection of Steel

	Building Columns			
	Aisle Width			
	4 ft (1.2 m)		8 ft (2.4 m)	
Commodity Classification	gpm/ft^2	(L/min)/m^2	gpm/ft^2	(L/min)/m^2
Class I	0.37	15.1	0.33	13.5
Class II	0.44	17.9	0.37	15.1
Class III	0.49	20	0.42	17.1
Class IV and Plastics	0.68	27.7	0.57	23.2

Source: NFPA 13, 2002, Table 12.3.1.7.1.

systems, using standard spray, large drop, and ESFR sprinkler systems at the ceiling. In some cases in-rack sprinklers are required and in others they are not. Annex C in NFPA 13 includes an explanation of test data and procedures that led to the development of sprinkler system discharge criteria for rack storage applications. The information in Annex C is referenced to particular sections in NFPA 13, Chapter 12, for rack storage. These explanations are helpful for understanding where some of the NFPA 13 rack storage requirements originated.

Rack storage facility owners generally do not want in-rack sprinklers and are likely to be willing to pay more money for a ceiling-only sprinkler system if they can eliminate the loss of business associated with broken pipes and sprinklers that can occur with an in-rack sprinkler system. It would be a source of embarrassment if a designer specifies ceiling and in-rack sprinklers for a large rack storage facility and the contractor follows the specifications using NFPA 13, installs the in-rack sprinklers, and then the owner finds out that by using an ESFR system or by using a higher density at the ceiling, the in-rack sprinklers would not be required. The owner's preferences are an important consideration in sprinkler system design.

A good way to get a feel for what is involved in determining the appropriate sprinkler protection for a rack storage area is to work through an example. Example 13.1 includes a determination of ceiling and in-rack storage protection using standard spray sprinklers at the ceiling. Later examples will include large drop and early suppression fast-response (ESFR) sprinklers for a comparison.

EXAMPLE 13.1 Using NFPA 13 Tables for Rack Storage Design

Example 13.1 focuses on storage of a Class III commodity with a top of storage height of 19 ft (5.79 m) on double-row racks in a building that has a roof height of 25 ft (7.6 m).

The designer starts by referencing Table 13.8, which is Table 12.3.2.1.2 from NFPA 13, beginning on the left side of the table with the column labeled "Height" and then following the table columns to the right at each step.

For this example, the height to the top of storage (not to the top of the rack structure) is 19 ft (5.79 m), which fits in the "over 12 ft (3.7 m) up to and including 20 ft (6.1 m) row."

The next column is labeled "Commodity Class." The commodity classification was stated to be Class III; however, the designer should verify that there are not isolated areas that will contain flammable or combustible liquids, aerosols, or other materials that are more "hazardous" than the Class III commodities to be stored. Areas with more hazardous materials or materials with a higher fuel load may change the protection requirements for a portion or all of the rack storage area. The designer should also consider the type of commodities that may be stored in the facility in the future. The Commodity Class "III" row is used.

The third column is labeled "Encapsulated." Encapsulation is a method of packaging consisting of a plastic sheet completely enclosing the sides and top of a pallet load containing a combustible commodity or a combustible package or a group of combustible commodities or combustible packages. Combustible commodities individually wrapped in plastic sheeting and stored exposed in a pallet load also are considered encapsulated. Totally noncombustible commodities on wood pallets enclosed only by a plastic sheet as described are not covered under this definition. Banding (i.e., stretch-wrapping around only the sides of a pallet load) is not considered to be encapsulation.

TABLE 13.8

Single- or Double-Row Racks—Storage Height Up to and Including 25 ft (7.6 m)
Without Solid Shelves

Height	Commodity Class	Encapsulated	Aisles*		Sprinklers Mandatory In-Rack	
			ft	m		
Over 12 ft (3.7 m), up to and including 20 ft (6.1 m)	I	No	4	1.2	No	
			8	2.4		
		Yes	4	1.2	No	
			8	2.4		
	II	No	4	1.2	No	
			8	2.4		
		Yes	4	1.2	No	
			8	2.4		
	III	No	4	1.2	No	
			8	2.4		
		Yes	4	1.2	1 level	
			8	2.4		
	IV	No	4	1.2	No	
			8	2.4		
		Yes	4	1.2	1 level	
			8	2.4		
Over 20 ft (6.1 m), up to and including 22 ft (6.7 m)	I	No	4	1.2	No	
			8	2.4		
		Yes	4	1.2	1 level	
			8	2.4		
	II	No	4	1.2	No	
			8	2.4		
		Yes	4	1.2	1 level	
			8	2.4		
	III	No	4	1.2	No	
			8	2.4		
		Yes	4	1.2	1 level	
			8	2.4		
	IV	No	4	1.2	No	
			8	2.4		
		Yes	4	1.2	1 level	
			8	2.4		
Over 22 ft (6.7 m), up to and including 25 ft (7.6 m)	I	No	4	1.2	No	
			8	2.4		
		Yes	4	1.2	1 level	
			8	2.4		
	II	No	4	1.2	No	
			8	2.4		
		Yes	4	1.2	1 level	
			8	2.4		

| | Ceiling Sprinkler Water Demand | | | | | |
| | With In-Rack Sprinklers | | | Without In-Rack Sprinklers | | |
	Figure	Curves	Apply Figure 12.3.2.1.5.1	Figure	Curves	Apply Figure 12.3.2.1.5.1
	12.3.2.1.2(a)	C and D	Yes	12.3.2.1.2(a)	G and H	Yes
		A and B			E and F	
	12.3.2.1.2(e)	C and D		12.3.2.1.2(e)	G and H	Yes
		A and B			E and F	
	12.3.2.1.2(b)	C and D		12.3.2.1.2(b)	G and H	Yes
		A and B			E and F	
	12.3.2.1.2(e)	C and D		12.3.2.1.2(e)	G and H	Yes
		A and B			E and F	
	12.3.2.1.2(c)	C and D		12.3.2.1.2(c)	G and H	Yes
		A and B			E and F	
	12.3.2.1.2(f)	C and D		—	—	—
		A and B				
	12.3.2.1.2(d)	C and D		12.3.2.1.2(d)	G and H	Yes
		A and B			E and F	
	12.3.2.1.2(g)	C and D		—	—	—
		A and B				
	12.3.2.1.2(a)	C and D	No	12.3.2.1.2(a)	F and H	Yes
		A and B			E and G	
	12.3.2.1.2(e)	C and D		—	—	—
		A and B				
	12.3.2.1.2(b)	C and D		12.3.2.1.2(b)	G and H	Yes
		A and B			E and F	
	12.3.2.1.2(e)	C and D		—	—	—
		A and B				
	12.3.2.1.2(c)	C and D		12.3.2.1.2(c)	G and H	Yes
		A and B			E and F	
	12.3.2.1.2(f)	C and D		—	—	—
		A and B				
	12.3.2.1.2(d)	C and D		12.3.2.1.2(d)	G and H	Yes
		A and B			E and F	
	12.3.2.1.2(g)	C and D		—	—	—
		A and B				
	12.3.2.1.2(a)	C and D	No	12.3.2.1.2(a)	F and H	Yes
		A and B			E and G	
	12.3.2.1.2(e)	C and D		—	—	—
		A and B				
	12.3.2.1.2(b)	C and D		12.3.2.1.2(b)	G and H	Yes
		A and B			E and F	
	12.3.2.1.2(e)	C and D		—	—	—
		A and B				

continues

TABLE 13.8

(Continued)

Height	Commodity Class	Encapsulated	Aisles*		Sprinklers Mandatory In-Rack	
			ft	m		
Over 22 ft (6.7 m), up to and including 25 ft (7.6 m)	III	No	4	1.2	No	
			8	2.4		
		Yes	4	1.2	1 level	
			8	2.4		
	IV	No	4	1.2	1 level	
			8	2.4		
		Yes	4	1.2		
			8	2.4		

*See NFPA 13, 12.3.2.1.2.1 for interpolation of aisle widths.

Source: NFPA 13, 2002, Table 12.3.2.1.2.

Where there are holes or voids in the plastic or waterproof cover on the top of the carton that exceed more than half of the area of the cover, the term *encapsulated* does not apply. The term also does not apply to plastic-enclosed products or packages inside a large, nonplastic, enclosed container.

For this example, the pallet loads are considered to be encapsulated, which will come under the row labeled "Yes." The densities for encapsulated storage are higher than for nonencapsulated because the plastic wrap used to encapsulate commodities tends to repel water and not absorb it.

The fourth column from the left is labeled "Aisles" and includes the choice of a 4-ft (1.2-m) or an 8-ft (2.4-m) aisle space between the face of the rack structures. If the aisle space is between 4 ft and 8 ft (1.2 m to 2.4 m), the designer may make a direct linear interpolation between the design densities indicated in the "Ceiling Sprinkler Water Demand" columns. If the aisles are over 8 ft (2.4 m), then the densities for 8-ft aisles are to be used. If the aisles are less than 4 ft down to 3½ ft (1.2 m to 1.1 m), the densities for 4 ft (1.2 m) may be used. If the aisles are less than 3½ ft (1.1 m), then the rack configuration is considered a multiple-row rack and the designer must use the tables and criteria for multiple-row racks in lieu of this table. A 6-ft (1.6-m) aisle space will be used in this example and the interpolation of densities will be discussed in the next few paragraphs.

The fifth column is labeled "Sprinklers Mandatory In-Rack." This column indicates whether in-rack sprinklers are required in addition to the ceiling sprinklers or not and also indicates how many levels of in-rack sprinklers are required. Since the commodities in this example are encapsulated, this column indicates that one level of in-rack sprinklers is required. If the commodities had not been encapsulated, a choice of whether to include the use of in-rack sprinklers versus ceiling sprinklers only is allowed. The ceiling sprinkler system density required is significantly higher without the use of in-rack sprinklers.

A "Figure" reference is indicated in the sixth column of the table and a "Curves" reference is indicated in the seventh column. Both columns are under the heading of "Ceiling Sprinkler Water Demand" and "With In-Rack Sprinklers." For this example, Figure 12.3.2.1.2(f) is specified, and is shown as Figure 13.21. Figure 13.21 shows design curves similar to the design curves

		Ceiling Sprinkler Water Demand				
	With In-Rack Sprinklers			Without In-Rack Sprinklers		
Figure	Curves	Apply Figure 12.3.2.1.5.1	Figure	Curves	Apply Figure 12.3.2.1.5.1	
12.3.2.1.2(c)	C and D	No	12.3.2.1.2(c)	G and H	Yes	
	A and B			E and F		
12.3.2.1.2(f)	C and D		—	—	—	
	A and B					
12.3.2.1.2(d)	C and D		—	—	—	
	A and B					
12.3.2.1.2(g)	C and D		—	—	—	
	A and B					

indicated for nonrack storage designs with light, ordinary, and extra hazard occupancies. The design density and design area indicated in Figure 13.21 are for determining the design density and design area for the ceiling sprinkler system. A description for each curve and the sprinkler temperature ratings associated with each curve are indicated to the right of the curves. If 8-ft (2.4-m) aisles are to be used, then curves A and B apply; if 4-ft (1.2-m) aisles are to be used, then curves C and D apply.

When referring to Figure 13.21, the designer must determine what temperature rating will be used for the ceiling sprinklers. Specifying 286°F (141°C) rated ceiling sprinklers will result in a design density that is less than that for 165°F (74°C) ceiling sprinklers. Getting back to the example, if 286°F (141°C) rated ceiling sprinklers are used, with 6-ft (1.8-m) aisles between the rack units, then a direct linear interpolation can be made to determine the required design density.

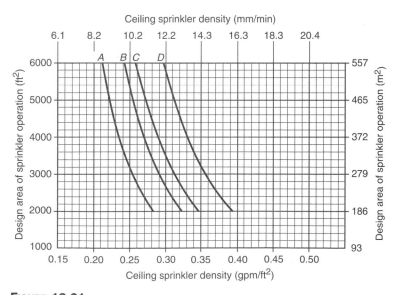

Curve	Legend
A —	8-ft (2.44-m) aisles with 286°F (141°C) ceiling sprinklers and 165°F (74°C) in-rack sprinklers
B —	8-ft (2.44-m) aisles with 165°F (74°C) ceiling sprinklers and 165°F (74°C) in-rack sprinklers

Curve	Legend
C —	4-ft (1.22-m) aisles with 288°F (141°C) ceiling sprinklers and 165°F (74°C) in-rack sprinklers
D —	4-ft (1.22-m) aisles with 165°F (74°C) ceiling sprinklers and 165°F (74°C) in-rack sprinklers

FIGURE 13.21

Single- or Double-Row Racks—20-ft (6.1-m)-High Rack Storage—Sprinkler System Design Curves—Class III Encapsulated Commodities—Conventional Pallets.

(Source: NFPA 13, 2002, Figure 12.3.2.1.2(f))

At this point the designer should discuss the future use of the rack storage units with the owner to determine whether at some future date the aisle space may change. As the aisles become narrower, the design density for the ceiling sprinkler system goes up. If, in the future, the owner wants to change to narrower aisles between the rack units, the ceiling sprinkler system may need to be revised significantly or additional in-rack sprinklers may be required. It may be of some advantage to design the ceiling sprinkler system to accommodate a heavier design density at this time rather than make a more costly change in the future. A similar discussion should occur with regard to future commodity classifications, storage heights, and possible shelving in the rack storage units.

Referring to curve "A" from Figure 13.21 for 8-ft (2.4-m) aisles, an appropriate density would be 0.28 gpm per square foot over a design area of 2000 ft² (186 m²) (the lowest point on the "A" curve). Referring to curve "C" from Figure 13.21 for 4-ft (1.2-m) aisles, an appropriate density would be 0.345 gpm per square foot over a design area of 2000 ft² (186 m²) (the lowest point on the "C" curve). Interpolating between the two densities gives a density of 0.32 gpm per square foot over a design area of 2000 ft² (186 m²) for a rack configuration with 6-ft (1.8-m) aisles.

The eighth column labeled "Apply Figure 12.3.2.1.5.1" indicates that the design density may be adjusted for storage height. Figure 12.3.2.1.5.1 from NFPA 13 is reproduced here (see Figure 13.22). Since the "height of storage" is 19 ft (5.79 m), the design density is reduced by locating the point corresponding to 19 ft (up from the bottom of the figure) along the curve in Figure 13.22 and then finding the corresponding number for the "Percent of design curve density" on the left side of the figure. The percent of the design curve density indicated on the figure is 92. Therefore, the required design density at the ceiling sprinkler system is 0.92 × 0.32 gpm/ft² (13 mm/min) = 0.29 gpm/ft² (12 mm/min) over a design area of 2000 ft² (186 m²). This is the final design density and design area required for the example described. The ceiling sprinkler must be a standard response sprinkler with a K-factor of at least 8.0 because the design density is between 0.20 gpm/ft² (8.15 mm/min) and

FIGURE 13.22

Ceiling Sprinkler Density vs. Storage Height.

(Source: NFPA 13, 2002, Figure 12.3.2.1.5.1)

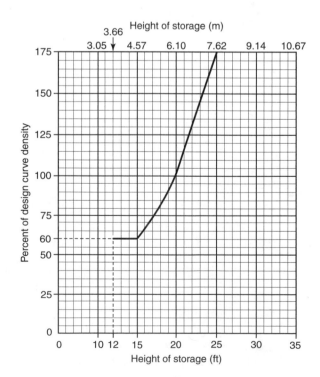

TABLE 13.9

Adjustment to Ceiling Sprinkler Density for Storage Height and In-Rack Sprinklers

Storage Height	In-Rack Sprinklers	Apply Figure 12.3.2.1.5.1 for Storage Height Adjustment	Permitted Ceiling Sprinklers Density Adjustments Where In-Rack Sprinklers are Installed
Over 12 ft (3.7 m) through 25 ft (7.6 m)	None	Yes	None
Over 12 ft (3.7 m) through 20 ft (6.1 m)	Minimum required	Yes	None
	More than minimum, but not in every tier	Yes	Reduce density 20% from that of minimum in-rack sprinklers
	In every tier	Yes	Reduce density 40% from that of minimum in-rack sprinklers
Over 20 ft (6.1 m) through 24 ft (7.5 m)	Minimum required	No	None
	More than minimum, but not in every tier	No	Reduce density 20% from that of minimum in-rack sprinklers
	In every tier	No	Reduce density 40% from that of minimum in-rack sprinklers

Source: NFPA 13, 2002, Table 12.3.2.1.5.3.

0.34 gpm/ft^2 (14 mm/min). Note that the protection scheme and design density determined are based upon no solid shelves being used in the rack units.

Reduction in Ceiling Sprinkler Density

Although not used in Example 13.1, the designer should be aware that in some cases NFPA 13 allows additional reductions to the ceiling sprinkler density. See NFPA 13 section 12.3.2.1.5.1. If additional levels of in-rack sprinklers are installed, the ceiling design density may be allowed to be adjusted an additional 20 or 40 percent above the reduction allowed by Figure 13.22. See Table 13.9.

EXAMPLE 13.2 USING NFPA 13 TO REDUCE CEILING SPRINKLER DENSITY

If in Example 13.1 the designer decides to install sprinklers at every tier of storage instead of the minimum required of just one level of in-rack sprinklers, the ceiling sprinkler density can be reduced by 40 percent, in addition to the reduction allowed by Figure 13.22. The final design curve density indicated in the example with one level of in-rack sprinklers is 0.294 gpm (12 mm/min) per square foot over a design area of 2000 ft^2 (186 m^2). If the design density is reduced by an additional 40 percent, the required design density at the ceiling sprinkler system would be 0.294 − (0.294 × 0.40) gpm per square foot = 0.18 gpm per square foot (7.4 mm/min) over a design area of 2000 ft^2 (186 m^2). This ceiling sprinkler design density is considerably less with an additional 40 percent reduction to the design density, but the designer should consider whether having the additional levels of in-rack sprinklers will create additional loss of business associated with broken pipes and sprinklers.

FIGURE 13.23

Adjustment of Design Area of Sprinkler Operation for Clearance from Top of Storage to Ceiling.

(Source: NFPA 13, 2002, Figure 12.3.2.1.5.7)

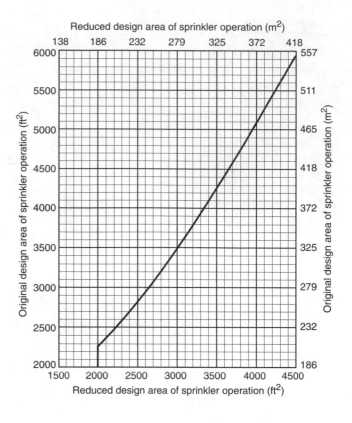

Reduction in Design Area

When rack storage areas are protected with sprinklers installed only at the ceiling, a reduction in the ceiling design area is allowed by NFPA 13, paragraph 12.3.2.1.5.7, with certain restrictions. Per NFPA 13: "Where clearance from ceiling to top of storage is less than 4½ ft (1.37 m), the sprinkler operating area indicated in curves E, F, G and H in Figure 12.3.2.1.2(a) through Figure 12.3.2.1.2(e) shall be permitted to be reduced as indicated in Figure 12.3.2.1.5.7 but shall not be reduced to less than 2000 ft² (186 m²)." Figure 12.3.2.1.5.7 is shown here as Figure 13.23.

EXAMPLE 13.3 USING NFPA 13 TO REDUCE DESIGN AREA

For Example 13.3, assuming storage of Class IV nonencapsulated commodities with a top-of-storage height of 22 ft (6.7 m) on double-row racks, with 8-ft (2.4 m) aisles, in a building that has a roof height of 25 ft (7.6 m), ceiling sprinklers rated 286°F (141°C) will be used at the ceiling only. Per Table 13.8 (Table 12.3.2.1.2 from NFPA 13, 2002 edition, Single- or Double-Row Racks—Storage Height Up to and Including 25 ft (7.6 m) Without Solid Shelves), Figure 12.3.2.1.2(d) (referenced here as Figure 13.24), Curve E, is the appropriate curve to use.

Using Figure 13.24, for a 3000-ft² (279-m²) design area for the ceiling sprinkler system from Curve E that corresponds to a design density of 0.45 gpm/ft² (18.3 mm/min), the size of the design area can be reduced in accordance with Figure 13.23 since the clearance from the ceiling to the top of storage is less than 4½ ft (1.37 m). Using Figure 13.23 and starting with a

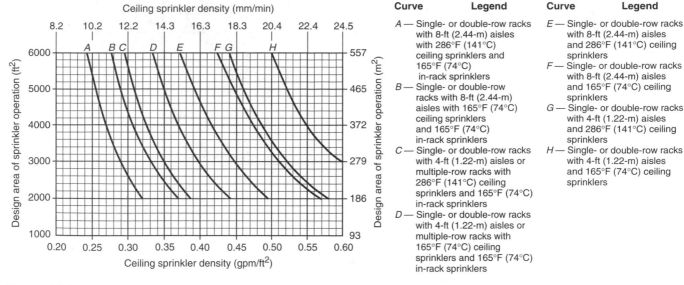

Curve | **Legend**

A — Single- or double-row racks with 8-ft (2.44-m) aisles with 286°F (141°C) ceiling sprinklers and 165°F (74°C) in-rack sprinklers

B — Single- or double-row racks with 8-ft (2.44-m) aisles with 165°F (74°C) ceiling sprinklers and 165°F (74°C) in-rack sprinklers

C — Single- or double-row racks with 4-ft (1.22-m) aisles or multiple-row racks with 286°F (141°C) ceiling sprinklers and 165°F (74°C) in-rack sprinklers

D — Single- or double-row racks with 4-ft (1.22-m) aisles or multiple-row racks with 165°F (74°C) ceiling sprinklers and 165°F (74°C) in-rack sprinklers

Curve | **Legend**

E — Single- or double-row racks with 8-ft (2.44-m) aisles and 286°F (141°C) ceiling sprinklers

F — Single- or double-row racks with 8-ft (2.44-m) aisles and 165°F (74°C) ceiling sprinklers

G — Single- or double-row racks with 4-ft (1.22-m) aisles and 286°F (141°C) ceiling sprinklers

H — Single- or double-row racks with 4-ft (1.22-m) aisles and 165°F (74°C) ceiling sprinklers

FIGURE 13.24

Sprinkler System Design Curves—20-ft (6.1-m)-High Rack Storage—Class IV Nonencapsulated Commodities—Conventional Pallets.

(Source: NFPA 13, 2002, Figure 12.3.2.1.2(d))

3000-ft² (279-m²) area, the design area can be reduced to 2650 ft² (246.2 m²). When using Figure 13.23 the design area cannot be reduced to less than 2000 ft² (186 m²).

In-Rack Sprinkler Location

The location of in-rack sprinklers depends on the height of the storage and whether a flue space is provided. Using NFPA 13 requirements, the designer must determine the required number of levels of in-rack sprinklers and specify them in the design documents.

In-Rack Sprinkler Location for Storage up to 20 ft (6.1 m). NFPA 13 states that the elevation of in-rack sprinkler deflectors with respect to storage is not a consideration in single- and double-row rack storage up to and including 20 ft high. Although a specific location is not indicated, it is good practice to locate the sprinkler deflectors at least 6 in. (150 mm) above the top of the storage tier. NFPA 13 states that if in-rack sprinklers are not located above the top of a storage tier, it is important that they be located at transverse flues. With multiple-row racks, the sprinkler deflectors are required to be at least 6 in. (150 mm) above the top of the storage tier.

In-Rack Sprinkler Location for Storage up to 25 ft (7.6 m). Where only one level of in-rack sprinklers is required for storage up to 25 ft (7.6 m), the sprinklers are to be located at the first tier level at or above one-half of the storage height. See Figure 13.25. Where two levels of in-rack sprinklers are required for storage up to 25 ft (7.6 m), the sprinklers are to be located at the first tier level at or above one-third and two-thirds of the storage height, as shown in Figure 13.26.

Rack horizontal members, sometimes referred to as "load beams," are not permitted to obstruct in-rack sprinklers.

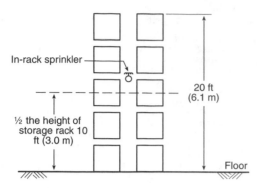

FIGURE 13.25

Positioning of One Level of In-Rack Sprinklers for Storage Heights Under 25 ft (7.6 m).

(Source: *Automatic Sprinkler Systems Handbook*, NFPA, 2002, Exhibit 12.7)

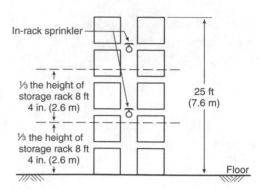

FIGURE 13.26

Positioning of Two Levels of In-Rack Sprinklers for Storage Heights Under 25 ft (7.6 m).

(Source: *Automatic Sprinkler Systems Handbook*, NFPA, 2002, Exhibit 12.8)

In-Rack Sprinkler Horizontal Spacing

Single- and Double-Row Racks. For single- and double-row racks, the horizontal spacing for in-rack sprinklers can be determined by reviewing NFPA 13 Table 12.3.2.4.2.1 (referenced here as Table 13.10). The designer needs to know the aisle width and the commodity classification. Where the commodities are encapsulated, the maximum horizontal spacing is 8 ft (2.4 m). NFPA 13, Section C.15, states that tests were not conducted with aisles wider than 8 ft (2.4 m) or narrower than 4 ft (1.2 m). It is, therefore, not possible to determine whether lower ceiling densities should be used for aisle widths greater than 8 ft (2.4 m).

Multiple-Row Racks. For multiple-row racks with encapsulated or nonencapsulated storage, NFPA 13 indicates the maximum horizontal spacing for in-rack sprinklers may not exceed 12 ft (3.7 m) for Class I, II, or III commodities, and 8 ft (2.4 m) for Class IV commodities. There is also a maximum area of coverage restriction of 100 ft² (9.3 m²) to consider for storage of Class I, II or III commodities, and of 80 ft² (7.4 m²) for Class IV commodities. The area of coverage does not include area in the aisle space. See NFPA 13 paragraph 12.3.2.4.2.2. The area of coverage is determined by using the plan view of the rack units. For example, if Class III commodities are stored in multiple-row

TABLE 13.10

In-Rack Sprinkler Spacing for Class I, II, III, and IV Commodities Stored up to 25 ft (7.6 m) in Height

Aisle Widths		Commodity Class					
		I and II		III		IV	
ft	m	ft	m	ft	m	ft	m
8	2.4	12	3.7	12	3.7	8	2.4
4	1.2	12	3.7	8	2.4	8	2.4

Source: NFPA 13, 2002, Table 12.3.2.4.2.1.

racks that are 13 ft (4.0 m) wide from outside face to outside face, the sprinklers could not be spaced more than 7 ft, 8 in. (2.34 m) apart horizontally even though the maximum allowed is 12 ft (3.7 m).

Rack Upright Obstructions. NFPA 13 does not restrict the placement of sprinklers behind rack uprights but recommends that they be placed away from rack uprights. NFPA 13, Section C.17, states that in a 20-ft (6.1-m)-high rack storage fire test with the sprinkler obstructed, the rack upright had no effect on sprinkler performance. However, it goes on to say that in tests with 30-ft (9.1-m)-high rack storage with sprinklers obstructed by rack uprights, the tests produced unsatisfactory results. When the sprinklers were placed at least 2 ft away from the uprights, the results improved.

In-Rack Sprinkler Water Flow Demand

The hydraulic calculations for in-rack sprinklers are handled differently from the density/area method. The number of sprinklers to be included in the hydraulic calculation is indicated in NFPA 13 and depends on the commodity classification and the number of levels of in-rack sprinklers required. Minimum in-rack sprinkler discharge pressure is 15 psi (1 bar).

Number of In-Rack Sprinklers to Be Calculated. The number of sprinklers to be hydraulically calculated for Class I, II, or III commodities with one level of in-rack sprinklers is 6. The number of sprinklers to be hydraulically calculated for a Class IV commodity with one level of in-rack sprinklers is 8. The number of sprinklers to be hydraulically calculated for Class I, II, or III commodities with more than one level of in-rack sprinklers is 10 (5 on each two top levels). The number of sprinklers to be hydraulically calculated for a Class IV commodity with more than one level of in-rack sprinklers is 14 (7 on each two top levels).

Pressure Balancing. The in-rack sprinkler system pressure and flow demand has to be balanced with the demand of the ceiling sprinkler system at the point where the two water flows combine in the hydraulic calculation. This may cause more water to flow from either the in-rack sprinklers or the ceiling sprinklers than was originally expected. It is not a good design for the in-rack sprinkler system to create an overdischarge in the ceiling sprinkler system. Where overdischarge of a system occurs, the designer modifies pipe sizes to minimize the pressure imbalance and resultant overdischarge.

EXAMPLE 13.4 Pressure Balancing

Assume a ceiling sprinkler system demand of 120 psi (8.2 bar) at a flow rate of 1050 gpm (3980 L/min) at the point where the in-rack sprinkler system is connected to the main riser. Also assume a Class IV commodity with one level of in-rack sprinklers. If eight 5.60 K-factor sprinklers flow at a minimum of 15 psi (1 bar), then the minimum flow rate will be $Q = K \times p^{1/2} = 5.6 \times 15^{1/2} =$ 21.7 gpm (82 L/min) per sprinkler. The minimum total flow rate for the eight sprinklers would be $8 \times 21.69 = 173.5$ gpm (657 L/min). The actual flow rate will be higher when friction losses and pressure balancing are taken into consideration. So an in-rack sprinkler system flow rate of 200 gpm (11 L/min) will be assumed. With an end head minimum pressure of 15 psi (1 bar), friction losses in the pipe and fittings, and pressure losses due to elevation, a demand pressure of 74 psi (5 bar) will be assumed. At the point where the ceiling sprinkler system and the in-rack sprinkler system join, there is a pressure difference of $120 - 74 = 46$ psi (3.1 bar). Since the two demands have to be balanced to the higher pressure, a K-factor for the lower-pressure system (in-rack system) is determined by the equation $K = Q/p^{1/2} = 200/74^{1/2} = 23.3$. Using the new K-factor and the ceiling pressure of 120 psi (8.2 bar), the required flow rate for the in-rack sprinkler system is $Q = K \times p^{1/2} = 23.3 \times 120^{1/2} = 255.2$ gpm (966 L/min). Therefore, the total flow rate to be carried back to the water supply source from the point where the ceiling and in-rack sprinkler system join is 1050 gpm (3980 L/min) (ceiling flow rate) plus 255.2 gpm (966 L/min) (balanced in-rack flow rate), which is 1305.2 (4946 L/min) gpm.

In addition to the ceiling and in-rack sprinkler demand, hose allowance must be added to the calculation. The total combined inside and outside hose allowance required for storage of Class I, II, and III commodities over 12 ft (3.7 m) and up to 25 ft (7.6 m) is 500 gpm (1892 L/min) for a duration of 90 minutes. Class IV commodities require a duration of 120 min.

Hydraulic calculation software can be used to perform these calculations. However the designer may need to perform some rudimentary calculations in order to determine whether a fire pump is required and what the rated flow of the fire pump should be when considering ceiling and in-rack sprinkler systems.

EXAMPLE 13.5 Large Drop Sprinklers

Now consider the same scenario as presented in rack storage Example 13.1, of Class III commodities with a top of storage height of 19 ft (5.8 m) on double-row racks in a building that has a roof height of 25 ft (7.6 m) using large drop sprinklers. Large drop sprinklers are specifically listed sprinklers with K-factors of 11.2. Manufacturer's catalog cutsheets will specifically indicate that a sprinkler is a large drop type. Just because a sprinkler has a K-factor of 11.2 or larger does not qualify a sprinkler as a large drop sprinkler. There are very few large drop sprinklers on the market. Large drop sprinklers are control mode sprinklers with large water droplets that have the ability to penetrate a high-velocity fire plume to reach a fire. See Figure 13.27. Note that the deflector of the large drop sprinkler is designed to create larger drops that are more likely to penetrate a fire plume and coat a combustible, such as rolled paper. Standard spray sprinklers create smaller drops that are more likely to evaporate and less likely to penetrate a plume and provide a coating of combustibles. A designer evaluates a commodity and determines which performance objective, coating or evaporation, is most amenable to control of a fire.

To start the example, see Table 13.11. Start on the left side of the table with the column labeled "Commodity Class" and then proceed through the columns to the right at each step.

Since a Class III commodity is being stored, either the second or third row down in the table applies for a Commodity Class of "I, II, III." Since the maximum storage height for this example is 20 ft (6.1 m), the designer must use the second row down. The third column indicates the maximum storage height of 20 ft (6.1 m). The roof height for this example is 25 ft (7.6 m), which is under

Figure 13.27
Large Drop Sprinkler.

(Source: *Automatic Sprinkler Systems Handbook*, NFPA, 2002, Exhibit 3.19; courtesy of Viking Corporation)

the maximum ceiling/roof height of 30 ft (9.1 m) indicated in the fourth column. With reference to the fifth column, the type of system will be "Wet." The sixth column, labeled "Number of Design Sprinklers/Minimum Pressure," indicates that the pipe-sizing hydraulic calculations must include 15 sprinklers flowing at a minimum pressure of 25 psi (1.7 bar). Table 13.11 does not indicate that in-rack sprinklers are required for this scenario. A hose stream demand of 500 gpm (1892 L/min) must also be included in the calculations. The water supply duration must be 1½ hours, as indicated in the last column.

The end result is not a true density/area calculation, but rather a number of sprinklers to calculate at a given minimum pressure. When performing the hydraulic calculations, the input to the program will need to include the minimum flow and pressure for each sprinkler instead of the usual density and area input. A restriction for large drop sprinklers is that the maximum discharge pressure at the hydraulically remote sprinkler cannot be over 95 psi (6.6 bar). At pressures over 95 psi (6.6 bar), the water droplet size is smaller than would be considered effective and the water spray pattern becomes sporadic and unreliable. Even though large drop sprinklers do not follow the design/area hydraulic calculation method, the design area is still required to be a rectangular area having a dimension parallel to the branchlines of at least 1.2 times the square root of the design area. Any fractional sprinkler is to be included in the design area.

A benefit of using large drop sprinklers is that structural steel column sprinkler protection is not required. As a side note, if a dry pipe system is used with large drop sprinklers, the size of the hydraulically calculated area does not need to be increased by 30 percent as with other sprinkler types. For a dry pipe system with large drop sprinklers, the number of design sprinklers is indicated in Table 13.11. For the previous example, 25 design sprinklers would be required, as indicated in the sixth column and second row of the table.

TABLE 13.11

Large Drop Sprinkler Design Criteria for Single-, Double-, and Multiple-Row Racks without Solid Shelves of Class I Through Class IV Commodities Stored Up to and Including 25 ft (7.6 m) in Height

Commodity Class	Nominal K-Factor	Maximum Storage Height		Maximum Ceiling/Roof Height		Type of System	
		ft	m	ft	m		
I, II	11.2	25	7.6	30	9.1	Wet	
						Dry	
I, II, III	11.2	20	6.1	30	9.1	Wet	
						Dry	
I, II, III	11.2	25	7.6	35	10.7	Wet	
						Dry	
IV	11.2	20	6.1	25	7.6	Wet	
						Dry	
IV	11.2	20	6.1	30	9.1	Wet	
						Dry	
IV	11.2	20	6.1	30	9.1	Wet	
						Dry	
IV	11.2	25	7.6	30	9.1	Wet	
						Dry	
IV	11.2	25	7.6	35	10.7	Wet	
						Dry	
IV	11.2	25	7.6	35	10.7	Wet	
						Dry	

Source: NFPA 13, 2002, Table 12.3.2.2.1(a).

EXAMPLE 13.6 Early Suppression Fast-Response (ESFR) Sprinklers

ESFR sprinklers are suppression mode sprinklers that operate with high water flow rates and have the largest K-factors (therefore the largest orifice) of any sprinklers manufactured. The sprinkler orifice is as large as 1 in. in diameter and the sprinkler itself is quite large when compared to standard sprinklers. ESFR sprinklers are intended for high-piled storage up to 40 ft (12.2 m) in buildings up to 45 ft (13.7 m) high. They are designed to react quickly to growing fires and deliver a heavy discharge. ESFR sprinklers are favored by building owners because they may remove the requirement for in-rack sprinklers. Early ESFR sprinklers had nominal K-factors of 14 and high end-head pressure requirements. Then came the ESFR sprinklers with larger orifices and nominal K-factors of 11, 17, 22, and 25. These larger orifice sprinklers allowed similar protection of rack storage than the previous K-14 ESFR sprinklers did, only

Number of Design Sprinklers/Minimum Pressure		Hose Stream Demand		Water Supply Duration (hours)
/psi	/bar	gpm	L/min	
20/25	20/1.7	500	1900	1½
30/25	30/1.7	500	1900	1½
15/25	15/1.7	500	1900	1½
25/25	25/1.7	N/A	N/A	1½
15/25 + 1 level of in-rack	15/1.7 + 1 level of in-rack	500	1900	1½
25/25 + 1 level of in-rack	25/1.7 + 1 level of in-rack	500	1900	1½
15/50	15/3.4	500	1900	2
N/A	N/A	N/A	N/A	N/A
20/50	20/3.4	500	1900	2
N/A	N/A	N/A	N/A	N/A
15/75	15/5.2	500	1900	2
N/A	N/A	N/A	N/A	N/A
15/50 + 1 level of in-rack	15/3.4 + 1 level of in-rack	500	1900	2
N/A	N/A	N/A	N/A	N/A
20/50 + 1 level of in-rack	20/3.4 + 1 level of in-rack	500	1900	2
N/A	N/A	N/A	N/A	N/A
15/75 + 1 level of in-rack	15/5.2 + 1 level of in-rack	500	1900	2
N/A	N/A	N/A	N/A	N/A

without the high end-head pressures that sometimes required a fire pump. Figures 13.28 and 13.29 show various ESFR sprinklers.

Per NFPA 13, ESFR protection does not apply to the protection of racks involving solid shelves or protection of rack storage involving combustible, open-top cartons or containers. The minimum operating pressure for ESFR sprinklers is 15 psi (1 bar), but the designer must follow the minimum pressures indicated in NFPA 13 and the manufacturer's catalog cutsheets. Longitudinal and transverse flue spaces are required regardless of the storage height.

Now consider the same rack storage example of Class III commodities with a top of storage height of 19 ft (5.59 m) on double-row racks in a building that has a roof height of 25 ft (7.6 m) using ESFR sprinklers. For the example, refer to Table 13.12, which is Table 12.3.2.3.1 from NFPA 13.

FIGURE 13.28
Pendent-Type ESFR
Sprinklers.

(Source: *Automatic Sprinkler Systems
Handbook*, NFPA, 2002, Exhibit 3.15)

FIGURE 13.29
Upright-Type ESFR Sprinkler.

(Source: *Automatic Sprinkler Systems
Handbook*, NFPA, 2002, Exhibit 3.16;
courtesy of Viking Corporation)

Start on the left side of Table 13.12 with the column labeled "Storage Arrangement" and then proceed through the columns to the right at each step. The first choice that can be made for this example is in the third column, labeled "Maximum Storage Height," which with a maximum storage height of 19 ft (5.79 m) the row for 20 ft (6.1 m) is selected. The maximum ceiling/roof height in the example is 25 ft (7.6 m), which matches with the first row under the fourth column labeled "Maximum Ceiling/Roof Height." At this point the designer has to decide what K-factor sprinkler to use that corresponds to a Minimum Operation Pressure (seventh column) that varies from 15 psi to 50 psi. In-rack sprinklers (eighth column) are not required for a maximum storage height of 20 ft (6.1 m) until the maximum roof height reaches 45 ft (13.7 m), so that is not a consideration in this example for the choice of sprinkler. The determining factor in the choice of sprinkler is primarily how much pressure is available at the facility and whether a fire pump is required. Larger K-factor sprinklers such as the 25.2 and 16.8 were designed in part to lower end-head pressures and, in many cases, remove the need for a fire pump when installing an ESFR suppression mode sprinkler system.

When performing hydraulic calculations for ESFR sprinkler systems, the design area must include the 12 most hydraulically demanding sprinklers using 4 sprinklers on each of three branchlines. The design area cannot include less than 960 ft² (89 m²). With 12 ESFR sprinklers spaced at the minimum spacing of 80 ft² (7.4 m²), the design area should always include at least 960 ft² (89 m²).

The column labeled "Hose Stream Demand" indicates that the required hose stream demand added into the hydraulic calculations is 250 gpm (946 L/min). The water supply duration must be 1 hour, as indicated in the last column.

As a special note, there are also ESFR sprinklers with specific listings that allow storage heights and building heights that exceed those indicated in NFPA 13. The designer should review manufacturer's data and listing requirements for ESFR sprinklers including the pertinent design data for each ESFR sprinkler to verify that the design criteria indicated in the manufacturer's data are being followed.

Summary of Examples 13.4, 13.5, and 13.6

It is not readily evident which of the three examples might be the best for a given storage facility using a wet sprinkler system. However, the fact that the large drop and ESFR sprinkler examples do not require in-rack sprinklers or column protection is a strong motivation to use one of those types of sprinklers. The designer will likely need to perform some rudimentary calculations in order to determine which of the two systems is the most cost-effective and whether a fire pump may be needed to provide the required sprinkler system pressure and flow rate.

In closing the discussion with regard to these examples, it is again important to emphasize that there are a large number of variables, which include storage height, ceiling height, rack storage arrangement, commodity classification and encapsulation, shelving, aisle width, sprinkler type, system type, sprinkler temperature rating, sprinkler K-factor, and minimum sprinkler operating pressure, that impact the type of sprinkler system and the design criteria for the protection of commodities stored on racks. The designer should make decisions that keep the long-term use of the facility in mind. If the lowest-cost system is installed, the owner must be made aware of the impact on sprinkler protection if there is a future change in the storage configuration and the commodities being stored.

Protection for Rack Storage of Class I through IV Commodities Over 25 ft (7.6 m) in Height

Although the method of determining in-rack sprinkler protection for commodities stored over 25 ft (7.6 m) is similar to the storage up to 25 ft, the end result may be significantly different. As storage heights get higher, the level of protection needs to increase accordingly.

In-Rack Sprinkler Water Flow Demand

The number of sprinklers to be included in the hydraulic calculations is the same as with rack storage below 25 ft (7.6 m) except that the minimum in-rack sprinkler discharge pressure is now 30 psi (2 bar).

Number of In-Rack Sprinklers to be Calculated

The number of in-rack sprinklers to be hydraulically calculated for Class I, II, or III commodities with one level of in-rack sprinklers is 6. The number of in-rack sprinklers to be hydraulically calculated for a Class IV commodity with one level of in-rack sprinklers is 8. The number of in-rack sprinklers to be hydraulically calculated for Class I, II, or III commodities with more than one level of in-rack sprinklers is 10 (5 on each two top levels). The number of in-rack sprinklers to be hydraulically calculated for a Class IV commodity with more than one level of in-rack sprinklers is 14 (7 on each two top levels).

TABLE 13.12

ESFR Protection of Rack Storage Without Solid Shelves of Class I Through Class IV
Commodities Stored Up to and Including 25 ft (7.6 m) in Height

Storage Arrangement	Commodity	Maximum Storage Height		Maximum Ceiling/Roof Height		Nominal K-Factor	Orientation	
		ft	m	ft	m			
Single-row, double-row, and multiple-row rack (no open-top containers)	Class I, II, III, or IV, encapsulated or unencapsulated	20	6.1	25	7.6	11.2	Upright	
						14.0	Upright or pendent	
						16.8	Pendent	
						25.2	Pendent	
				30	9.1	14.0	Upright or pendent	
						16.8	Pendent	
						25.2	Pendent	
				35	10.7	14.0	Upright or pendent	
						16.8	Pendent	
						25.2	Pendent	
				40	12.2	14.0	Pendent	
						16.8	Pendent	
						25.2	Pendent	
				45	13.7	14.0	Pendent	
						16.8	Pendent	
						25.2	Pendent	
		25	7.6	30	9.1	14.0	Upright or pendent	
						16.8	Pendent	
						25.2	Pendent	
				32	9.8	14.0	Upright or pendent	
						16.8	Pendent	
				35	10.7	14.0	Upright or pendent	
						16.8	Pendent	
						25.2	Pendent	
				40	12.2	14.0	Pendent	
						16.8	Pendent	
						25.2	Pendent	
				45	13.7	14.0	Pendent	
						16.8	Pendent	
						25.2	Pendent	

Source: NFPA 13, 2002, Table 12.3.2.3.1.

EXAMPLE 13.7 Protection of Class I through Class IV Commodities Stored Over 25 ft (7.6 m) in Height

This example will include a determination of ceiling and in-rack sprinkler protection for storage of Class I through IV commodities using control mode density-area standard spray sprinkler protection per NFPA 13, 2002 edition. The

Minimum Operating Pressure			Hose Stream Demand		
psi	bar	In-Rack Sprinkler Requirements	gpm	L/min	Water Supply Duration (hours)
50	3.4	No	250	946	1
50	3.4	No			
35	2.4	No			
15	1.0	No			
50	3.4	No			
35	2.4	No			
15	1.0	No			
75	5.2	No			
52	3.6	No			
20	1.4	No			
75	5.2	No			
52	3.6	No			
25	1.7	No			
90	6.2	Yes			
64	4.4	Yes			
40	2.8	No			
50	3.4	No			
35	2.4	No			
15	1.0	No			
60	4.1	No			
42	2.9	No			
75	5.2	No			
52	3.6	No			
20	1.4	No			
75	5.2	No			
52	3.6	No			
25	1.7	No			
90	6.2	Yes			
64	4.4	Yes			
40	2.8	No			

example will focus on storage of encapsulated Class IV commodities with a top of storage height of 55 ft (16.8 m) on double-row racks in a building that has a roof height of 60 ft (18.3 m). The racks do not have solid shelves, the aisle width between racks is 10 ft (3.1 m), and the typical load height on the racks is 5 ft (1.5 m). The paragraph in NFPA 13 for the protection of Class I through IV commodities stored over 25 ft (7.6 m) in height is paragraph 12.3.4.

TABLE 13.13

Double-Row Racks Without Solid Shelves, of Class I Through Class IV Commodities Stored Over 25 ft (7.6 m) in Height, Aisles 4 ft (1.2 m) or Wider

Commodity Class	In-Rack Sprinklers Approximate Vertical Spacing at Tier Nearest the Vertical Distance and Maximum Horizontal Spacing[1,2,3]		Figure
	Longitudinal Flue[4]	Face[5,6]	
I	Vertical 20 ft (6.1 m) Horizontal 10 ft (3.1 m) under horizontal barriers	None	12.3.4.4.1.1(a)
	Vertical 20 ft (6.1 m) Horizontal 10 ft (3.1 m)	Vertical 20 ft (6.1 m) Horizontal 10 ft (3.1 m)	12.3.4.4.1.1(b)
I, II, III	Vertical 10 ft (3.1 m) or at 15 ft (4.6 m) and 25 ft (7.6 m)	None	12.3.4.4.1.1(c)
	Vertical 10 ft (3.1 m) Horizontal 10 ft (3.1 m)	Vertical 30 ft (9.1 m) Horizontal 10 ft (3.1 m)	12.3.4.4.1.1(d)
	Vertical 20 ft (6.1 m) Horizontal 10 ft (3.1 m)	Vertical 20 ft (6.1 m) Horizontal 5 ft (1.5 m)	12.3.4.4.1.1(e)
	Vertical 25 ft (7.6 m) Horizontal 5 ft (1.5 m)	Vertical 25 ft (7.6 m) Horizontal 5 ft (1.5 m)	12.3.4.4.1.1(f)
	Horizontal barriers at 20 ft (6.1 m) Vertical intervals—two lines of sprinklers under barriers—maximum horizontal spacing 10 ft (3.1 m), staggered		12.3.4.4.1.1(g)
I, II, III, IV	Vertical 15 ft (4.6 m) Horizontal 10 ft (3.1 m)	Vertical 20 ft (6.1 m) Horizontal 10 ft (3.1 m)	12.3.4.4.1.1(h)
	Vertical 20 ft (6.1 m) Horizontal 5 ft (1.5 m)	Vertical 20 ft (6.1 m) Horizontal 5 ft (1.5 m)	12.3.4.4.1.1(i)
	Horizontal barriers at 15 ft (4.6 m) Vertical intervals—two lines of sprinklers under barriers—maximum horizontal spacing 10 ft (3.1 m), staggered		12.3.4.4.1.1(j)

[1]Minimum in-rack sprinkler discharge, 30 gpm (114 L/min).
[2]Water shields required.
[3]All in-rack sprinkler spacing dimensions start from the floor.
[4]Install sprinklers at least 3 in. (76.2 mm) from uprights.
[5]Face sprinklers shall not be required for a Class I commodity consisting of noncombustible products on wood pallets (without combustible containers), except for arrays shown in Figure 12.3.4.4.1.1(g) and Figure 12.3.4.4.1.1(j).
[6]In Figure 12.3.4.4.1.1(a) through Figure 12.3.4.4.1.1(j), each square represents a storage cube that measures 4 ft to 5 ft (1.2 m to 1.5 m) on a side. Actual load heights can vary from approximately 18 in. to 10 ft (0.46 m to 3.1 m). Therefore, there can be one load to six or seven loads between in-rack sprinklers that are spaced 10 ft (3.1 m) apart vertically.
[7]For encapsulated commodity, increase density 25 percent.
[8]Clearance is distance between top of storage and ceiling.
[9]See A.12.3.1.12 for protection recommendations where clearance is greater than 10 ft (3.1 m).

Source: NFPA 13, 2002, Table 12.3.4.1.1.

Ceiling Sprinkler Design Criteria

The ceiling design criteria and in-rack sprinkler locations are indicated in NFPA 13, Table 12.3.4.1.1 (referenced here as Table 13.13).

Maximum Storage Height	Stagger	Ceiling Sprinkler Operating Area		Ceiling Sprinkler Density Clearance up to 10 ft (3.1 m)[7,8,9]			
				Ordinary Temperature		High Temperature	
		ft²	m²	gpm/ft²	mm/min	gpm/ft²	mm/min
30 ft (9.1 m)	No	2000	186	0.25	10.2	0.35	14.3
Higher than 25 ft (7.6 m)	Yes			0.25	10.2	0.35	14.3
30 ft (9.1 m)	Yes	2000	186	0.3	12.2	0.4	16.3
Higher than 25 ft. (7.6 m)	Yes			0.3	12.2	0.4	16.3
	Yes			0.3	12.2	0.4	16.3
	No			0.3	12.2	0.4	16.3
	Yes			0.3	12.2	0.4	16.3
Higher than 25 ft (7.6 m)	Yes	2000	186	0.35	14.3	0.45	18.3
	No			0.35	14.3	0.45	18.3
	Yes			0.35	14.3	0.45	18.3

EXAMPLE 13.8 Ceiling Design Density

To determine the required ceiling design density, start on the left side of Table 13.13 under the commodity Class I, II, III, IV. The ceiling criteria are indicated in the columns on the far right side of the table. If ordinary temperature–rated sprinklers are used at the ceiling, the ceiling density indicated for a Class IV commodity is 0.35 gpm/ft² (14.3 mm/min). If high temperature–rated sprinklers are used, the ceiling density indicated is 0.45 gpm/ft² (18.3 mm/min). In an effort to keep the ceiling sprinkler system water demand low, ordinary temperature–rated sprinklers and a design density of 0.35 gpm/ft² (14.3 mm/min) will be used in this example. The design density must be adjusted if the commodities are encapsulated. Footnote 7 under Table 13.13 states that for encapsulated commodities the design density is to be increased 25 percent. Therefore, the ceiling density for this example is increased from 0.35 gpm/ft² (14.3 mm/min) to 0.44 gpm/ft² (17.9 mm/min). The ceiling operating area indicated in the table is 2000 ft² (186 m²).

In-Rack Sprinkler Design Criteria

In-rack sprinkler arrangement figures. Table 13.13 references several figures that indicate options for the location of required in-rack sprinklers. See the table column headed "Figure." These figures apply to double-row racks without solid shelves and with aisles 4 ft (1.2 m) or wider. The figures referenced in Table 13.13 for this example are NFPA 13 Figures 12.3.4.4.1.1(h), 12.3.4.4.1.1(i) and 12.3.4.4.1.1(j), referenced here as Figures 13.30, 13.31, and 13.32. For the protection of double-row racks, the figures include three different options, depending on number of in-rack sprinklers and whether horizontal barriers are installed within the rack structure. Sprinkler locations are indicated in the second and third columns from the left under the table heading "In-Rack Sprinklers Approximate Vertical Spacing at Tier Nearest the Vertical Distance and Maximum Horizontal Spacing."

Horizontal barriers. Figure 13.32 includes the use of horizontal barriers. A horizontal barrier is a solid barrier in the horizontal position covering the entire rack, including all flue spaces, at certain height increments to prevent vertical fire spread. The shaded areas in the plan view indicate the extent of the horizontal barriers. Note that the racks against the wall have the horizontal

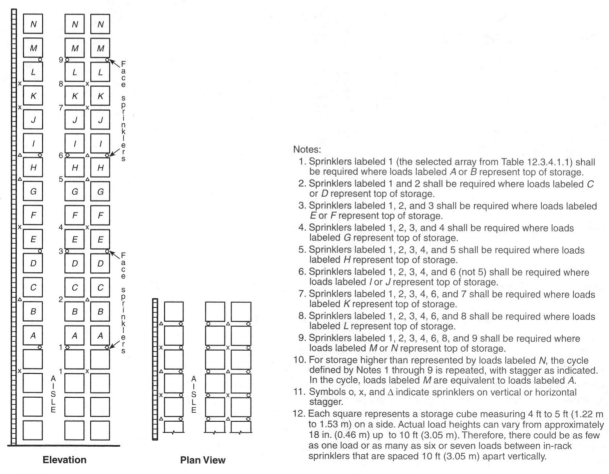

Elevation   **Plan View**

Notes:
1. Sprinklers labeled 1 (the selected array from Table 12.3.4.1.1) shall be required where loads labeled *A* or *B* represent top of storage.
2. Sprinklers labeled 1 and 2 shall be required where loads labeled *C* or *D* represent top of storage.
3. Sprinklers labeled 1, 2, and 3 shall be required where loads labeled *E* or *F* represent top of storage.
4. Sprinklers labeled 1, 2, 3, and 4 shall be required where loads labeled *G* represent top of storage.
5. Sprinklers labeled 1, 2, 3, 4, and 5 shall be required where loads labeled *H* represent top of storage.
6. Sprinklers labeled 1, 2, 3, 4, and 6 (not 5) shall be required where loads labeled *I* or *J* represent top of storage.
7. Sprinklers labeled 1, 2, 3, 4, 6, and 7 shall be required where loads labeled *K* represent top of storage.
8. Sprinklers labeled 1, 2, 3, 4, 6, and 8 shall be required where loads labeled *L* represent top of storage.
9. Sprinklers labeled 1, 2, 3, 4, 6, 8, and 9 shall be required where loads labeled *M* or *N* represent top of storage.
10. For storage higher than represented by loads labeled *N*, the cycle defined by Notes 1 through 9 is repeated, with stagger as indicated. In the cycle, loads labeled *M* are equivalent to loads labeled *A*.
11. Symbols o, x, and △ indicate sprinklers on vertical or horizontal stagger.
12. Each square represents a storage cube measuring 4 ft to 5 ft (1.22 m to 1.53 m) on a side. Actual load heights can vary from approximately 18 in. (0.46 m) up to 10 ft (3.05 m). Therefore, there could be as few as one load or as many as six or seven loads between in-rack sprinklers that are spaced 10 ft (3.05 m) apart vertically.

FIGURE 13.30
In-Rack Sprinkler Arrangement, Class I, Class II, Class III, or Class IV Commodities, Storage Height Over 25 ft (7.6 m)—Option 1.

(Source: NFPA 13, 2002, Figure 12.3.4.4.1.1(h))

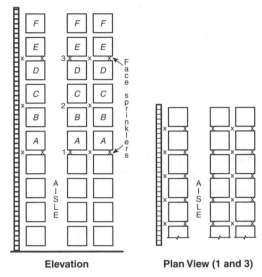

Figure 13.31

In-Rack Sprinkler Arrangement, Class I, Class II, Class III, or Class IV Commodities, Storage Height Over 25 ft (7.6 m)—Option 2.

(Source: NFPA 13, 2002, Figure 12.3.4.4.1.1(i))

Notes:

1. Sprinklers labeled 1 (the selected array from Table 12.3.4.1.1) shall be required where loads labeled *A* or *B* represent top of storage.
2. Sprinklers labeled 1 and 2 shall be required where loads labeled *C* or *D* represent top of storage.
3. Sprinklers labeled 1 and 3 shall be required where loads labeled *E* or *F* represent top of storage.
4. For storage higher than represented by loads labeled *F*, the cycle defined by Notes 2 and 3 is repeated.
5. Symbol x indicates face and in-rack sprinklers.
6. Each square represents a storage cube measuring 4 ft to 5 ft (1.22 m to 1.53 m) on a side. Actual load heights can vary from approximately 18 in. (0.46 m) up to 10 ft (3.05 m). Therefore, there could be as few as one load or as many as six or seven loads between in-rack sprinklers that are spaced 10 ft (3.05 m) apart vertically.

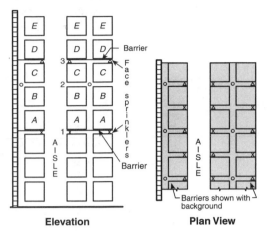

Figure 13.32

In-Rack Sprinkler Arrangement, Class I, Class II, Class III, or Class IV Commodities, Storage Height Over 25 ft (7.6 m)—Option 3.

(Source: NFPA 13, 2002, Figure 12.3.4.4.1.1(j))

Notes:

1. Sprinklers labeled 1 (the selected array from Table 12.3.4.1.1) shall be required where loads labeled *A* or *B* represent top of storage.
2. Sprinklers labeled 1 and 2 and barrier labeled 1 shall be required where loads labeled *C* represent top of storage.
3. Sprinklers and barriers labeled 1 and 3 shall be required where loads labeled *D* or *E* represent top of storage.
4. For storage higher than represented by loads labeled *E*, the cycle defined by Notes 2 and 3 is repeated.
5. Symbol Δ or x indicates sprinklers on vertical or horizontal stagger.
6. Symbol o indicates longitudinal flue space sprinklers.
7. Each square represents a storage cube measuring 4 ft to 5 ft (1.22 m to 1.53 m) on a side. Actual load heights can vary from approximately 18 in. (0.46 m) up to 10 ft (3.05 m). Therefore, there could be as few as one load or as many as six or seven loads between in-rack sprinklers that are spaced 10 ft (3.05 m) apart vertically.

barrier extending all the way to the wall. The horizontal barriers do not extend into the aisle spaces, but do cover both the transverse and longitudinal flue spaces to prevent vertical spread of fire.

Figure notes. The notes under each figure are important and must be reviewed by the designer to correctly apply the requirements indicated in the figures. The last note under each of the referenced figures states that each square represents a storage cube of 4 ft to 5 ft (1.22 m to 1.53 m) on a side. Therefore, sprinklers indicated in the figures are approximately 8 ft to 10 ft (2.44 m to 3.05 m) apart where two cubes are shown between the levels of sprinklers. The note also indicates that load heights can vary from approximately 18 in. to 10 ft (0.46 m to 3.05 m) and therefore, there could be as few as one load or as many as six or seven loads between in-rack sprinklers that are spaced 10 ft (3.05 m) apart vertically. Since the load height in this example is 5 ft, there will be two loads between in-rack sprinklers that are spaced 10 ft (3.05 m) apart vertically. If the load height in the example were 2 ft (0.6 m) high, there would be five loads between in-rack sprinklers that are spaced 10 ft (3.05 m) apart vertically.

In-Rack Sprinklers. In-rack sprinklers are primarily indicated by an "x," however, an "o," triangle, and diamond-shaped symbols are also used to indicate sprinklers in particular locations. The different symbols do not represent different types of sprinklers or sprinklers that are connected to individual piping systems. The sprinklers are to be located in the horizontal space nearest the vertical intervals shown in the figures. A vertical clear space of at least 6 in. (0.15 m) is to be maintained between the top of a tier of storage and the sprinkler deflector.

Face sprinklers. Face sprinklers cannot be located out in the aisle space and must be located within 18 in. of the face of the racks. Face sprinklers are to be located a minimum of 3 in. away from rack uprights to minimize obstruction to the spray pattern. Face sprinklers reduce the spread of fire vertically in the rack unit.

Option 1—Figure 13.30. The "o" and diamond symbols indicated on Figure 13.30 represent the location of face sprinklers that are staggered in the horizontal and vertical positions. The face sprinklers are staggered vertically every fourth cube, and horizontally every other cube. The "x" and triangle symbols represent longitudinal flue space sprinklers that are also staggered in the horizontal and vertical positions. If each cube in the elevation view of Figure 13.30 is 5 ft (1.5 m), then the top of storage at 55 ft (16.8 m) will be at the top side of the cubes labeled G. Footnote 4 under Figure 13.30 states that sprinklers labeled 1, 2, 3, and 4 shall be required where loads labeled G represent the top of storage. Since there are two sprinkler levels labeled with a 1, a total of five in-rack sprinkler levels are required for option 1. The plan view indicates that sprinklers are required at the face of each transverse flue space and also in the longitudinal flue space where it intersects each transverse flue space.

Option 2—Figure 13.31. The "x" symbol indicated in Figure 13.31 represents the location of face and flue space sprinklers. The face sprinklers are located vertically every fourth cube, and horizontally at every cube. The longitudinal flue space sprinklers are spaced vertically at every other cube. If each cube in the elevation view of Figure 13.31 is 5 ft (1.5 m), then the top of storage at 55 ft (16.8 m) will be located above the top cubes labeled F. Footnote 3 under Figure 13.31 states that sprinklers labeled 1 and 3 shall be required where loads labeled E or F represent the top of storage. Footnote 4 states that for storage higher than represented by loads labeled F, the cycle defined by Notes 2 and 3 is repeated. The total number of in-rack sprinkler levels required is three for this option: two levels as indicted by labels 1 and 3, and one additional level located 10 ft (3.05 m) above the sprinklers labeled 3. The top level is required to be the same arrangement as the sprinklers labeled 2. The plan view indicates that sprinklers are required at the face of each transverse flue space and also in the longitudinal flue space where it intersects each transverse flue space.

Option 3—Figure 13.32. The "o" symbol indicated in Figure 13.32 represents the location of longitudinal flue space sprinklers. The "x" and triangle symbols indicate face sprinklers that are staggered in the horizontal and vertical positions. If each cube in the elevation view of Figure 13.32 is 5 ft (1.5 m), then the top of storage at 55 ft (16.8 m) will be located above the top cubes labeled E. Footnote 3 under Figure 13.32 states that sprinklers and barriers labeled 1 and 3 shall be required where loads labeled D and E represent the top of storage. Footnote 4 states that for storage higher than represented by loads labeled E, the cycle defined by Notes 2 and 3 is repeated. The top of storage of 55 ft (16.8 m) is three cubes higher than the elevation shows. A third barrier and a level of sprinklers under that barrier are required at a level 45 ft (13.7 m) above the floor. The total number of barriers required is three and the number of in-rack sprinkler levels required is also three (one level under each of three barriers) for this option.

The plan view indicates that sprinklers are required at the face of each transverse flue space and also in the longitudinal flue space where it intersects every other transverse flue space.

Number of Sprinklers to Be Calculated. The hydraulic calculations for in-rack sprinklers are handled differently from the density/area method. The number of sprinklers to be included in the hydraulic calculation is indicated in NFPA 13 in paragraph 12.3.4.4.3 and depends on the number of levels of in-rack sprinklers required. For the example, 14 sprinklers (7 on each two top levels) are required where more than one level is installed in racks with Class IV commodities.

Sprinkler Discharge. The minimum in-rack sprinkler discharge rate for storage of all classes of commodities over 25 ft (7.6 m) is 30 gpm (113.6 L/min).

Pressure Balancing. The in-rack sprinkler system pressure and flow demand has to be balanced with the demand of the ceiling sprinkler system at the point where the two water flows join in the hydraulic calculation. This may cause more water to flow from either the in-rack sprinklers or the ceiling sprinklers than was originally expected. It is not a good design for the in-rack sprinkler system to create an overdischarge in the ceiling sprinkler system. Where overdischarge of a system occurs, the designer modifies pipe sizes to minimize the pressure imbalance and resultant overdischarge.

Hose Stream and Water Supply Duration. The hose stream demand for storage of Class I through IV commodities is indicated in NFPA 13 Table 12.3.4.1.5, shown here as Table 13.14. For the example, the total combined inside and outside hose is 500 gpm (1900 L/min) and the water supply duration is 120 minutes.

TABLE 13.14

Hose Stream Demand and Water Supply Duration Requirements for Rack Storage of Class I through Class IV Commodities Stored Above 25 ft (7.6 m) in Height

Commodity Classification	Storage Height		Inside Hose		Total Combined Inside and Outside Hose		Duration (minutes)
	ft	m	gpm	L/min	gpm	L/min	
Class I, II, and III	>25	>7.6	0, 50, or 100	0, 190, 380	500	1900	90
Class IV	>25	>7.6	0, 50, or 100	0, 190, 380	500	1900	120

Source: NFPA 13, 2002, Table 12.3.4.1.5.

Summary of Options. The designer must consider each option indicated in Table 13.14 and in Figures 13.30, 13.31, and 13.32. To make a simple comparison of the three options, a count of the number of in-rack sprinklers included within four transverse flue space sections was made. Option 1 requires a total of 14 sprinklers and no horizontal barriers, option 2 requires a total of 28 sprinklers and no horizontal barriers, and option 3 requires a total of 12 sprinklers and three horizontal barriers. Options 1 and 3 have significantly fewer sprinklers than option 2, and option 1 has another advantage in that it does not require horizontal barriers. The number of sprinklers is only one of several considerations, such as differences in cost, water demand, sprinkler location, horizontal barrier location, piping arrangement, and protection of sprinklers from material handling equipment. Each of the options has the same requirement for the ceiling design density of 0.44 gpm/ft^2 (17.9 mm/min).

ESFR and Large Drop Sprinklers. There are no large drop sprinkler options for the protection of rack storage of Class I through IV commodities over 30 ft (9.1 m). The options for the use of ESFR sprinklers are available only up to storage heights of 40 ft (12.2 m). Since the example uses a storage height of 55 ft (16.8 m), neither large drop or ESFR sprinkler systems are allowed per NFPA 13.

Protection for Rack Storage of Plastics Commodities

Plastics present a greater fire protection challenge because their combustion can produce about 1.5 to 3 times as much heat per unit of weight as wood or paper. In addition, plastics can burn at a much faster rate, resulting in a very challenging, high-heat-release-rate fire [3]. The protection of plastics is determined by the use of a decision tree, as shown in Figure 13.33.

In the decision tree, plastics are classified into three groups: Group A, Group B, and Group C. These groups contain the following materials:

Group A

ABS (acrylonitrile-butadiene-styrene copolymer)

Acetal (polyformaldehyde)

Acrylic (polymethyl methacrylate)

Butyl rubber

FIGURE 13.33
Decision Tree.
(Source: NFPA 13, 2002, Figure 12.3.1.1)

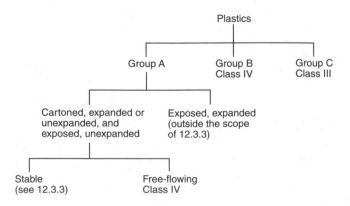

EPDM (ethylene-propylene rubber)

FRP (fiberglass-reinforced polyester)

Natural rubber (if expanded)

Nitrile-rubber (acrylonitrile-butadiene-rubber)

PET (thermoplastic polyester)

Polybutadiene

Polycarbonate

Polyester elastomer

Polyethylene

Polypropylene

Polystyrene

Polyurethane

PVC (polyvinyl chloride—highly plasticized, with plasticizer content greater than 20 percent; rarely found)

SAN (styrene acrylonitrile)

SBR (styrene-butadiene rubber)

Group B

Cellulosics (cellulose acetate, cellulose acetate butyrate, ethyl cellulose)

Chloroprene rubber

Fluoroplastics (ECTFE—ethylene-chlorotrifluoro-ethylene copolymer; ETFE—ethylene-tetrafluoroethylene-copolymer; FEP—fluorinated ethylene-propylene copolymer)

Natural rubber (not expanded)

Nylon (nylon 6, nylon 6/6)

Silicone rubber

Group C

Fluoroplastics (PCTFE—polychlorotrifluoroethylene; PTFE—polytetrafluoroethylene)

Melamine (melamine formaldehyde)

Phenolic

PVC (polyvinyl chloride—flexible PVCs with plasticizer content up to 20 percent)

PVDC (polyvinylidene chloride)

PVDF (polyvinylidene fluoride)

PVF (polyvinyl fluoride)

Urea (urea formaldehyde)

By using the decision tree the designer can determine whether the storage of some types of plastics can be protected the same as Class III or Class IV commodities. Starting with the first level on the decision tree, it indicates that

Group B plastics are protected the same as Class IV commodities and Group C plastics are protected the same as Class III commodities. Group A plastics are divided into two groups on the second level of the decision tree. Where the chart lists Stable, it refers to Group A plastic commodities where collapse, spillage of contents, or leaning of stacks across flue spaces is not likely to occur soon after initial fire development. Exposed, expanded plastics are indicated to be outside the scope of this chapter in NFPA 13.

Those plastics not in packaging or coverings that absorb water or otherwise appreciably retard the burning hazard of the commodity are considered exposed Group A plastic commodities. (Paper wrapped or encapsulated, or both, should be considered exposed.)

Expanded (foamed or cellular) plastics are those plastics the density of which is reduced by the presence of numerous small cavities (cells), interconnecting or not, dispersed throughout their mass.

Cartoned, expanded or unexpanded, and exposed, unexpanded plastics are further divided on the third level of the decision tree. Free-flowing Group A plastics are to be protected the same as Class IV commodities. Free-flowing plastic materials are those plastics that fall out of their containers during a fire, fill flue spaces, and create a smothering effect on the fire. Examples include powder, pellets, flakes, or random-packed small objects [e.g., razor blade dispensers, 1-oz to 2-oz (28-g to 57-g) bottles].

Again, NFPA 13 uses multiple tables and figures to break up particular storage heights and type of plastics commodity classifications. The tables also reference the use of large drop, ESFR, and standard spray sprinklers. The use of overhead large drop and ESFR sprinkler systems provides protection for a wide variety of commodities without the use of in-rack sprinklers if installed in accordance with NFPA 13 criteria. Using a large drop or ESFR sprinkler system may make discussions with the owner and the future use of the facility a little easier. However, the use of solid shelves or the storage of higher hazard commodities such as flammable and combustible liquids must be recognized and the appropriate protection provided.

Example 13.9 Protection for Rack Storage of Plastics Commodities Over 25 ft (7.6 m) in Height

This example will include a determination of ceiling and in-rack storage protection of plastic commodities using control mode density-area standard spray sprinkler protection per NFPA 13. The example will focus on storage of Group A plastic commodities with a top of storage height of 55 ft (16.76 m) on double-row racks in a building that has a roof height of 60 ft (18.3 m). The typical load height on the racks is 5 ft (1.5 m).

Ceiling Sprinkler Density. The required ceiling design density varies based on the height of commodities stored above the highest level of in-rack sprinklers. If the storage height above the top level of in-rack sprinklers in a rack storage unit is 5 ft (1.5 m) or less than a density of 0.30 gpm/ft² (12.2 mm/min) over a design area of 2000 ft² (186 m²) is required. If the storage height above the top level of in-rack sprinklers is 5 ft to 10 ft (1.5 to 3.1 m), then a density of 0.45 gpm/ft² (18.3 mm/min) is required. This is indicated in Table 13.15, which is Table 12.3.5.1.1 in NFPA 13. Where a clearance of over 10 ft (3.1 m)

TABLE 13.15

Control Mode Density-Area Sprinkler Discharge Criteria for Single-, Double-, and Multiple-Row Racks of Plastics Commodities with Storage Over 25 ft (7.6 m) in Height

Storage Height above Top Level In-Rack Sprinklers	Ceiling Sprinklers Density (gpm/ft^2)
5 ft or less	0.30/2000
Over 5 ft up to 10 ft	0.45/2000

Source: NFPA 13, 2002, Table 12.3.5.1.1.

exists, another layer of in-rack sprinklers is required to bring the clearance to below 10 ft (3.1 m).

Figures. NFPA 13 references several figures to indicate the required location of in-rack sprinklers. The figures for the protection of double-row racks include two different protection options, depending on whether horizontal barriers are installed within the rack structure or not. As defined previously, a horizontal barrier is a solid barrier in the horizontal position covering the entire rack, including all flue spaces, at certain height increments to prevent vertical fire spread. The figures for reference in this example are Figures 13.34 and 13.35. These figures apply to double-row racks without solid shelves and with a maximum of 10 ft (3.1 m) between the top of storage and the ceiling.

Horizontal Barriers. Figure 13.34 includes the use of horizontal barriers. The shaded areas in the plan view indicate the extent of the horizontal barriers. Note that the racks against the wall have the horizontal barrier extending all the way to the wall. The horizontal barriers do not extend into the aisle

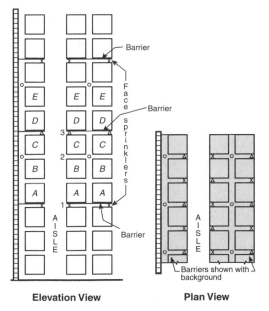

Notes:
1. Sprinklers and barriers labeled 1 shall be required where loads labeled *A* or *B* represent top of storage.
2. Sprinklers labeled 1 and 2 and barriers labeled 1 shall be required where loads labeled *C* represent top of storage.
3. Sprinklers and barriers labeled 1 and 3 shall be required where loads labeled *D* or *E* represent top of storage.
4. For storage higher than represented by loads labeled *E*, the cycle defined by Notes 2 and 3 is repeated.
5. Symbol Δ or x indicates face sprinklers on vertical or horizontal stagger.
6. Symbol o indicates longitudinal flue space sprinklers.
7. Each square represents a storage cube measuring 4 ft to 5 ft (1.22 m to 1.53 m) on a side. Actual load heights can vary from approximately 18 in. (0.46 m) up to 10 ft (3.05 m). Therefore, there could be as few as one load or as many as six or seven loads between in-rack sprinklers that are spaced 10 ft (3.05 m) apart vertically.

Elevation View   **Plan View**

FIGURE 13.34
In-Rack Sprinkler Arrangement, Group A Plastic Commodities, Storage Height Over 25 ft (7.6 m)—Option 1.

(Source: NFPA 13, 2002, Figure 12.3.5.1.2(a))

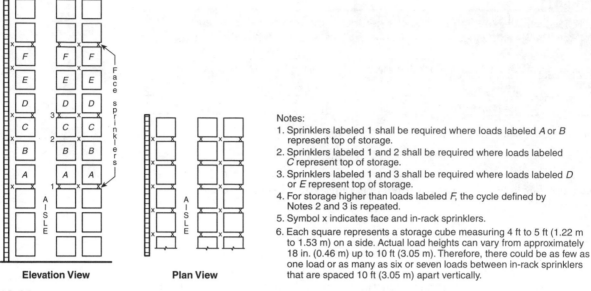

Elevation View   **Plan View**

Notes:
1. Sprinklers labeled 1 shall be required where loads labeled *A* or *B* represent top of storage.
2. Sprinklers labeled 1 and 2 shall be required where loads labeled *C* represent top of storage.
3. Sprinklers labeled 1 and 3 shall be required where loads labeled *D* or *E* represent top of storage.
4. For storage higher than loads labeled *F*, the cycle defined by Notes 2 and 3 is repeated.
5. Symbol x indicates face and in-rack sprinklers.
6. Each square represents a storage cube measuring 4 ft to 5 ft (1.22 m to 1.53 m) on a side. Actual load heights can vary from approximately 18 in. (0.46 m) up to 10 ft (3.05 m). Therefore, there could be as few as one load or as many as six or seven loads between in-rack sprinklers that are spaced 10 ft (3.05 m) apart vertically.

FIGURE 13.35
In-Rack Sprinkler Arrangement, Group A Plastics Commodities, Storage Height Over 25 ft (7.6 m)—Option 2.
(Source: NFPA 13, 2002, Figure 12.3.5.1.2(b))

spaces, but do cover both the transverse and longitudinal flue spaces to prevent vertical spread of fire.

Figure Notes. The notes under each of these figures are important and must be reviewed by the designer to correctly apply the requirements indicated in the figures. The last note under each of the referenced figures states that each square represents a storage cube of 4 ft to 5 ft (1.22 m to 1.53 m) on a side. Therefore, sprinklers indicated in the figures are approximately 8 ft to 10 ft (2.4 m to 3.1 m) apart where two cubes are shown between the levels of sprinklers. The note also indicates that load heights can vary from approximately 18 inches (0.5 m) to 10 ft (3.1 m) and therefore, there could be as few as one load or as many as six or seven loads between in-rack sprinklers that are spaced 10 ft (3.1 m) apart vertically. Since the load height in this example is 5 ft (1.5 m), there will be two loads between in-rack sprinklers that are spaced 10 ft (3.1 m) apart vertically. If the load height in the example were 2 ft (0.6 m) high, there would be five loads between in-rack sprinklers that are spaced 10 ft (3.1 m) apart vertically.

Sprinklers. The sprinklers are primarily indicated by an "x"; however, an "o" or a triangle-shaped symbol are also used to indicate sprinklers in particular locations. The different symbols do not represent different types of sprinklers. The "o" indicated on Figure 13.34 represents the location of longitudinal flue space sprinklers, and the triangle indicates sprinklers that are staggered in the horizontal position. The triangles labeled as number 3 in Figure 13.34 elevation view are staggered horizontally. The elevation view shows sprinklers labeled 1 (x's) and sprinklers labeled 3 (triangles) at every third cube. Referencing the plan view, it is evident that the sprinklers represented by x's and triangles are staggered every other cube horizontally. The sprinklers are to be located in the horizontal space nearest the vertical intervals shown in the figures.

A vertical clear space of at least 6 in. (150 mm) is to be maintained between the top of a tier of storage and the sprinkler deflector.

ESFR and Large Drop Sprinklers. There are no large drop sprinkler options for the protection of rack storage of plastics over 25 ft (7.6 m). The options for the use of ESFR sprinklers are available only up to storage heights of 40 ft (12.2 m). Since the example uses a storage height of 55 ft (16.76 m), neither a large drop nor an ESFR sprinkler system is allowed.

Summary of Example. The designer needs to consider both options. Option 1 that includes horizontal barriers has significantly fewer in-rack sprinklers, especially in the longitudinal flue space. The cost of the barriers versus the cost of the additional in-rack sprinklers required in option 2 should be considered. Damage to the in-rack sprinklers in both options must be considered, but option 2 has more in-rack sprinklers.

For both options face sprinklers are required. Face sprinklers cannot be located out in the aisle space, and must be located within 18 in. (1.5 m) of the face of the racks. Face sprinklers are to be located a minimum of 3 in. (7.6 cm) away from rack uprights to minimize obstruction of the spray pattern. Face sprinklers reduce the spread of fire vertically in the rack unit.

In-Rack Sprinkler Water Flow Demand

The hydraulic calculations for in-rack sprinklers are handled differently from the density/area method. The number of sprinklers to be included in the hydraulic calculation is indicated in NFPA 13 and depends on the number of levels of in-rack sprinklers required. The minimum in-rack sprinkler discharge rate for storage over 25 ft (7.6 m) is 30 gpm (113.5 L/min).

Number of Sprinklers to Be Calculated

The number of sprinklers to be hydraulically calculated for plastics commodities with one level of in-rack sprinklers is 8. Where more than one level of in-rack sprinklers is required, such as in the example for either option 1 or 2, the number of sprinklers to be hydraulically calculated is 14 (7 on each top two levels).

Pressure Balancing

The in-rack sprinkler system pressure and flow demand has to be balanced, after adjusting pipe sizes in an effort to minimize the pressure imbalance, with the demand of the ceiling sprinkler system at the point where the two water flows join in the hydraulic calculation. This may cause more water to flow from either the in-rack sprinklers or the ceiling sprinklers than was originally expected. It is not a good design for the in-rack sprinkler system to create an overdischarge in the ceiling sprinkler system. Where overdischarge of a system occurs, the designer modifies pipe sizes to minimize the pressure imbalance and resultant overdischarge.

The hose stream demand for storage of plastics over 25 ft (7.6 m) is 500 gpm (1892 L/min) and the water supply duration is 120 minutes.

EXAMPLE 13.10 Protection for Rack Storage of Plastics Commodities up to 25 ft (7.6 m) in Height

This example will include a determination of ceiling and in-rack storage protection using standard spray sprinklers at the ceiling of a heated building. Further examples will include specific application control mode and early suppression fast-response (ESFR) sprinklers for comparison. The examples will focus on storage of stable, cartoned, unexpanded, encapsulated Group A plastics commodities with a top of storage height of 25 ft (7.6 m) on double-row racks in a building that has a ceiling/roof height of 30 ft (9.1 m). The aisles between the racks are 10 ft (3.1 m) wide. Typical pallet load height is 5 ft (1.53 m) stored in racks without solid shelves.

Sprinkler protection for rack storage of plastics up to 25 ft (7.6 m) is located in NFPA 13, subsection 12.3.3, "Protection Criteria for Rack Storage of Plastics Commodities Stored Up to and Including 25 ft (7.6 m) in Height."

EXAMPLE 13.11 Standard Spray Sprinklers

Ceiling Sprinkler Water Demand. The ceiling sprinkler water demand is determined by choosing the appropriate figure from NFPA 13, Figures 12.3.3.1.5(a) through 12.3.3.1.5(f). Figure 12.3.3.1.5(f) applies to this example because it is for storage up to 25 ft (7.6 m) and for ceiling clearances between 5 to 10 ft (1.5 to 3.1 m). See NFPA 13 Figure 12.3.3.1.5(f) (referenced here as Figure 13.36). The ceiling clearance for this example is 5 ft (1.5 m) [30 ft (9.1 m) less 25 ft (7.6 m)]. The text at the top of Figure 13.36 indicates a required design density at the ceiling of 0.30 gpm/ft^2 per 2000 ft^2 (12.2 mm/min per 186 m^2).

Figure Notes. The notes under each NFPA 13 figure are important and must be reviewed by the designer to correctly apply the requirements indicated in the figures. Note 1 indicates design criteria for in-rack sprinklers where required. Note 2 at the bottom of Figure 13.36 indicates that "ceiling-only" protection shall not be permitted for this storage configuration. Note 3 indicates an option to use a K-factor 16.8 spray sprinkler that is specifically listed for storage use where the ceiling height in the protected area does not exceed 30 ft (9.1 m). The ceiling height in this example is 30 ft (9.1 m), so the K-factor 16.8 "ceiling-only" design criterion is allowed. This application will be reviewed in the next example. Note 4 states that each square in the figure represents a storage cube of 4 ft to 5 ft (1.22 m to 1.53 m) on a side. Therefore, sprinklers indicated in the figure are approximately 8 ft to 10 ft apart where two cubes are shown between the levels of sprinklers. Note 4 also indicates that load heights can vary from approximately 18 in. (0.46 m) up to 10 ft (3.05 m). Therefore, there could be as few as one load or as many as six or seven loads between in-rack sprinklers that are spaced 10 ft apart vertically. Since the load height in this example is 5 ft (1.53 m), there will be two loads between in-rack sprinklers that are spaced 10 ft (3.05 m) apart vertically. If the load height in the example were 2 ft (0.61 m) high, there would be five loads between in-rack sprinklers that are spaced 10 ft (3.05 m) apart vertically.

Figure Elevation and Plan View. Figure 13.36 includes a plan view and an elevation view. In each of these views the figure indicates sprinkler positions for a single-row rack, double-row rack, and 3-deep and 4-deep multiple-row racks.

0.30 gpm/ft² per 2000 ft²
(12.2 mm/min per 186 m²)

5 ft to 10 ft (1.5 m to 3.1 m) ceiling clearance
See Notes 1, 2, and 3

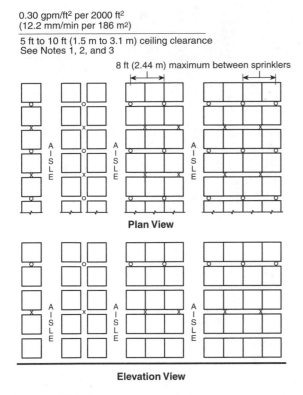

Plan View

Elevation View

Notes:

1. Two levels of in-rack sprinklers [½ in. or ¹⁷⁄₃₂ in. (12.7 mm or 13.5 mm) operating at 15 psi (1.03 bar) minimum] installed on 8 ft to 10 ft (2.5 m to 3.12 m) spacings located as indicated and staggered in the transverse flue space.

2. Ceiling-only protection shall not be permitted for this storage configuration.

3. Where K-16.8 spray sprinklers listed for storage use are installed at the ceiling, the in-rack sprinklers shall not be required, provided the ceiling sprinklers criteria is increased to 0.8 gpm/ft² over 2000 ft² (32.6 mm/min over 186 m²) for wet systems and 4500 ft² (419 m²) for dry systems and the ceiling height in the protected area does not exceed 30 ft (9.1 m).

4. Each square represents a storage cube measuring 4 ft to 5 ft (1.22 m to 1.53 m) on a side. Actual load heights can vary from approximately 18 in. (0.46 m) up to 10 ft (3.05 m). Therefore, there could be as few as one load or as many as six or seven loads between in-rack sprinklers that are spaced 10 ft (3.05 m) apart vertically.

FIGURE 13.36
25-ft (7.6-m) Storage; 5-ft to 10-ft (1.5-m to 3.1-m) Ceiling Clearance.

(Source: NFPA 13, 2002, Figure 12.3.3.1.5(f))

This example uses the portion of the figure that indicates a double-row rack (second group of racks from the left). The elevation view indicates one level of sprinklers at approximately 10 ft (3.1 m) high and a second level of sprinklers at approximately 20 ft (6.2 m) high. The plan view indicates that a lower-level sprinkler "x" is staggered with an upper-level sprinkler "o" at every other pallet load. All the sprinklers are located in the longitudinal flue space.

In-Rack Sprinklers. The requirements for in-rack sprinklers begin in NFPA 13, subsection 12.3.3.4. This section includes criteria for in-rack sprinkler spacing, clearance from storage, water demand, and discharge pressure for rack storage of plastic commodities stored up to and including 25 ft (7.6 m) in height.

In-Rack Sprinkler Spacing. The sprinkler spacing is required to follow the spacing indicated in NFPA Figures 12.3.3.1.5(a) through 12.3.3.1.5(f). Note 1 in Figure 13.36 indicates that two levels of in-rack sprinklers are required. The sprinklers can be either ½-in. or ¹⁷⁄₃₂-in. (12.7 mm or 13.5 mm) sprinklers operating at 15 psi (1.03 bar) minimum. The sprinklers must be installed on 8 ft to 10 ft (2.5 m or 3.12 m) spacings and located as indicated in the figure and staggered in the transverse flue space (3- and 4-deep multiple-row racks only). The sprinklers are indicated by an "x" or an "o"-shaped symbol. The different symbols do not represent different types of sprinklers. The "o" indicated on Figure 13.36 represents the location of upper-level in-rack sprinklers, and the "x" indicates lower-level in-rack sprinklers.

TABLE 13.16

Hose Stream Demand and Water Supply Duration Requirements for Rack Storage of Plastics Commodities Stored Up to and Including 25 ft (7.6 m) in Height

Commodity Classification	Storage Height		Inside Hose		Total Combined Inside and Outside Hose		Duration (minutes)
	ft	m	gpm	L/min	gpm	L/min	
Plastic	>5 up to 20	>1.5 up to 6.1	0, 50, or 100	0, 190, or 380	500	1900	120
	>20 up to 25	>6.1 up to 7.6	0, 50, or 100	0, 190, or 380	500	1900	150

Source: NFPA 13, 2002, Table 12.3.3.1.11.

Clearance from Storage. NFPA 13, paragraph 12.3.3.4.2.1, calls for a minimum of 6-in. (152.4 mm) vertical clear space to be maintained between the in-rack sprinkler deflectors and the top of a tier of storage. This is to allow a portion of the water from a discharging sprinkler to spray over the top of the pallet load.

Water Demand. NFPA 13, paragraph 12.3.3.4.3, states that the water demand for sprinklers installed in racks shall be based on simultaneous operation of the most hydraulically remote 8 sprinklers where only one level is installed in racks. Where more than one level is installed in racks, the operation of the most hydraulically remote 14 sprinklers (7 on each top two levels) must be used. For this example, the water demand for the simultaneous operation of the most hydraulically remote 14 sprinklers (7 on each of the two levels) is to be included in the pipe-sizing hydraulic calculations.

Sprinkler Discharge Pressure. NFPA 13, paragraph 12.3.3.4.4, states that sprinklers in racks shall discharge at not less than 15 psi (1 bar) for all classes of commodities.

Hose Stream Demand and Water Supply Duration. NFPA 13 Table 12.3.3.1.11 (referenced here as Table 13.16) indicates the hose stream demand and water supply duration requirements. For this example, the total combined inside and outside hose is 500 gpm (1900 L/min) and the duration is 150 min.

Pressure Balancing. The in-rack sprinkler system pressure and flow demand has to be balanced in an effort to minimize the pressure imbalance with the demand of the ceiling sprinkler system at the point where the two water flows join in the hydraulic calculation. This may cause more water to flow from either the in-rack sprinklers or the ceiling sprinklers than was originally expected. It is not a good design for the in-rack sprinkler system to create an overdischarge in the ceiling sprinkler system. Where overdischarge of a system occurs, the designer modifies pipe sizes to minimize the pressure imbalance and resultant overdischarge.

EXAMPLE 13.12 Specific Application Control Mode Sprinklers

The same rack storage example will be considered of Group A plastics commodities with a top of storage height of 25 ft (7.6 m) on double-row racks in a building that has a roof height of 30 ft (9.1 m) using specific application

TABLE 13.17

Specific Application Control Mode (16.8 K-factor) Sprinkler Design Criteria for Single-, Double-, and Multiple-Row Racks Without Solid Shelves of Plastics Commodities Stored Up to and Including 25 ft (7.6 m) in Height

Commodity Class	Maximum Storage Height		Maximum Building Height		Type of System	Number of Design Sprinklers by Minimum Operating Pressure		Hose Stream Demand		Water Supply Duration (hours)
	ft	m	ft	m		10 psi (0.7 bar)	22 psi (1.5 bar)	gpm	L/min	
Cartoned or exposed unexpanded plastics	25	7.6	30	9.1	Wet	—	15	500	1900	2

Source: NFPA 13, 2002, 12.3.3.2.1(b).

control mode sprinklers. Specific application control mode sprinklers are specifically listed sprinklers with K-factors of 16.8. These control mode sprinklers have large water droplets that have the ability to penetrate a high-velocity fire plume to reach a fire. To start the example, reference NFPA 13 Table 12.3.3.2.1(b) (referenced here as Table 13.17).

The use of specific application control mode sprinklers for the protection of plastics is only allowed with wet sprinkler systems and for storage up to 25 ft (7.6 m) in a building with a maximum ceiling/roof height of 30 ft (9.1 m).

The heading "Number of Design Sprinklers by Minimum Operating Pressure" indicates that the pipe-sizing hydraulic calculations must include 15 sprinklers flowing at a minimum pressure of 22 psi (1.5 bar). In-rack sprinklers are not required. A hose stream demand of 500 gpm must be included in the calculations. The required water supply duration is 2 hours.

The end result is not a true density/area calculation, but rather a number of sprinklers to calculate at a given pressure. When performing the hydraulics calculations, the input to the program will need to include the minimum flow and pressure for each sprinkler instead of the usual density and area input. A restriction is that the maximum discharge pressure at the hydraulically remote sprinkler cannot be over 95 psi (6.5 bar). At pressures over 95 psi (6.5 bar), the water droplet size and pattern are negatively impacted. Even though specific application control mode sprinklers do not follow the design/area hydraulic calculation method, the design area is still required to be a rectangular area having a dimension parallel to the branchlines of at least 1.2 times the square root of the design area. Any fractional sprinkler is to be included in the design area. A benefit of using specific application control mode sprinklers in this application is that building steel protection is not required.

EXAMPLE 13.13 Early Suppression Fast-Response (ESFR) Sprinklers

ESFR sprinklers are suppression mode sprinklers that operate with high water flow rates and have the largest K-factors (therefore, the largest orifice) of any sprinklers manufactured. The sprinkler orifice is as large as 1 in. in diameter and the sprinkler itself is quite large when compared to standard sprinklers. ESFR sprinklers are intended for high-piled storage up to 40 ft (12.2 m) in buildings up to 45 (13.7 m) feet high. They are designed to react quickly to growing fires and deliver a heavy discharge. ESFR sprinklers are favored by building owners because they may remove the requirement for in-rack sprinklers.

Per NFPA 13, ESFR protection does not apply to the protection of racks with solid shelves or to protection of rack storage of combustible, open-top cartons or containers. The minimum operating pressure for ESFR sprinklers is 15 psi (1 bar), but the designer must follow the minimum pressures indicated in NFPA 13 and the manufacturer's catalog cutsheets. Longitudinal and transverse flue spaces are required regardless of the storage height.

The same rack storage example will be used of plastics commodities with a top of storage height of 25 ft (7.6 m) on double-row racks in a building that

TABLE 13.18

ESFR Protection of Rack Storage Without Solid Shelves of Plastics Commodities Stored Up to and Including 25 ft (7.6 m) in Height

Storage Arrangement	Commodity	Maximum Storage Height		Maximum Ceiling/Roof Height		Nominal K-Factor	Orientation
		ft	m	ft	m		
Single-row, double-row and multiple-row rack (no open-top containers)	Cartoned unexpanded	20	6.1	25	7.6	11.2	Upright
						14.0	Upright or pendent
						16.8	Pendent
						25.2	Pendent
				30	9.1	14.0	Upright or pendent
						16.8	Pendent
						25.2	Pendent
				35	10.7	14.0	Upright or pendent
						16.8	Pendent
						25.2	Pendent
				40	12.2	14.0	Pendent
						16.8	Pendent
						25.2	Pendent
				45	13.7	14.0	Pendent
						16.8	Pendent
						25.2	Pendent
		25	7.6	30	9.1	14.0	Upright or pendent
						16.8	Pendent
						25.2	Pendent
				32	9.8	14.0	Upright or pendent
						16.8	Pendent
				35	10.7	14.0	Upright or pendent
						16.8	Pendent
						25.2	Pendent
				40	12.2	14.0	Pendent
						16.8	Pendent

has a roof height of 30 ft (9.1 m) using ESFR sprinklers. This example refers to NFPA 13, Table 12.3.3.3.1, referenced here as Table 13.18.

Starting on the left side of the table with the column labeled "Storage Arrangement" and then proceeding through the columns to the right at each step, the first choice that can be made for this example is in the second column labeled "Commodity." This example is using a cartoned, unexpanded commodity. In the third column the maximum storage height is 25 ft (7.6 m), and in the fourth column the maximum ceiling/roof height is 30 ft (9.1 m). At this point, the designer has to decide what K-factor (fifth column) ESFR sprinkler to use that corresponds to a minimum operating pressure (seventh column) that varies from 15 psi to 50 psi (1.0 bar to 3.4 bar). The determining

Minimum Operating Pressure			Hose Stream Demand		
psi	bar	In-Rack Sprinkler Requirements	gpm	L/min	Water Supply Duration (hours)
50	3.4	No	250	946	1
50	3.4	No			
35	2.4	No			
15	1.0	No			
50	3.4	No			
35	2.4	No			
15	1.0	No			
75	5.2	No			
52	3.6	No			
20	1.4	No			
75	5.2	No			
52	3.6	No			
25	1.7	No			
90	6.2	Yes			
63	4.3	Yes			
40	2.8	No			
50	3.4	No			
35	2.4	No			
15	1.0	No			
60	4.1	No			
42	2.9	No			
75	5.2	No			
52	3.6	No			
20	1.4	No			
75	5.2	No			
52	3.6	No			

continues

TABLE 13.18

(Continued)

Storage Arrangement	Commodity	Maximum Storage Height		Maximum Ceiling/ Roof Height		Nominal K-Factor	Orientation	
		ft	m	ft	m			
Single-row, double-row and multiple-row rack (no open-top containers)						25.2	Pendent	
				45	13.7	14.0	Pendent	
						16.8	Pendent	
						25.2	Pendent	
	Exposed unexpanded	20	6.1	25	7.6	14.0	Pendent	
						16.8	Pendent	
				30	9.1	14.0	Pendent	
						16.8	Pendent	
				35	10.7	14.0	Pendent	
						16.8	Pendent	
				40	12.2	14.0	Pendent	
						16.8	Pendent	
				45	13.7	14.0	Pendent	
						16.8	Pendent	
		25	7.6	30	9.1	14	Pendent	
						16.8	Pendent	
				32	9.8	14.0	Pendent	
						16.8	Pendent	
				35	10.7	14.0	Pendent	
						16.8	Pendent	
				40	12.2	14.0	Pendent	
						16.8	Pendent	
				45	13.7	14.0	Pendent	
						16.8	Pendent	
	Cartoned expanded	20	6.1	25	7.6	14.0	Upright or pendent	
						16.8	Pendent	
				30	9.1	14.0	Upright or pendent	
						16.8	Pendent	
		25	7.6	30	9.1	14.0	Upright or pendent	
						16.8	Pendent	
				32	9.8	14.0	Pendent	
						16.8	Pendent	

Source: NFPA 13, 2002, Table 12.3.3.3.1.

	Minimum Operating Pressure			Hose Stream Demand		Water Supply Duration (hours)
	psi	bar	In-Rack Sprinkler Requirements	gpm	L/min	
	25	1.7	No	250	946	1
	90	6.2	Yes			
	63	4.3	Yes			
	40	2.8	No			
	50	3.4	No			
	35	2.4	No			
	50	3.4	No			
	35	2.4	No			
	75	5.2	No			
	52	3.6	No			
	75	5.2	No			
	52	3.6	No			
	90	6.2	Yes			
	63	4.3	Yes			
	50	3.4	No			
	35	2.4	No			
	60	4.1	No			
	42	2.9	No			
	75	5.2	No			
	52	3.6	No			
	75	5.2	No			
	52	3.6	No			
	90	6.2	Yes			
	63	4.3	Yes			
	50	3.4	No			
	35	2.4	No			
	50	3.4	No			
	35	2.4	No			
	50	3.4	No			
	35	2.4	No			
	60	4.1	No			
	42	2.9	No			

factor in the choice of sprinkler is primarily how much pressure is available at the facility and whether a fire pump is required. Larger K-factor sprinklers such as the 25.2 and 16.8 were designed in part to lower end-head pressures and, in many cases, remove the need for a fire pump when installing an ESFR suppression mode sprinkler system. In-rack sprinklers (eighth column) are not required for a maximum storage height of 25 ft (7.6 m) until the maximum roof height reaches 45 ft (13.7 m), so that is not a consideration in this example for the choice of sprinkler.

When performing hydraulic calculations for ESFR sprinkler systems, the design area must include the 12 most hydraulically demanding sprinklers using 4 sprinklers on each of three branchlines. The design area can not include less than 960 ft². With 12 ESFR sprinklers spaced at the minimum spacing of 80 ft², the design area should always include at least 960 ft².

The column labeled "Hose Stream Demand" indicates that the required hose stream demand added into the hydraulic calculations is 250 gpm (946 L/min). The water supply duration must be 1 hour, as indicated in the last column.

Summary of Examples for Storage of Plastics Commodities up to 25 ft (7.6 m)

It is not readily evident which of the three examples might be the best for a given storage facility using a wet sprinkler system. However, the fact that the specific application control mode and ESFR sprinkler examples do not require in-rack sprinklers or steel protection is a strong motivation to use one of those types of sprinklers. The designer will likely need to perform some rudimentary hydraulic calculations in order to determine which of the two systems is the most cost-effective and whether a fire pump may be needed to provide the required sprinkler system pressure and flow rate.

Protection for Rack Storage of Rubber Tires, Baled Cotton, and Roll Paper

There are specific sections and tables in NFPA 13 for rack protection of rubber tire storage, baled cotton storage, and roll paper storage. This section provides an introduction to the sprinkler protection concepts for storage of these commodities.

Rubber Tire Storage

Rubber tires stored in racks present a unique challenge because their interior surfaces burn vigorously and are not easily reached by water from sprinkler systems. Because tires are not packaged, they contribute their own circular air space to the storage, forming considerable horizontal or vertical flues [3].

NFPA 13 protection requirements depend in part on the method in which the tires are stored. Storage methods include tires stored on-side, on-tread, and laced. On-side tires are tires stored horizontally or flat. On-tread tire storage is defined as tires stored vertically or on their treads. Laced tire storage is defined as tires stored where the sides of the tires overlap, creating a woven or laced appearance. See Figures 13.37, 13.38, and 13.39. The protection criteria in NFPA 13 apply to buildings with ceiling slopes not exceeding 2 in 12 (16.7 percent).

FIGURE 13.37
On-Tread Storage of Tires in
Open, Portable Racks.

(Courtesy of Goodyear)

FIGURE 13.38
On-Side Storage of Rubber Tires in Palletized Racks.

(Courtesy of Goodyear)

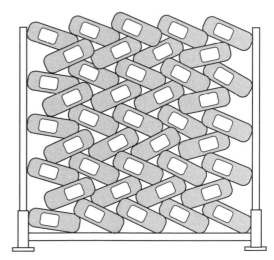

FIGURE 13.39
Laced Tire Configuration.

High-Expansion Foam. The design densities indicated in NFPA 13 for rack protection of rubber tires are high, and the criteria include high-expansion foam for some of the design options. Where high-expansion foam is used in accordance with NFPA 11A, *Standard for Medium- and High-Expansion Foam Systems* (now incorporated in NFPA 11, *Standard for Low-, Medium-, and High-Expansion Foam*), a reduction in the required sprinkler discharge density is allowed. High-expansion foam is defined by NFPA 11A as an aggregation of bubbles resulting from the mechanical expansion of a foam solution by air or other gases with a foam-to-solution volume ratio ranging from 200:1 to approximately 1000:1. These foams provide a unique agent for transporting water to inaccessible places; for total flooding of confined spaces; and for volumetric displacement of vapor, heat, and smoke. Tests have shown that, under certain circumstances, high-expansion foam, when used in conjunction with water sprinklers, will provide more positive control and extinguishment than either extinguishment system by itself.

Large Drop and Early Suppression Fast-Response Sprinklers. Protection criteria for storage of rubber tires with large drop and early suppression fast-response (ESFR) sprinklers was introduced into NFPA 231D, *Standard for Storage of Rubber Tires*, in 1996. NFPA 13, Chapter 12, contain tables indicating the design criteria for large drop, specific application control mode, and ESFR sprinklers that were previously in NFPA 231D.

Baled Cotton Storage

The protection requirements for baled cotton storage were not covered in NFPA 13 until the 1999 edition. Baled cotton is defined by NFPA 13 as a natural seed fiber wrapped and secured in industry-accepted materials, usually consisting of burlap, woven polypropylene, or sheet polyethylene, and secured with steel, synthetic, or wire bands or wire. It can also include linters (lint removed from the cottonseed) and motes (residual materials from the ginning process).

The concepts for the protection of baled cotton are based on the use of control mode density-area sprinkler protection using design areas over 3000 sq ft (279 m²) for wet systems and 3900 square ft (362 m²) for dry pipe systems. The criteria in NFPA 13 pertain to storage heights of up to and including 15 ft (4.6 m). The water supply must be able to sustain the sprinkler system demand plus a minimum of 500 gpm (1892 L/min) for hose streams for a minimum 2-hour time period. See Table 13.19.

There are specific requirements in NFPA 13 for the type of sprinkler that can be used. Although it is not specifically mentioned in the baled cotton

TABLE 13.19

Baled Cotton Storage Up to and Including 15 ft (4.6 m)

System Type	Baled Cotton Storage Up to and Including 15 ft		
	Tiered Storage	Rack Storage	Untiered Storage
Wet	0.25/3000	0.33/3000	0.15/3000
Dry	0.25/3900	0.33/3900	0.15/3900

Source: NFPA 13, 2002, Table 12.5.2.1.

requirements, the designer needs to keep in mind that standard response upright or pendent spray sprinklers with K-factors of at least 8.0 are required for design densities between 0.20 gpm/ft^2 and 0.34 gpm/ft^2 (8.1 mm/min to 14.1 mm/min). For design densities over 0.34 gpm/ft^2 (14.1 mm/min), standard response upright or pendent spray sprinklers with K-factors of at least 11.2 that are listed for storage applications are required.

Roll Paper Storage

Roll paper is classified in detail in Chapter 5 of NFPA 13, 2002 edition. In the classification there are three major categories, as follows:

- Heavyweight—Basis weight of 20 lb/1000 ft^2 (9.1 kg/92.9 m^2) or greater
- Medium weight—Basis weight of 10 lb to 20 lb/1000 ft^2 (4.5 kg to 9.1 kg/92.9 m^2)
- Lightweight—Basis weight of less than 10 lb/1000 ft^2 (4.5 kg/92.9 m^2) and tissues regardless of basis weight

Tissues are defined by NFPA to include the broad range of papers of characteristic gauzy texture, which, in some cases, are fairly transparent. The definition also includes soft absorbent-type, regardless of basis weight—specifically, crepe wadding and the sanitary class including facial tissue, paper napkins, bathroom tissue, and toweling. Roll tissue creates a high-challenge fire condition because of the rapid flame spread across the surface of the rolls.

The protection requirements in NFPA 13 are based on fire test results by Factory Mutual Research, some of which are presented in Chapter 12 and also in a summary in Annex C of the 2002 edition. Fire tests conducted on roll paper storage indicate that storage of lightweight and medium weight rolled paper, with heavyweight paper used for a protective wrapper, performs better. Therefore, the requirements allow medium weight rolled paper to be protected as heavyweight paper where wrapped completely on the sides and both ends or where wrapped on the sides only with steel bands. The requirements similarly allow lightweight paper to be protected as medium weight paper where wrapped completely on the sides and both ends or where wrapped on the sides only with steel bands.

The options for the protection of rolled paper include the use of control mode density-area sprinkler systems, large drop sprinkler systems, and ESFR sprinkler systems. NFPA 13 has several tables that indicate the specific requirements depending on the type of sprinkler, storage height, weight of the paper, type of array (standard, open, or closed), and whether the rolls are banded or unbanded.

A standard array is defined by NFPA 13 as a vertical storage arrangement in which the distance between columns in one direction is short (1 in.) and is in excess of 2 in. in the other direction. A closed array is defined by NFPA 13 as a vertical storage arrangement in which the distances between columns in both directions is short (not more than 2 in. in one direction and 1 in. in the other). An open array is defined as a vertical storage arrangement in which the distances between columns in both directions are lengthy (all vertical arrays other than closed or standard). When a roll of paper is banded, it is provided with a circumferential steel strap (⅜ in. or wider) at each end of the roll.

The requirements for new installations of control mode density-area sprinkler systems are limited to storage heights of 25 ft (7.6 m). Large drop sprinklers are limited to storage heights of 26 ft (7.9 m), and ESFR sprinkler systems for building heights up to 45 ft (13.7 m). When using control mode density-area sprinkler systems, roll paper storage over 15 ft (4.6 m) requires high-temperature rated sprinklers, and the protection area per sprinkler cannot exceed 100 ft^2 (9.2 m^2) or be less than 70 ft^2 (6.5 m^2).

For control mode density-area and large drop sprinkler systems, the water supply duration is 2 hours and at least 500 gpm (1892 L/min) of hose stream allowance must be added to the sprinkler demand. For ESFR sprinkler systems, the water supply duration is 1 hour and at least 250 gpm (946 L/min) of hose stream allowance must be added to the sprinkler demand.

High-expansion foam systems may also be used for the protection of rolled paper. A reduction in the design density and design area are allowed when used with medium- and heavyweight class storage areas.

Summary

The protection of stored commodities is complicated and requires a thorough understanding of the protection terminology, requirements, and various options for each method of protection. The designer needs to consider not only the protection of the storage at the time of design but also the future use of the space, and should discuss options with the building owner.

BIBLIOGRAPHY

References Cited

1. *Automatic Sprinkler Systems Handbook*, 9th edition, NFPA, Quincy, MA, 2002.
2. Jensen, R., "A Brief History of Sprinklers, Sprinkler Systems, and the NFPA Sprinkler Standards," in *Automatic Sprinkler Systems Handbook*, 9th edition, NFPA, Quincy, MA, 2002.
3. Golinveaux, J. E., and Hankins, J. B., "Sprinkler Systems for Storage Facilities," in Cote, A. E., ed., *Fire Protection Handbook*, 19th edition, NFPA, Quincy, MA, 2003, p. 10-213.

NFPA Codes, Standards and Recommended Practices

NFPA Publications National Fire Protection Association, 1 Batterymarch Park, Quincy, MA 02169-7471

The following is a list of NFPA codes, standards, and recommended practices cited in this chapter. See the latest version of the *NFPA Catalog* for availability of current editions of these documents.

NFPA 11A, *Standard for Medium- and High-Expansion Foam Systems* (now incorporated in NFPA 11, *Standard for Low-, Medium-, and High-Expansion Foam*)
NFPA 13, *Standard for the Installation of Sprinkler Systems*
NFPA 25, *Standard for the Inspection, Testing, and Maintenance of Water-Based Fire Protection Systems*
NFPA 230, *Standard for the Fire Protection of Storage*

14

Specification Writing

Cindy Gier, PE

Specifications are a written component of the contract documents which include provisions for all systems and components in the building. Beginning with architectural items and ending with engineered systems in the building, the specifications indicate all approved materials and methods that are to be used for the construction, installation, and testing of the systems contained therein. The written collection of specifications accompanies the construction plans for the building and is an important component of the construction documents. Too often, specifications are not given the necessary attention to adequately explain the building systems and sometimes fail to give bidding and installing contractors the necessary instructions for proper design and installation of building systems.

Fire protection requirements are included with the engineering portions in different locations of the specification, depending on the format chosen by the specification writers. The main purpose of the sections is to specify exactly which materials are permitted, to outline installation and testing methods, and to provide a general fire protection system description. Written system specifications are a very important part of the contract document package and require attention and effort in their creation to alleviate misinformation, conflicts, or incomplete descriptions of the fire sprinkler systems.

It is important to use a separate specification section for each type of system included in the project. Fire sprinklers, piping, wet or dry standpipes, diesel and electric fire pumps, and specialty systems should all have their own specification sections. If separate sections are used, then it is much more clear to the bidding contractor and confusing repetition can be eliminated.

Types of Contract Documents

Contract documents include project specifications and contract drawings and are prepared by the architect and engineer of record.

Contract documents are used as bid documents for contractors who are competing for work on a project, and are usually selected on the basis of having submitted the lowest bid. Contract documents are also submitted to the authority having jurisdiction (AHJ), the governmental agency that will review and approve the project for conformance to adopted code and ordinances for their jurisdiction and will ultimately issue a building permit allowing construction to commence on the project. The contractors are then allowed to use the construction documents to build the project. The contractors become very familiar with the project specifications during the bidding and construction process and will submit requests for information if any of the contract documents are not clear or present a conflict.

Specifications are required to be submitted with the application for permit in the administrative section of *NFPA 5000®, Building Construction and Safety Code®*, 2003 edition. Paragraph 1.7.6.3 requires that specifications be submitted with the sets of plans and all applicable calculations prior to the issuance of a building permit. For some buildings—including high-rise, covered mall buildings and buildings containing atria—a fire protection narrative is also required to explain the scope of the systems. The description must include the basic concepts used for the design of the systems and their integration with other systems or components of the building, and should be included as part of the fire protection specifications and permanent plan set.

Engineers usually start with some standard specification format that is edited for each project. It is important to note that each building is unique and requires its own specifications. Too often, the engineer may be inclined to quickly adapt the specifications from the last project or use a standard specification that is not edited for a particular project. Many Requests for Information (RFI) result when the specifications do not appropriately address a particular project.

For example, in a large museum remodeling and retrofit project, the specifications called for a single-interlock sprinkler system, but the notes on the plans indicated that the system was a dry double-interlock sprinkler system. The calculations submitted were for a wet-type sprinkler system that would be acceptable if the system were single-interlock. The owners thought they were getting a double-interlock system tied into the fire alarm system for the release of water. Many RFIs resulted from this error. The problems could have been easily eliminated if the plans and specifications had been more carefully reviewed to make sure that the requirements were clearly coordinated and explained.

Shop Drawings

Shop drawings, on the other hand, are usually prepared by the contractor for specific systems in the building, such as the fire alarm or fire suppression system. Refer to Chapter 15, "Shop Drawings for Automatic Sprinkler Systems." The project specifications will instruct the contractor on how to prepare shop drawings, what information is required on them, how many copies must be submitted, to whom the shop drawings will be submitted, and sometimes will include supplemental information such as reports or prior approvals or variances received from the AHJ. It is important that the drawings and specifications are coordinated and do not have conflicts. The specifications are a written document that "specifies" requirements for materials, manufacturers, installation, and testing methods for all components of the fire sprinkler system. When the plans and specifications have different information, conflicts can result in a potentially chaotic fire protection system coordination scenario with other trades and a needlessly and excessively expensive fire protection system installation.

Purpose and Types of Specifications

Purpose of Specifications

The purpose of specifications is to meet the needs of the owner for use of a given building and to communicate the results of decisions concerning the project that have been made between the owner or user of a building and the design engineer. It is a necessary function of a set of specifications to point out the resolution of any complex issues and provide a thorough system explanation relative to performance objective and function. Many potential conflicts can be resolved when the design intent of the system is fully explained to the parties that are bidding on or attempting to install the system.

Specifications also are provided to meet the requirements of the AHJ and address any significant code issues that may arise during system design or installation, and coordinate requirements of the AHJ with other requirements applicable to the building. For example, the AHJ or the design engineer may require a factor of safety for the sprinkler calculations. Project specifications should be used to indicate requirements above those listed in the code and standards and to address installation and material requirements for the particular project.

It is important to coordinate the physical juxtaposition of the fire protection system with other systems in the building, especially with respect to mechanical duct work and piping. Concerns of limited space above ceilings or in vertical shafts should be resolved in the specifications so the bidding contractor is aware of these issues.

It is a further role of a set of specifications to functionally integrate the fire protection, particularly electric components installed on the fire protection system, with other building systems, such as security, elevator control, smoke control, and fire alarm and detection. Specifications can help to render the fire protection system a functional building component and not an apparent afterthought.

Types of Specifications

There are several different specification types but most fit into broad categories depending on how much detail they outline for the bidding contractor or designers. However, the main types of specifications are detailed specifications and performance specifications.

Detailed Specifications. Detailed specifications indicate specific requirements for every component of the system, from tubing, fittings, and valves through installation, testing, and commissioning. Everything is controlled by the specification writer, and all equipment manufacturers that are approved are included in lists. Alternative materials or manufacturers usually have to be approved prior to bid for this type of specification, or may not be approved at all. The system must include the precise components specified and must perform precisely as stated in the project specifications, and potential cost-saving methods or materials may not always be considered. So a disadvantage is that the most economical or efficient methods permitted by NFPA 13, *Standard for the Installation of Sprinkler Systems*, are not always allowed by detailed specifications.

Performance Specifications. Performance specifications allow the contractor greater latitude by requiring conformance with adopted codes and that the installation be performed in accordance with the appropriate standards and allowing the latest cost-efficient technologies to be considered. Especially as the building industry moves more towards design/build construction projects, there is a growing consensus that performance specifications have the potential to save time and money on behalf of the engineer, contractor, and owner.

Sources of Specifications

Specifications are available from computer specification databases or can be generated in-house. Regardless of which type of specification is used, it is important to make sure that the database used is appropriate and relevant and carefully edited for each project. Very few projects are identical in scope and nature, and therefore very few specifications should be identical.

Computer Specification Database

There are computer specification databases available for a fee to engineers and architects. The most common is called *MASTERSPEC*®, a product of the American Institute of Architects [1].

The MASTERSPEC Mechanical/Electrical (M/E) Library provides a set of over 250 specifications and reference materials from Divisions 15 and 16 as well as selected sections from Divisions 1–14. (See "Specification Organization," later, for a discussion of divisions.) Periodically reviewed by representatives of the American Consulting Engineers Council (ACEC) and the National Society of Professional Engineers (NSPE), *MASTERSPEC* is produced by a staff of engineering specification writers and consulting engineers.

Specifications are also available from the Unified Facilities Guide Specifications (UFGS) for use in specifying construction for the military services and are a joint effort of the U.S. Army Corps of Engineers (USACE), the Naval Facilities Engineering Command (NAVFAC), and the Air Force Civil Engineer Support Agency (AFCESA). Many corporations have specifications that they have adapted as their "standard" for use by consultants employed to design their facilities.

In-House Specifications

Many engineering firms write their own specifications "in-house." This can be a problem if the firm is not familiar with all the components, equipment, and materials of the systems. This can be a problem in such firms as a mechanical, electrical and plumbing (MEP) design firm that also includes fire protection design in its scope of services. Because most fire protection equipment is specialized and the systems are complex, many mechanical or plumbing engineers do not have sufficient knowledge to adequately address some of these systems, and in such cases it is important to hire or train fire protection engineering professionals.

Also a concern with in-house specifications is the need that they be edited or updated to address new technology, or changes to codes and standards. Old specifications may tend to perpetuate old ideas (such as the old requirement for universally using Schedule 40 pipe on all systems) and the specification of pipe schedule systems whose use is severely restricted by NFPA 13 for new system design. Many in-house specifications still reference standards such as NFPA 231C, D, E, and F, which were merged into NFPA 13, *Standard for the Installation of Sprinkler Systems*, in 1999. Some older methods are not always justified on a cost basis versus newer, more potentially cost-efficient materials and methods. New technologies need to be studied carefully and sufficient empirical and functional data need to be collected, in addition to being subjected to all relevant testing and listing protocol, before they are permitted on a project. An example is flexible sprinkler tubing that allows easy and quick modifications in a lay-in ceiling by relocating lighting and sprinklers on their flex connectors. On the other hand, engineers should be vigilant on behalf of the owner of the building in permitting cost-efficient new technologies and possibly allowing them initially for specific projects or tenant finishes once their cost and advantages are clearly known.

Specification Organization

Specifications are organized into separate divisions that reflect broad categories of building component groupings, dealing with general requirements, architectural requirements, structural requirements, and mechanical and electrical systems. Detailed requirements for each trade are contained within the various divisions. Specialty services and systems like fire protection, lightning and cathodic protection, security, and water storage tanks are located in Division 13 on Special Construction. The following section numbers refer to the fire protection sections of documents produced by the American Institute of Architects (AIA).

Specifications with 16 Divisions

Most standard specifications have 16 divisions. The divisions start with requirements for general issues, outside work, and structural and building materials. The latter divisions (Divisions 13 through 16) address fire protection, mechanical, and electrical work. Table 14.1 summarizes all the typical divisions as they relate to fire protection.

Relocation of Fire Protection Requirements

Until recently, fire sprinkler requirements were included in the Mechanical Section (Division 15), and fire alarm requirements were included in the Electrical Section (Division 16). The sprinkler contractor predominately received either a copy of Division 15 or the applicable section in Division 15, and the electrician received a copy of Division 16 to share with the fire alarm contractor. This almost guaranteed that the sprinkler contractor and electrician did not coordinate fire protection components that required an electrical connection.

TABLE 14.1

Summary of Divisions

Division	Title	Contents Related to Fire Protection
0	Instructions to Bidders	Scheduling Submittals Construction Testing Addresses and phone numbers of important individuals to contact
1	General Requirements	Scope Requirements for coordination meetings Requirements for liquidated damages
2	Site Work	Underground fire protection piping Cathodic protection
3	Concrete	May apply for underground vaults or pump houses Required fire ratings
4	Masonry	May apply for underground vaults or pump houses Required fire ratings
5	Metals	May apply for underground vaults or pump houses Required fire ratings
6	Wood and Plaster	May apply for underground vaults or pump houses Required fire ratings
7	Thermal and Moisture Protection	Firestop systems
8	Doors and Windows	Access doors and frames
9	Finishes	Painting
10	Specialties	Fire extinguishers, louvers and vents
11	Equipment	Some equipment may require specialized fire protection
12	Furnishings	Flammability of furnishings may be relevant to modeled sprinkler design
13	Special Construction	Now used for fire protection and fire alarm systems
14	Conveying Systems	Coordination may be needed if conveying systems are protected by sprinklers
15	Mechanical	Specifications for fire suppression systems were formerly located here.
16	Electrical	Specifications for fire alarm systems were formerly located here

Unfortunately, the fire protection contractor may not request or be given all of the relevant specification sections prior to bid or during the completion of the contract. As shown on Table 14.1, each division may contain important information for the fire protection contractor, so it is important that these sections be shared with them. Also, a big problem is that most contractors get or request only the section relating to fire protection, and this can lead to an expensive lack of coordination, where it may not be clear who installs the underground fire main outside the building or who provides and installs the fire alarm supervisory devices. For this reason, standardized specification formats collect and organize the relevant fire suppression and detection specifications into a more central location in the specification, in Division 13. It should therefore be easier to tie each specification section to all other applicable specification sections throughout the project.

With the new specification format, all relevant and interdependent fire protection requirements are located and cross-referenced within Division 13, enhancing the probability for much better coordination. The sprinkler and electrical contractors can be more responsive to each other, and the possibility is open for them to be subcontracted to each other in order to complete the project more expeditiously and cost-efficiently.

Specification Division 13

Master Specification Division 13—Miscellaneous Items. By utilizing Specification Division 13 for fire protection and specialty systems, better coordination is also seen for Section 13700, Security Access and Surveillance, including intrusion detection and security alarms. These systems are sometimes integrated with fire alarm systems, and many contractors have the technical expertise to install both types of systems. Building Automation and Control, Division 13800, concerning elevator control and the relationship to fire alarm and suppression systems, also is well located for its logical ties to the fire alarm system and similarity of contractors.

Specification Division 13—Fire Alarm. Division 13850, Detection and Alarm, encompasses fire alarm systems, fire detection systems, and fire alarm control panels. Since most fire suppression systems are required to be monitored, supervised, or integrated with a fire alarm system, this coordination is critical to the success of the overall fire protection design.

Specification Division 13—Fire Suppression Systems. Table 14.2 presents typical fire suppression division subsections within Division 13.

TABLE 14.2

Fire Suppression Requirements within Division 13

Section	Title	Contents
13915	Fire Suppression Piping	Piping criteria Fittings Installation methods for standpipe systems, wet pipe, dry pipe, preaction, and deluge sprinkler systems
13920	Fire Pumps	Basic information on fire pumps
13921	Electric-Drive, Centrifugal Fire Pumps	Basic information and details on end-suction and in-line, centrifugal, split-case pumps and accessories
13922	Diesel-Drive, Centrifugal Fire Pumps	Basic information and details on end-suction, centrifugal, split-case pumps and accessories
13926	Electric-Drive, Vertical-Turbine Fire Pumps	Basic information and details on vertical-turbine pumps, controllers, and accessories
13927	Diesel-Drive, Vertical-Turbine Fire Pumps	Basic information and details on vertical-turbine pumps, controllers, and accessories
13955	Foam Fire Extinguishing	Suppression, fixed, low-expansion, and high-expansion systems for flammable liquid fires
13967	Clean-Agent Extinguishing Systems	Suppression agents and approved substitutes for Halon 1301 systems

Other Specification Considerations

Format and Content

A standard format is recommended for all specifications. They are often read or skimmed quickly and it is important that designers and bidders be able to quickly focus on project scope and any irregularities of the project or design.

TABLE 14.3

Typical Section Contents

Title	Contents
General Description	A general description of the types of systems encountered and an outline of the extent of the contract.
Related Work	Nonsprinkler work, such as electrical wiring or outside work is described and the appropriate sections are referenced.
System Design, Calculation, and Coordination	*Hydraulic calculations:* hydraulic densities, remote areas, classification of occupancies, hose demand, and special situations are described. *Water flow test:* Complete water flow test information is provided, and a copy of the water flow test report should be provided as an attachment. This is an especially important, often-neglected item. *Coordination:* Requirements for coordination with other trades, coordination with reflected ceiling plans, and other items are detailed. Qualifications and quality assurance.
Installation Experience	The specifying engineer may provide minimum experience criteria for work of an unusual or complex nature.
Layout Credentials	It is recommended that minimum NICET credentials be established for the layout, supervision, and checking of the system. Usually, NICET Level II is sufficient for layout on residential systems and NICET Level III is required by many AHJs for layout of most sprinkler systems. Specialized systems, complex systems, or systems designed for crucial or sensitive installations may warrant heightened layout credentials or length of experience.
Submittals	*Catalog cuts:* Requirements for submittal of catalog cuts are outlined. *Shop drawings:* Shop drawing submittal requirements, including minimum scale and any specialized requirements are provided. *Hydraulic calculations:* Requirements for submittal of hydraulic calculations are outlined. *Certificates:* Submittal of certificates related to NICET-level verification, material and test certificates, or other forms of required documentation are outlined. *As-built drawings:* Any submittal requirements for as-built drawings and as-built hydraulic calculations of the completed system are listed. *Operation and maintenance manuals:* Submittal requirements for operation and maintenance manuals, if required, are outlined in this section.
Specification Considerations	Applicable design and installation documents. A list of documents that apply to the design of a fire protection system is provided.
Products	Detailed requirements for components, such as piping, valves, sprinklers, fire department connection, switches and electrical wiring, hangers and supports, sleeves, and signs are provided.
Installation	Specialized requirements for system installation are provided.
Testing	The types of tests and notification requirements for the tests are detailed. Commissioning of the system and training of any building maintenance personnel are included in this section.

The length of specification sections should be no longer than necessary to convey important information. Some engineering firms have what they call a "long spec" and a "short spec." The short specification format is especially useful in small "standard" projects with typical fire protection systems. There is no need to go to great length to explain and specify a small wet pipe sprinkler system that is to be strictly in conformance with *NFPA 5000*® and with NFPA 13. Table 14.3 presents sample items that may be included in a specification section that applies to a fire protection system.

The Need for Specification Editing

Potential problems arise when specifications are not carefully edited for each project. The type of facility and fire protection systems should be clearly indicated early in the specification sections, then each item and section should clearly agree with the systems and facilities indicated. Sometimes, a warehouse can look suspiciously like a hospital in a poorly edited specification, or a dry standpipe system may be called out for a fully sprinklered building. Again, confusion and requests-for-information can be caused by quick and sloppy editing of the documents.

A tremendous problem results when the specifications do not match drawings. A final quality control check, preferably by a different individual than the one who wrote the specification, should be performed to coordinate all the construction documents and make sure that all plan notes, drawings, and specifications are in agreement with the specified systems. Inconsistencies between the documents may ultimately result in extras to the contract, and field problems may arise as systems are installed at a delayed time in the project. Contractors may take advantage of inconsistencies, and the engineer who caused the confusion may be responsible for change orders or increased costs.

It is necessary to carefully evaluate a specification with respect to a drawing, matching the document line by line with what is shown on plans and details. The engineer must look at a specification with "fresh eyes," as if it had not been seen before, and look for any inconsistencies. Firms should have an in-house quality control department in which engineers are trained to review each other's work and look for confusing or conflicting information. Regardless, it is a good idea to require that the specifications and drawings be reviewed by at least two engineers involved in the project before they are submitted.

Summary

Specifications fulfill an important function in the design, installation, and testing of any fire protection system and should be completed by staff with the knowledge and experience to be familiar with all parts of the system and the operation of the system components. Too often, the specification writing is left to the newest and least-experienced designers or provided as an afterthought that is edited shortly before the project goes out to bid. Specifications, ideally, should be completed first, before the design is started, and the design should follow provisions that the specifications outline. With this methodology, the plans, specifications, other trades, and the installation of the systems will always be coordinated.

BIBLIOGRAPHY

Reference Cited

1. MASTERSPEC Specification System Copyright 2004, American Institute of Architects, Washington, DC. MASTERSPEC is published by ARCOM, 332 East 500 South, Salt Lake City, UT 84111-3309.

NFPA Codes, Standards, and Recommended Practices

NFPA Publications National Fire Protection Association, 1 Batterymarch Park, Quincy, MA 02169-7471

The following is a list of NFPA codes, standards, and recommended practices cited in this chapter. See the latest version of the *NFPA Catalog* for availability of current editions of these documents.

NFPA 13, *Standard for the Installation of Sprinkler Systems*
NFPA 5000®, *Building Construction and Safety Code*®

15

Shop Drawings for Automatic Sprinkler Systems

Cindy Gier, PE

Shop drawings are an integral part of every automatic sprinkler system design. They are the visual representation of the proposed fire sprinkler system. A shop drawing shows all fire sprinkler piping, lengths of each pipe on the system, fittings, sprinkler locations, control valves and appurtenances, and much more information about the system, related systems, and the building. These drawings are usually produced by the fire sprinkler contractor or a company hired by the contractor. A fire protection engineer can also perform shop drawings in accordance with the SFPE Position Statement "The Engineer and the Technician: Designing Fire Protection Systems," [1] as described in Chapter 19, "Professional Issues in Fire Protection Engineering Design."

This chapter explores shop drawing requirements and the attributes that define a good, correct, and complete set of shop drawings. NFPA 13, *Standard for the Installation of Sprinkler Systems*, includes a complete list of items required to be included on shop drawings, and a competent reviewer of shop drawings will also look for conflicts with building structure or systems, underground fire main entry information, and architectural design. The review also includes the checking of hydraulic calculations for inaccuracies and comparing the calculation reference points to what is shown on the sprinkler plans.

The completed fire sprinkler shop drawings will be reviewed by the authority having jurisdiction (AHJ), normally the state, local, city, or fire department plans examiner, by the insurance agency, and by the architect and fire protection or mechanical engineer. Each one of these entities is looking for slightly different information. This chapter will also explore the concerns that each of these "stakeholders" looks for in the fire sprinkler shop drawings.

Responsibility for Shop Drawings

The Design Team—the Architect and Engineer of Record

The mechanical or fire protection engineer specifies the design documents, as discussed in previous chapters. The design is based on the hazard protected, storage arrangements, encapsulation, and many other conditions such as occupancy, design method, and use of specially listed sprinkler technologies. The engineer of record also prepares design specifications, as discussed in detail in Chapter 14, "Specification Writing." *NFPA 5000®*, *Building Construction and Safety Code®*, does not provide detailed sprinkler requirements for specification writing, but some general guidance may be ascertained from this document.

This information should be provided by the engineer on the design documents that are submitted to the jurisdiction for plan review. The engineer is required to specify the scope of the fire sprinkler system, the applicable codes and standards, the occupancy type and hazard classification, the density, water flow, and pressure requirements, confirmation of the available water supply data, preliminary hydraulic calculations, and any areas of the building that require special attention. The intent of providing this information on bid documents is to give guidance for the sprinkler contractors so that the bids will have the same scope and include similar quality of materials, and therefore they should be more equally competitive. Also, the specification of the design criteria for a fire sprinkler system is part of the practice of engineering and should be completed by a licensed engineer. The contractor can then use the design documents to complete shop drawings for review and approval.

The Authority Having Jurisdiction

The design criteria are reviewed by the AHJ for conformity to standards and adopted codes. An AHJ can be a state or local governmental official or engineer, a fire protection engineer or the engineer of record, an insurance company representative, or other individual or entity recognized by the contract documents as having jurisdiction over the design and installation of the sprinkler system.

This review is usually started when the design documents are submitted by the engineer of record to the plans examiner for initial review. At this time, a general contractor and a sprinkler contractor may not yet be hired. It is important, therefore, for the design criteria to be reviewed and any nonconformance to adopted codes or standards be identified before the future contractor starts shop drawing preparation.

The reviewing agency looks for conformance to sprinkler design requirements contained in NFPA 13 and the adopted building and fire codes for the jurisdiction. Building and fire codes are adopted by a jurisdiction to regulate construction materials and methods by establishing the minimum regulations for building systems. Common code documents include NFPA 1, *Uniform Fire Code*™, NFPA 5000, and the *International Building Code*.

The Sprinkler Contractor

After a project is awarded or negotiated with a general contractor, the owner and owner's representatives choose a sprinkler contractor to perform the work. The selection is usually based on low-bid conditions from prices submitted earlier as part of the general contractor's total bid price. Sometimes the selection of a sprinkler contractor, or any subcontractor, is based on length and quality of experience on a specific type of project, or past experience on joint projects with the general contractor.

Included in the scope of services provided by the sprinkler contractor is the production of fire sprinkler working shop drawings, which shall be based on the design documents prepared by the engineer. Many sprinkler contractors have an "in-house" design team that works with the installers and fabricators for the sprinkler company. Some sprinkler installation firms hire outside engineering or design firms to produce their sprinkler shop drawings.

Preparation of Working Shop Drawings

Regardless of who produces the drawings, the same basic information is required in each submittal package for review by the authorities. There are several components to the sprinkler shop-drawing submittal. The basic components include the following:

- Floor plans indicating sprinkler pipe routing, sprinkler locations, and equipment associated with the sprinkler system
- Hydraulic calculations for each remote or hazard area
- Material safety data sheets for each component
- Water flow information, based on a recent water flow test for the building

Collecting Basic Information

The review of fire sprinkler working shop drawings includes the collection of many pieces of information. The sprinkler contractor will submit a package that must be complete to commence the review of the system. The engineer or reviewing authority may have previously reviewed specifications and contract drawings to gain a full understanding of the impact of the systems.

Complete Building Plans

Complete building plans are required for the adequate design or review of the fire sprinkler systems in a building. Architectural plans, details, and building sections clarify the three-dimensional aspects of the building design. Mechanical, electrical, plumbing and reflected ceiling plans are necessary for coordination. Structural plans, schedules, and sections are also vitally important when routing sprinkler piping around obstructions and checking clearances to exposed structure. Only with the complete plans can all architectural features of a facility be addressed. Electronic AutoCAD files facilitate the overlaying of systems to foresee any potential conflicts between systems. Common issues arise between locations of lighting fixtures, HVAC diffusers, and automatic sprinklers. Although it may seem easier to relocate the sprinklers rather than the other systems, the positions of sprinklers and piping are required to remain within specified parameters to achieve their performance objective. Much time can be saved when complete final plans are reviewed against the working fire sprinkler shop drawings and these conflicts can be addressed before the systems are installed.

Water Supply Information

Water supply information must be collected before the calculations can be evaluated. Hydrant flow data is available from most water purveyors. Private companies can perform hydrant flow tests in conformance with NFPA 24,

Standard for the Installation of Private Fire Service Mains and Their Appurtenances, or NFPA 291, *Recommended Practice for Fire Flow Testing and Marking of Hydrants*. Most AHJs require a recent flow test, with 1 year being the predominant span of time permitted between conduct of the test and submission of design documentation.

The hydrant flow test is used by hydraulic calculation programs to compare the system demand to the water supply. The results are reflected in a graph included with the calculations. Information required by subsection 14.2.1 of NFPA 13, 2002 edition, includes location of hydrants, hydrant pressure readings, date, time, personnel present at test, and a listing of other sources of water supply that can influence water supply availability. Also see Chapter 6, "Evaluation of Water Supply," for a rigorous discussion of that topic.

Preparing Plans and Calculations

Plans and calculations are required to be prepared in accordance with Chapter 14, "Plans and Calculations," of NFPA 13. Shop drawings, or "working plans" as NFPA 13 calls them, are required to "be submitted for approval to the authority having jurisdiction before any equipment is installed or remodeled." Working plans are also submitted to the architect and engineer of record for review in accordance with the contract documents.

Basic information required on the plans includes walls, ceiling heights, sections, and all sprinkler design and installation instructions. Subsection 14.1.3 of NFPA 13 provides a list of 44 required items that pertain to the design of the system. Plans are required to be drawn to scale, on uniform-size sheets (usually 24 in. by 36 in.), with floor plans of each floor. The complete list of subsection 14.1.3 requirements appears in Box 15.1.

BOX 15.1

Plan Requirements from NFPA 13, 2002 Edition

14.1.3 Working plans shall be drawn to an indicated scale, on sheets of uniform size, with a plan of each floor, and shall show those items from the following list that pertain to the design of the system:

(1) Name of owner and occupant

(2) Location, including street address

(3) Point of compass

(4) Full height cross section, or schematic diagram, including structural member information if required for clarity and including ceiling construction and method of protection for nonmetallic piping

(5) Location of partitions

(6) Location of fire walls

(7) Occupancy class of each area or room

(8) Location and size of concealed spaces, closets, attics, and bathrooms

(9) Any small enclosures in which no sprinklers are to be installed

(10) Size of city main in street and whether dead end or circulating; if dead end, direction and distance to nearest circulating main; and city main test results and system elevation relative to test hydrant (*see A.15.1.8*).

(11) Other sources of water supply, with pressure or elevation

(12) Make, type, model, and nominal K-factor of sprinklers including sprinkler identification number

(13) Temperature rating and location of high-temperature sprinklers

(14) Total area protected by each system on each floor

(15) Number of sprinklers on each riser per floor

(16) Total number of sprinklers on each dry pipe system, preaction system, combined dry pipe-preaction system, or deluge system

(17) Approximate capacity in gallons of each dry pipe system

(18) Pipe type and schedule of wall thickness

(19) Nominal pipe size and cutting lengths of pipe (or center-to-center dimensions). Where typical branch lines prevail, it shall be necessary to size only one typical line

(20) Location and size of riser nipples

(21) Type of fittings and joints and location of all welds and bends. The contractor shall specify on drawing any sections to be shop welded and the type of fittings or formations to be used

(22) Type and locations of hangers, sleeves, braces, and methods of securing sprinklers when applicable

(23) All control valves, check valves, drain pipes, and test connections

(24) Make, type, model, and size of alarm or dry pipe valve

(25) Make, type, model, and size of preaction or deluge valve

(26) Kind and location of alarm bells

(27) Size and location of standpipe risers, hose outlets, hand hose, monitor nozzles, and related equipment

(28) Private fire service main sizes, lengths, locations, weights, materials, point of connection to city main; the sizes, types and locations of valves, valve indicators, regulators, meters, and valve pits; and the depth that the top of the pipe is laid below grade

(29) Piping provisions for flushing

(30) Where the equipment is to be installed as an addition to an existing system, enough of the existing system indicated on the plans to make all conditions clear

(31) For hydraulically designed systems, the information on the hydraulic data nameplate

(32) A graphic representation of the scale used on all plans

(33) Name and address of contractor

(34) Hydraulic reference points shown on the plan that correspond with comparable reference points on the hydraulic calculation sheets

(35) The minimum rate of water application (density), the design area of water application, in-rack sprinkler demand, and the water required for hose streams both inside and outside

(36) The total quantity of water and the pressure required noted at a common reference point for each system

(37) Relative elevations of sprinklers, junction points, and supply or reference points

(38) If room design method is used, all unprotected wall openings throughout the floor protected

(39) Calculation of loads for sizing and details of sway bracing

(40) The setting for pressure-reducing valves

(41) Information about backflow preventers (manufacturer, size, type)

(42) Information about antifreeze solution used (type and amount)

(43) Size and location of hydrants, showing size and number of outlets and if outlets are to be equipped with independent gate valves. Whether hose houses and equipment are to be provided, and by whom, shall be indicated. Static and residual hydrants that were used in flow tests shall be shown

(44) Size, location, and piping arrangement of fire department connections

The information in Box 15.1 is presented as a minimum list of items required for a complete submittal for most buildings. The designer or reviewer should be cognizant of the fact that much additional information is required for each unique situation. The quality of the review and design may depend on items not included in this list. Therefore, the list can be used as a basis for starting the design, or as a final checklist, but not as the only information required in a complete submittal.

The Review of Shop Drawings

Shop drawings are reviewed by several parties including the AHJ, architects, the project engineers, fire protection engineers, and/or property insurance agencies. Reviewing entities have specific concerns for which they are responsible, relative to verifying specific features shown on the shop drawings.

The specifications prepared by the engineer usually require the fire sprinkler contractor to prepare and submit shop drawings for review by the engineer. Because the engineer's review has several important implications for the project team, it is important for the engineer to know exactly what to look for on working plans and hydraulic calculations.

Shop Drawing Review by the Engineer of Record

The main responsibility of the engineer of record in the review of shop drawings is to verify conformance to the design documents that were part of the contract documents. The engineer must have the technical knowledge, education, and experience to review fire sprinkler hydraulic calculations and a working knowledge of the requirements of NFPA 13, *Standard for the Installation of Sprinkler Systems*, to note any inconsistencies with the requirements and the engineer's design.

The engineer specifies the information that describes the system in conformance with the adopted codes and standards for the jurisdiction and the project's design goals. The engineer's review should include a review of the following four basic design criteria:

- Conformance to general and specific details
- Conformance to design documents which establish the design goals and objectives

- Conformance with local ordinances and building code requirements
- Presence of current and complete sets of manufacturer's annotated catalog cut sheets

A complete review of the fire sprinkler plans performed by the engineer of record should include at a minimum the following items:

- The location of the fire protection service entrance and sprinkler riser to verify adequate space is provided in the room for sprinkler apparatus, backflow preventers, control panels, and fire pumps, and that the location corresponds with that shown on civil and architectural drawings.
- Fire sprinkler control valves, backflow preventers, piping, fittings and other appurtenances on the fire sprinkler riser for conformance to specifications and listing requirements. These should be indicated in the engineer's design documents. Specific manufacturers may be listed in the specifications or substitutions may need to be reviewed and approved.
- Indications for the orifice, K-factor, orientation, finish, color, and model of sprinklers on the plans. The sprinkler types and finishes should be coordinated with the ceiling types. The reflected ceiling plans should be compared to the sprinkler plans and differences in ceiling types noted. Specifications may indicate specific sprinkler types and finishes if they are determined to be important to the installation.
- The presence of such architectural features of the building as ceiling peaks, skylights, materials—anything architectural and aesthetic that may impact the installation or coverage area of sprinklers. Special attention should be paid to atria, glazing locations, vaulted ceilings, and special uses in a building. Auditoriums, conference rooms, executive offices, and many other areas frequently have special ceiling treatments, and therefore special fire sprinklers may have been specified for the location.
- Coordination of such special features as maximum building slope and clearance from structural elements with requirements for early suppression fast response (ESFR) sprinklers, the presence of draft curtains and heat vents and their effect on sprinkler performance, and special storage requirements that may affect sprinkler location.

The engineer of record should also verify conformance to details shown on the design documents. The sprinkler contractor usually provides more complete details and dimensions. The engineer may have used standard details that did not exactly reflect field conditions. It is important that the engineer carefully review and understand the sprinkler details presented by the sprinkler contractor's shop drawings. This way, the engineer can check that plan details are followed or appropriately modified for field conditions. The engineer should also

- Verify that special conditions are met, elevation differences are reflected, and that recommended pipe routing and notations are followed
- Check fire pump room layouts, elevations, distances
- Verify coordination with other building systems, such as the fire alarm system and smoke management system, to ensure that all design performance objectives are met

Most importantly, it is essential to verify that nothing is excluded from the shop drawings that were required by the design documents. The specifications should provide any applicable requirements for pressure reducing valves, fire pumps, or backflow preventers, any of which would be expensive items to add at a later time in the project. Any notes or modifications can be clarified or added to the plans in handwritten notes to the designers and installers. The contractor is required to maintain an approved set of plans on the site, but the engineer will want to take the marked-up set to the field for construction administration site investigations.

A complete review of the fire sprinkler hydraulic calculations performed by the engineer of record should include the following items:

- Complete information on fire pumps, controllers, jockey pumps, relief valves or other regulating means, pressure reducing valves, and pressure ratings for all components served by the fire pump.

- Conformity to specified densities, calculated operating areas, hose stream allowances, and safety factors. The engineer carefully evaluates the calculations to verify that the correct water flow data is used.

- Evaluation and verification of protection areas for each sprinkler, design areas used in calculations, and system protection areas to ensure that all comply with NFPA 13 requirements and the engineer's design documents.

The review of fire sprinkler plans and calculations on the working shop drawings is an important function of the engineer of record, as the representative of the design team. Many important items can be identified on the plans that could prevent field conflicts or installation problems during construction. Time and care should be taken during this time in the process to verify all information.

Subsection 14.3.3 of NFPA 13 provides a checklist of items that are required on the hydraulic calculation worksheets or computer printouts for a complete submittal (see Box 15.2).

BOX 15.2

Shop Drawing Requirements from NFPA 13, 2002 Edition

(Note that the *italics* are not part of NFPA 13)

14.3.3 Detailed Worksheets. Detailed worksheets or computer printout sheets shall contain the following information:

(1) Sheet number

(2) Sprinkler description and discharge constant (K.) [*This information is required to be given for every sprinkler on the project.*]

(3) Hydraulic reference points. [*These should be checked against the plans, with the location of reference point verified, and every reference point for every juncture shown in calculations.*]

(4) Flow in gpm (L/min) [*for each sprinkler and cumulative flow for the system.*]

(5) Pipe size [*verified for each type of pipe material, inside diameter, with a check to ensure conformance to the pipe sizes shown on the plans*]

(6) Pipe lengths, center-to-center of fittings [*verified against plans and including elevation distances*]

(7) Equivalent pipe lengths for fittings and devices [*shown for each fitting and compared to plans*]

(8) Friction loss in psi/ft (bar/m) of pipe [*compared to friction loss tables*]

(9) Total friction loss between reference points [*Check the math between nodes.*]

(10) In-rack sprinkler demand balanced to ceiling demand

(11) Elevation head in psi (bar) between reference points [*in addition to friction loss*]

(12) Required pressure in psi (bar) at each reference point [*totaled on the calculations*]

(13) Velocity pressure and normal pressure if included in calculations

(14) Notes to indicate starting points or reference to other sheets or to clarify data shown

(15) Diagram to accompany gridded system calculations, to indicate flow quantities and directions for lines, with sprinklers operating in the remote area [*Gridded systems cannot be reviewed without this diagram.*]

(16) Combined K-factor calculations for sprinklers on drops, armovers, or sprigs where calculations do not begin at the sprinkler [*It is important to verify this calculation because most or all of the sprinklers may have calculated K-factors.*]

A graph sheet, usually computer generated, accompanies the hydraulic calculations and provides a pictoral summary of water supply and system demand information, as required by NFPA 13, subsection 14.3.4. The graph should be verified against the calculations and the cover sheet to make sure that all three reflect the same information. Unfortunately, it is common for differences to occur between the sheets in the same set of calculations. Some AHJs require hand-drawn graphs to ensure conformance between hydraulic results.

The graphic representation should be plotted on semi-exponential graph paper ($Q^{1.85}$). The following should be included and plotted correctly on the graph, as required by NFPA 13, subsection 14.3.4:

- Water supply curve
- Sprinkler system demand
- Hose demand (where applicable)
- In-rack sprinkler demand (where applicable)

Shop Drawing Review by the AHJ

The primary responsibility of the AHJ is to verify that plans and calculations are in conformance with all codes, standards, and regulations adopted by the jurisdiction. The AHJ also needs to verify that the design criteria selected is appropriate and adequate for the hazard described. This may involve familiarity with the facility and its operations, knowledge of similar occupancies in the jurisdiction, and typical storage methods or any assortment of operational issues concerning adequate protection of the facility.

Most AHJs expect sprinkler plans to be prepared by a design technician certified by the National Institute of Certification Engineering Technologies (NICET) or a professional engineer. Such qualification provides assurance that the sprinkler contractor is associated with someone who should have the knowledge and ability to produce a code-compliant, workable sprinkler installation.

The AHJ needs to maintain a permanent record of the facility plans, shop drawings, and specifications, as required by Section 4.1 of NFPA 25, "Responsibility of the Owner or Occupant." As part of the permanent record, there should be documentation, including a plan review letter and a checklist for the sprinkler shop drawing review. The 44 items indicated earlier are a good start on the plan review checklist, but it should be clear, to both the engineer and the AHJ, that there is much more information that is needed to complete a thorough shop drawing review.

The Review of Shop Drawings for Performance-Based Sprinkler Systems

Some commonalties exist between performance-based and prescriptive-based design review, but there are some differences, which are covered in this chapter.

As described more fully in Chapter 17, "The Performance-Based Design Process and Automatic Sprinkler System Design," performance-based design (PBD) is a method of evaluating a building or occupancy and assigning performance requirements for systems or designs rather than applying standard code requirements. For PBD to be effective, the code official should ideally be involved in PBD projects from inception through the life of the building. By the time shop drawings are received in the code official's office, the official should have already reviewed many documents related to the design and be very familiar with the requirements related to the fire suppression system. The code official plays a critical role in the progress of the PBD project. By the time sprinkler shop drawings are received, it is too late to point out that the design is not what the code official expected from the design team.

The designer should review the *Code Official's Guide to Performance-Based Design Review* [2] for general guidance on sprinkler criteria and a review of sprinkler systems in performance-based projects. The review of sprinkler systems in PBD may exceed the requirements of NFPA 13 in density, spacing, type, or many other factors, as agreed by the design team and other stakeholders. The review of these systems by both the AHJ and the design team is extremely critical. It is necessary to verify that any stated goals or objectives are met through the design of the sprinkler system.

The reviewing code official needs to verify that proper and complete information is submitted, that all necessary calculations are correctly completed, and that adequate information is provided to analyze the performance of the systems involved. The code official does not take on responsibility for the design by performing the review. The responsibility for the design lies with the design professional, who provides a Verification of Compliance to ensure an indication of involvement in the construction process and that compliance with the design intent is met.

Sprinklers may be used in combination with vents, draft curtains, and smoke removal or pressurization systems to provide a specified level of performance of the systems in a building. The plans reviewer needs to be familiar with all aspects of the design and be aware of any agreements between stakeholders in a project. The reviewer needs to have an understanding of the engineering concepts and the knowledge and ability to work through detailed engineering calculations to verify all information.

The code official may be reviewing sprinkler hydraulic calculations, response time indices of fire sprinklers, fire model output data, egress time studies, or any combination of these and other methods. Of all the reviews, the input data review is as critical, or more so, than the output information. The review of the limitations of any calculation method is necessary to verify that the correct model or method is being used. Finally, the code official will witness system testing and commissioning of the systems. The acceptance criteria should be indicated in the drawings, specifications, and supporting information.

Summary

Many fire protection engineers, especially in the consulting field, are actively involved in the design of automatic fire sprinkler systems and most fire protection engineers are responsible for reviewing sprinkler system shop drawings created by others. This job function should be taken seriously and performed methodically to identify serious errors that could compromise system function and the safety of human lives. This chapter can be used to formulate a procedure for conducting a rigorous review of sprinkler system shop drawings to ensure compliance with the national consensus and owners' specifications.

BIBLIOGRAPHY

References Cited

1. SFPE Position Statement, "The Engineer and the Technician: Designing Fire Protection Systems," Society of Fire Protection Engineers, Bethesda, MD, 2005. Available from www.SFPE.org.
2. *Code Official's Guide to Performance-Based Design Review*, Society of Fire Protection Engineers and International Code Council, 2003.

NFPA Codes, Standards, and Recommended Practices

NFPA Publications National Fire Protection Association, 1 Batterymarch Park, Quincy, MA 02169-7471

The following is a list of NFPA codes, standards, and recommended practices cited in this chapter. See the latest version of the *NFPA Catalog* for availability of current editions of these documents.

NFPA 1, *Uniform Fire Code*™
NFPA 13, *Standard for the Installation of Sprinkler Systems*
NFPA 24, *Standard for the Installation of Private Fire Service Mains and Their Appurtenances*
NFPA 291, *Recommended Practice for Fire Flow Testing and Marking of Hydrants*
NFPA 5000®, *Building Construction and Safety Code*®

PART 3

Regulation and Professional Responsibility

16

Fire Sprinkler Codes and Standards

Grant Cherkas

This chapter reviews the codes, standards, and recommended practices applicable to fire sprinkler systems installed in the United States and many other countries. Local building and fire codes typically mandate the installation of fire sprinklers based on features of a building such as size, height, or occupancy. Building codes are adopted by legislative acts, which in turn reference installation standards for fire sprinkler systems. Installation standards which govern the design of fire sprinkler systems such as NFPA 13, *Standard for the Installation of Sprinkler Systems*, are separate from building codes, and do not mandate when sprinklers are required, only how to install them once a code determines they are needed.

Overview of Codes, Standards, and Recommended Practices

Installation standards and recommended practices are not codes. Codes are adopted by laws through acts of the legislature. Typically, a code directly references a standard and thereby becomes an enforceable requirement.

Building codes are adopted by state and local legislatures and are usually modified by amendments to suit local practice or local requirements. Legislative bodies write their own codes or adopt model building codes and model fire codes at their discretion, with or without amendments. When a code is adopted through a legislative act it becomes law and all private property owners within the jurisdiction are required to comply with the law.

Over time, codes change to suit new technologies, a better understanding of hazards, and changes in the level of risk society is willing to accept and for which it is willing to pay. Compliance with a building code is determined based on the date of construction of a building, often called the *code of construction*, and new editions of a building code are almost never retroactively applied to existing buildings unless a law is passed which recognizes an immediate threat to life or other unsafe condition, such as high rise building protection or fraternity house protection. New editions of a building code are applied to existing buildings when there is a change in occupancy, or a major building addition, alteration, or rehabilitation. Fire codes are generally applied to all buildings regardless of the date of construction. In simple terms, building codes instruct designers and builders how to construct a building and fire codes instruct owners how to operate and maintain the building.

Most building codes specify where and when fire sprinkler systems are to be installed and reference a specific standard, predominately NFPA 13, for their design and installation. Not all building codes simply reference external standards for design criteria. Some building codes also contain specific design information that overrides and modifies the design criteria of the referenced standard.

NFPA Codes, Standards, and Recommended Practices

Codes, Standards, and Recommended Practices Defined

Although the terms *codes, standards*, and *recommended practices* are sometimes used interchangeably, they are actually three different kinds of documents. The following definitions are from the *NFPA Glossary of Terms*.

Code. The *NFPA Glossary of Terms* defines a code as "a standard that is an extensive compilation of provisions covering broad subject matter or that is suitable for adoption into law independently of other codes and standards" [1]. NFPA 1, *Uniform Fire Code™*; NFPA *101®*, *Life Safety Code®*; and *NFPA 5000®* are among codes published by NFPA and adopted into law.

Standard. The *NFPA Glossary of Terms* defines a standard as "a document, the main text of which contains only mandatory provisions using the word 'shall' to indicate requirements and which is in a form generally suitable for mandatory reference by another standard of code or for adoption into law. Nonmandatory provisions shall be located in an appendix or annex, footnote, or fine-print note and are not to be considered a part of the requirements of a standard" [2]. NFPA 13 and other documents relating to sprinkler systems are examples of standards.

Recommended Practice. The *NFPA Glossary of Terms* defines a recommended practice as "a document that is similar in content and structure to a code or standard but that contains only nonmandatory provisions using the word 'should' to indicate recommendations in the body of the text" [3]. One example of a recommended practice is NFPA 77, *Recommended Practice on Static Electricity*.

The NFPA Standards-Making Process

NFPA codes and standards are widely adopted and are developed and periodically reviewed by more than 5000 volunteer committee members with a wide range of professional expertise. These volunteers serve on more than 200 technical committees and are overseen by the NFPA Board of Directors, which also appoints a 13-person Standards Council to administer the standards-making activities and regulations. All NFPA members are eligible to vote on the content of the standard.

The NFPA process also includes provisions to change codes and standards between editions. These provisions are called Tentative Interim Amendments (TIAs) and exist to address specific concerns or immediate wording that creates hazards and are then adopted into the code text on the next revision cycle. TIAs are binding and usually are worded using "shall."

Structure

NFPA codes, standards, and recommended practices generally consist of a main body and numerous annexes (previously called appendices) that provide supplemental information or references to external documents. The chapters in NFPA codes, standards, and recommended practices follow a uniform order so that, for example, Chapter 1 is always Administration, Chapter 2 is always Referenced Publications, and Chapter 3 is always Definitions.

In codes and standards, the main body of the text is mandatory provisions and utilizes the prescriptive wording "shall." Statements using the term "shall" are direct requirements that, unless prior approval is obtained from the local Authority Having Jurisdiction (AHJ), must be followed. NFPA standards follow a standardized layout of mandatory text with annexes that contain explanatory material or good practices. Where specific text is referenced in the annexes, then the text will be denoted by an asterisk, such as 7.3.4.5*, and the corresponding paragraph, A.7.3.4.5, can be found in Annex A.

Annexes typically provide additional information to assist the user in understanding why a requirement in the body of the standard exists, or to outline good practice on how to implement or to provide references from where the requirement was derived. Annexes are not part of the standard and, unless specifically adopted in the building code, are not binding. However, a municipal code, local amendment, or project specification may make an annex mandatory.

If a section in an NFPA document has an asterisk, it will have useful information in a corresponding annex. A vertical bar to the left of the text indicates it was recently revised during the last revision cycle.

Retroactivity

As with building codes, NFPA standards are usually not retroactively applied. NFPA 13, 2002 edition, contains the following statement about retroactivity:

> The provisions of this standard reflect a consensus of what is necessary to provide an acceptable degree of protection from the hazards addressed in this standard at the time the standard was issued. Unless otherwise specified, the provisions of this standard shall not apply to facilities, equipment, structures, or installations that existed or were approved for construction or installation prior to the effective date of this standard. Where specified, the provisions of this standard shall be retroactive. In those cases where the authority having jurisdiction determines that the existing situation presents an unacceptable degree of risk, the authority having jurisdiction shall be permitted to apply retroactively any portions of this standard deemed appropriate. [13-02:1.4]

Equivalency

It is important to highlight that there may be alternate means of achieving an equivalent level of safety and property protection through alternate means, equivalencies, or new technology. As an example, NFPA 13 advises users:

> Nothing in this standard is intended to prevent the use of systems, methods, or devices of equivalent or superior quality, strength, fire resistance, effectiveness, durability, and safety over those prescribed by this standard. Technical documentation shall be submitted to the authority having jurisdiction to demonstrate equivalency. The system, method, or device shall be approved for the intended purpose by the authority having jurisdiction. [13-02:1.5]

New Technology

NFPA documents encourage the use of new technology. NFPA 13 states, "Nothing in this standard shall be intended to restrict new technologies or alternate arrangements, provided the level of safety prescribed by this standard is not lowered" and "Materials or devices not specifically designated by this standard shall be utilized in complete accord with all conditions, requirements, and limitations of their listings."

NFPA Fire Sprinkler Related Standards

NFPA 13, *Standard for the Installation of Sprinkler Systems*

NFPA 13 traces its roots back to a document entitled *Rules and Regulations of the National Board of Fire Underwriters for Sprinkler Equipments, Automatic and Open Systems*, which was developed in the late 1800s in New England and adopted as a standard in 1896. On average, NFPA 13 has been revised and issued every 4 years since 1896, and at each revision it incorporated the latest technological and installation techniques and changes.

The stated scope of NFPA 13, 2002 edition, is "This standard shall provide the minimum requirements for the design and installation of automatic fire sprinkler systems and exposure protection sprinkler systems covered within this standard." The stated purpose in subsection 1.2.1 is "The purpose of this standard shall be to provide a reasonable degree of protection for life and property from fire through standardization of design, installation, and testing requirements for sprinkler systems, including private fire service mains, based on sound engineering principles, test data, and field experience." It is important to note that the standard only attests to provide a reasonable degree of protection. Despite the excellent record of fire sprinklers over the last century, the standard makes no warranty on the elimination of total loss of property damage or loss of life. Although automatic sprinklers are often successful in achieving fire suppression, the technology (other than a special class of sprinklers called Early Suppression Fast Response sprinklers) is based upon a fire control principle.

The design of fire sprinkler systems requires experience, education, and special knowledge, as stated in subsection 1.2.2, "Sprinkler systems and private fire service mains are specialized fire protection systems and shall require knowledgeable and experienced design and installation." As detailed in Chapter 14,

"Specification Writing," engineers may create criteria for minimum competency in the design and installation of sprinkler systems. NFPA 13 is intended to apply to the assessment of the character and adequacy of water supplies and selection of sprinklers, piping material, pipe fittings, valves and other appurtenances required to form a functional system.

In 1999, NFPA documents containing fire sprinkler system design and installation information underwent a major reorganization. Almost all sprinkler design criteria that were scattered across over a hundred separate codes, standards, and recommended practices were incorporated into the 1999 edition of NFPA 13. At that time, the scope of NFPA 13 was expanded to coordinate and address all sprinkler system applications, such as the installation of underground pipe from NFPA 24, and sprinkler system discharge criteria storage occupancies, which were moved from the NFPA 231 series of documents related to protection of storage. Sprinkler system information for special hazard protection from over 40 other NFPA documents was either reproduced in the 1999 edition or cross-referenced. This effort resulted in a more complete and better coordinated compendium of fire sprinkler design criteria in one document.

Changes to the 2002 edition included reference to the new sprinkler identification system, the designation of sprinkler sizes by nominal K-factors, new rules for the use of steel pipe in underground applications, and new requirements to address microbiologically influenced corrosion. Obstruction rules for specific sprinkler types and rules for locating sprinklers in concealed spaces were revised and limitations were placed on the sprinkler sizes in select storage applications. In addition, design rules for using K-25 orifice sprinklers and new rules for seismic applications were added.

The 2002 edition of NFPA 13 was once again substantially revised, this time in both style and technical content as a result of new technology and additional peer-reviewed fire testing. Chapter 13, in the 2002 edition of NFPA 13, now incorporates all design and installation requirements extracted from the various documents previously mentioned.

A companion document to NFPA 13 is the *Automatic Sprinkler Systems Handbook* [4]. The *Handbook* includes the complete text of NFPA 13 as well as expert commentary from NFPA staff and committee members. A unique feature of the *Handbook* is that it also covers NFPA 13D, *Standard for the Installation of Sprinkler Systems in One- and Two-Family Dwellings and Manufactured Homes*, and NFPA 13R, *Standard for the Installation of Sprinkler Systems in Residential Occupancies up to and Including Four Stories in Height*.

NFPA 13R, *Standard for the Installation of Sprinkler Systems in Residential Occupancies up to and Including Four Stories in Height*

NFPA 13R is the installation standard for fire sprinklers in residential occupancies up to and including four stories in height. There are a wide variety of occupancies that may exist within such buildings, such as commercial kitchens, mechanical rooms, lobbies, and assembly rooms, which fall under the scope of NFPA 13R.

NFPA 13R was first published in 1989 to promote the life safety aspects of automatic sprinkler systems while reducing the cost of installation by taking advantage of new residential fire sprinkler technology. The stated purpose of

NFPA 13R, 2002 edition, is "... to provide design and installation requirements for a sprinkler system to aid in the detection and control of fires in residential occupancies and thus provide improved protection against injury, life loss, and property damage. A sprinkler system designed and installed in accordance with this standard shall be expected to prevent flashover (total involvement) in the room of fire origin, where sprinklered, and to improve the chance for occupants to escape or be evacuated." The criteria for prevention of total involvement in the room of origin is an important goal in reducing the loss of life in residential fires, which was identified early in the development of residential sprinkler technology. It is important to note that the 2002 edition of NFPA 13R establishes a minimum design discharge density as a result of testing.

NFPA 13D, *Standard for the Installation of Sprinkler Systems in One- and Two-Family Dwellings and Manufactured Homes*

The scope of this standard is to address the design and installation of automatic sprinkler systems for protection against the fire hazards in one- and two-family dwellings and manufactured homes. Similar to the purpose of NFPA 13R, the purpose of NFPA 13D, 2002 edition, is to "... provide a sprinkler system that aids in the detection and control of residential fires and thus provides improved protection against injury, life loss, and property damage. A sprinkler system designed and installed in accordance with this standard shall be expected to prevent flashover (total involvement) in the room of fire origin, where sprinklered, and to improve the chance for occupants to escape or be evacuated."

The first edition of NFPA 13D was issued in 1975 as a result of recognition of the large annual loss of life resulting from fires in one- and two-family dwellings and manufactured homes. The 1980 edition revised the installation criteria to take advantage of new knowledge on sprinkler performance stemming from the residential sprinkler research program funded by the Factory Mutual Research Corporation and the Los Angeles City Fire Department [5].

The 1989 and 1991 editions established criteria for the use of antifreeze systems as well as installation criteria associated with specially listed pipe. The 1994 edition provided additional design criteria for nonmetallic pipe and introduced a design option to reduce the water storage volume required for limited-area homes.

The 1996 edition reintroduced the use of ½-in. pipe. The 1999 edition updated the installation requirements for multipurpose piping systems, and the exception for omitting sprinklers in attics and crawl spaces was modified. In the 2002 edition, a requirement for minimum design discharge density was added.

NFPA 14, *Standard for the Installation of Standpipe and Hose Systems*

NFPA 14 was first adopted as a standard in 1915 and has been periodically revised in the intervening years. Although this standard applies predominantly to standpipes, hydrants, and hose systems, some sprinkler systems utilized elements common to a standpipe system, such as water supply mains and risers, and therefore NFPA 14 should be consulted during the design of a sprinkler system for additional design criteria. In addition, the type of standpipe system installed in a building is influenced by the presence or absence of automatic sprinklers.

NFPA 20, *Standard for the Installation of Stationary Pumps for Fire Protection*

The scope of NFPA 20, 2002 edition, addresses the "... selection and installation of pumps supplying liquid for private fire protection. Items considered include liquid supplies; suction, discharge, and auxiliary equipment; power supplies; electric drive and control; diesel engine drive and control; steam turbine drive and control; and acceptance tests and operation. This standard does not cover system liquid supply capacity and pressure requirements, nor does it cover requirements for periodic inspection, testing, and maintenance of fire pump systems. This standard does not cover the requirements for installation wiring of fire pump units." The stated purpose of NFPA 20 is to "... provide a reasonable degree of protection for life and property from fire through installation requirements for stationary pumps for fire protection based upon sound engineering principles, test data, and field experience."

The first National Fire Protection Association standard for automatic sprinklers published in 1896 contained information on steam and rotary fire pumps. In 1899, a Committee on Fire Pumps was formed consisting of five members from insurance organizations. The first fire pumps were alternative supplies for automatic sprinkler systems or standpipe and hydrant systems and were started manually.

Fire pumps are now automatic-starting and are routinely utilized for high-rise and high hazard applications requiring an automatic water supply. In 1913, gasoline engine-driven pumps appeared but are no longer permitted. The permission to use gasoline engine-driven pumps has now evolved into the use of compression-ignition diesel engines, which are considered as reliable as electric-driven pumps.

Two publications provide guidance on the use of NFPA 20. These are the *Fire Pump Handbook* [6], the code handbook for the 1996 edition of NFPA 20, and the technical reference book *Pumps for Fire Protection Systems* [7]. Both books were jointly developed by NFPA and the National Fire Sprinkler Association (NFSA).

NFPA 22, *Standard for Water Tanks for Private Fire Protection*

In 1909, the NFPA Committee on Gravity Tanks developed the *Standard on Gravity Tanks*. The standard was adopted in 1914 and revised or amended approximately every 4 years since that time.

NFPA 22, 2003 edition, provides "... the minimum requirements for the design, construction, installation, and maintenance of tanks and accessory equipment that supply water for private fire protection, including the following: (1) Gravity tanks, suction tanks, pressure tanks, and embankment-supported coated fabric suction tanks, (2) Towers, (3) Foundations, (4) Pipe connections and fittings, (5) Valve enclosures, (6) Tank filling, (7) Protection against freezing."

The purpose of NFPA 22 "... is to provide a basis for the design, construction, operation, and maintenance of water tanks for private fire protection."

NFPA 22 is normally utilized when the water supply for a sprinkler system is required to be supplied by a private water storage tank or reservoirs. Typically, tanks or reservoirs are only used when an adequate automatic municipal water supply is unavailable to the building. NFPA 24 is normally used to design and install the piping leading from the reservoir or tank to the system.

NFPA 24, *Standard for the Installation of Private Fire Service Mains and Their Appurtenances*

The predecessor to NFPA 24 was published in 1903 by the NFPA Committee on Hose and Hydrants and was titled "*Specifications for Mill Yard Hose Houses.*" Numerous revisions and changes to the title of the document have occurred in the intervening years.

The purpose of NFPA 24, as stated in the 2002 edition, is to "... provide a reasonable degree of protection for life and property from fire through installation requirements for private fire service main systems based on sound engineering principles, test data, and field experience." The scope of NFPA 24 is to "... cover the minimum requirements for the installation of private fire service mains and their appurtenances supplying the following: (1) Automatic sprinkler systems, (2) Open sprinkler systems, (3) Water spray fixed systems, (4) Foam systems, (5) Private hydrants, (6) Monitor nozzles or standpipe systems with reference to water supplies, (7) Hose houses and combined service mains used to carry water for fire service and other uses."

NFPA 24 is used for the design and installation of private piping systems to supply water to sprinkler systems. Typically, the piping is installed underground from a tank, fire pump, or municipal supply.

NFPA 25, *Standard for the Inspection, Testing, and Maintenance of Water-Based Fire Protection Systems*

The purpose of NFPA 25 is to detail the inspection, testing, and maintenance requirements to ensure a reasonable degree of protection for life and property from fire through minimum inspection, testing, and maintenance methods for water-based fire protection systems.

NFPA 25 is applicable to systems such as water supplies, fire pumps, fire service mains, water storage tanks and valves, sprinklers, standpipes and hoses, fixed water sprays, and foam water. Included are the water supplies that are part of these systems, such as private fire service mains and appurtenances, fire pumps and water storage tanks, and valves that control system flow.

NFPA 25 details how to conduct inspection, testing, and maintenance activities and also details the frequency of the activities for water-based fire protection systems. Requirements are provided for impairment procedures, notification processes, and system restoration. NFPA 25 prescriptively details the maintenance activities, including periodic field observations and testing, required to ensure the availability and reliable operation of water-based fire protection systems when demanded.

A companion document to NFPA 25 is *Inspection, Testing, and Maintenance of Water-Based Fire Protection Systems: The NFPA 25 Handbook* [8]. In addition to the complete text of NFPA 25, this book includes commentary on the standard and Frequently Asked Questions.

Other Sprinkler System Standards

There are a number of other sprinkler system standards that may be applicable to an automatic fire sprinkler installation. The AHJ (including the property's insurance company) may have requirements that are unique to the occupancy or location that is being protected. Some insurance companies publish their own set of standards relating to automatic sprinkler system design and installation, which in some cases have significant deviations from the NFPA standards listed here.

Organizations such as the Department of Energy, FM Global, and Industrial Risk Insurers are a few of the groups who have specific design and installation requirements that deviate from the NFPA standards for automatic sprinkler systems. Prior to starting the design of an automatic sprinkler system, the various AHJs and stakeholders should be identified and contacted to ensure that the sometimes layered requirements of the local AHJ, insurance company, and owner are addressed.

Summary

The National Fire Protection Association develops several standards related to fire sprinkler systems, including NFPA 13, *Standard for the Installation of Sprinkler Systems*, and NFPA 14, *Standard for the Installation of Standpipe and Hose Systems*. NFPA standards are periodically reviewed and updated in cycles and become enforceable requirements when referenced by building codes that have been adopted by law. NFPA standards are usually not applicable retroactively and allow for alternative means of achieving an equivalent level of safety and property protection. Other organizations, such as insurance companies, may also specify their own requirements for the design and installation of sprinkler systems.

BIBLIOGRAPHY

References Cited

1. *NFPA Glossary of Terms*, NFPA, Quincy, MA, 2003 edition, p. 54.
2. *NFPA Glossary of Terms*, NFPA, Quincy, MA, 2003 edition, p. 319.
3. *NFPA Glossary of Terms*, NFPA, Quincy, MA, 2003 edition, p. 280.
4. *Automatic Sprinkler Systems Handbook*, 9th edition, NFPA, Quincy, MA, 2002.
5. Kung, H. C., Spaulding, R. D., Hill, E. E., Jr., and Symonds, A. P., "Technical Report Field Evaluation of Residential Prototype Sprinkler, Los Angeles Fire Test Program," Factory Mutual Research Corporation, Norwood, MA, February 1982.
6. Puchovsky, M. T., and Isman, K. E., *Fire Pump Handbook*, NFPA, Quincy, MA, 1998.
7. Isman, K. E., and Puchovsky, M. T., *Pumps for Fire Protection Systems*, NFPA, Quincy, MA, 2002.
8. *Inspection, Testing, and Maintenance of Water-Based Fire Protection Systems: The NFPA 25 Handbook*, NFPA, Quincy, MA, 2002.

NFPA Codes, Standards, and Recommended Practices

NFPA Publications National Fire Protection Association, 1 Batterymarch Park, Quincy, MA 02169-7471

The following is a list of NFPA codes, standards, and recommended practices cited in this chapter. See the latest version of the *NFPA Catalog* for availability of current editions of these documents.

NFPA 13, *Standard for the Installation of Sprinkler Systems*

NFPA 13D, *Standard for the Installation of Sprinkler Systems in One- and Two-Family Dwellings and Manufactured Homes*

NFPA 13R, *Standard for the Installation of Sprinkler Systems in Residential Occupancies up to and Including Four Stories in Height*

NFPA 14, *Standard for the Installation of Standpipe and Hose Systems*

NFPA 20, *Standard for the Installation of Stationary Pumps for Fire Protection*

NFPA 22, *Standard for Water Tanks for Private Fire Protection*

NFPA 24, *Standard for the Installation of Private Fire Service Mains and Their Appurtenances*

NFPA 25, *Standard for the Inspection, Testing, and Maintenance of Water-Based Fire Protection Systems*

17

The Performance-Based Design Process and Automatic Sprinkler System Design

Cindy Gier, PE

Performance-based design (PBD) is a concept developed from the design and code process that recognizes that not all buildings, systems, or processes fit into the confines of the adopted building and fire code requirements, many of which are prescriptive in nature. When a building is unusually large, unusually tall, or just unusual and challenging, it may be best to take a different approach to the design and review process for the structure. Performance-based design is a recognized method of design and analysis, used to focus on the manner in which a building is intended to perform for its occupants and emergency response personnel in the presence of hazards that may impact the building.

The traditional building design process, carried out in conformance with adopted codes and standards, is called *prescriptive* design. Prescriptive design dictates or mandates exactly how a building is to be designed, in a manner that may be compared to a recipe from a cookbook. A typical example of a code containing prescriptive criteria is *NFPA 5000®*, *Building Construction and Safety Code®*. The design process begins by classifying the building into prescribed occupancy use groups and typical construction types. Once this information is determined, the code applies requirements for area, height, number of stories, and other limiting conditions on the building. The length of exit access is prescribed. The need for a fire alarm or fire sprinkler system is determined by the text of the code depending on the size of the fire area, the particular hazards in the building, or the number of occupants. Requirements for interior finish and other building materials are specified. *NFPA 5000* also provides criteria for performance-based design options.

In classifying building use into typical prescriptive categories, generalizations must be made that may greatly impact the design of a building. For example, a nightclub and a school cafeteria may fall into the same occupancy classification since food and/or drink consumption take place in both facilities. Both uses vary dramatically in type of occupants, familiarity with their surroundings, noise levels, lighting levels, and general awareness. Prescriptive-based rules would require a fire alarm system if there are over 300 occupants. If a building has a floor area that is more than 5000 ft^2, an automatic sprinkler system is required. Performance-based design allows building designers to recognize the inherent differences in building use classifications and create a design that uniquely defines their client and facility.

Performance-based design is described in publications such as the *SFPE Engineering Guide to Performance-Based Fire Protection Analysis and Design of Buildings* [1]; NFPA 101®, *Life Safety Code*®; NFPA 5000, and the *ICC Performance Code for Buildings and Facilities* [2].

Prescriptive Codes Compared to Performance Codes

Prescriptive codes have been around since ancient times, and the method of "prescribing" exactly how a code requirement is to be met has historically been the basis for building design. The codes have allowed an equivalent method of design that can be approved through administrative channels. It is now recognized that all buildings do not fit into the tables, all numerical values in codes are not absolute, and the code is not the only acceptable method of compliance. This thought process has provided a way for fire protection engineers to design better and safer buildings that often exceed the minimum requirements laid out in the adopted codes of a jurisdiction. However, the fire protection engineer needs in-depth knowledge and good empirical data to analyze fire protection systems and their reliability, human response and behavior, and fire statistics. These ultimately create the input data that are used in the evaluation of the trial designs. The engineer also frequently becomes the "overseer" of the project through concept, design, plan review, site observation, and project completion. The site observation includes supervising the installation of systems, reviewing shop drawings, and commissioning or witnessing the testing of fire protection systems. It is important that the fire protection engineer carefully supervise the performance-based design and its implementation to ensure that all individuals responsible accomplish the intended goals and objectives of the project. Following the completion of the building construction, the fire protection engineer is involved with the oversight and development of "as-built" drawings and records, the "living documents" of the PBD.

Performance-based design often provides more flexibility in building design than prescriptive design. The design team needs to evaluate the appropriateness and applicability of the design, relative to the goals and objectives of the code, relative to life safety, property protection, mission continuity, and environmental protection. Further, responsible evaluation relative to the potential for failure of building systems and features and careful analysis to ensure that the design demonstrates compliance with the original performance goals and objectives are essential functions of the application of quality assurance to the performance-based design process.

Summary of the Performance-Based Design Process

Performance-based design is an engineering approach to fire protection design using the following principles as its basis:

- Established fire safety goals and objectives
- Deterministic and probabilistic analysis of fire scenarios
- Quantitative assessment of design alternatives against the fire safety goals and objectives using accepted engineering tools, methodologies, and performance criteria

Before a building design is started, the fire safety goals and objectives are established. These may be as basic as getting people out of the building in advance of the onset of untenable conditions, flashover, or building collapse; allowing firefighters adequate and safe access; or protecting the environment from fire exposure. Or they may be as complicated as keeping the smoke layer above the heads of exiting building occupants, preventing exposed steel from local or global collapse by minimizing heat production or heat transfer to the structural members, or providing fire protection systems in place of compartmentalization or fire-resistive construction. Methods to be discussed later in this chapter can be used in a performance-based analysis to determine the acceptable performance of proposed methods. The results are analyzed and the design refined to achieve the desired goals.

Team Concept

The design team serves an important role. Among their responsibilities are items such as maintaining agreement on the code-specified goals and objectives, settling questions on assumptions surrounding the proposed design, and ensuring that proper documentation is developed and available for review by the authority having jurisdiction (AHJ). A building design that relied heavily on automatic sprinkler systems would be expected to have an analysis centered on reliability and redundancy of system features. Members of the design team would have various points of view to consider, such as

- Reliability and redundancy of water supplies
- Reliability of electric motors (and electrical supplies)
- Diesel motors, if system operation depended on proper operation of a fire pump
- Expected sprinkler system performance in areas prone to seismic events
- Safeguards against tampering with sprinkler components

A necessary component of the performance-based design method requires the involvement of all members of the design team early in the design process, including the architect; the structural, mechanical, and electrical engineers; fire protection engineers; AHJs; the building and fire officials; property insurance companies; contractors; users; and owners. It is through the cooperative efforts of all these parties that the success of the performance-based design is achieved.

Ultimately, the AHJ's understanding of the design approach, including the preliminary design concepts, and expeditions approval of the final design, is essential to keep a project on track and ensure acceptance by all parties at the completion of the project. Although the design engineers and architects hold the professional responsibility for the design, in PBD it is all parties that must come to a consensus for the process to work.

The work of the PBD team does not stop at the start of construction. The team works as a unit to enforce agreements made during the design process and to ensure that the building will be inspected and maintained during its lifespan in a way that ensures conformance to the original PBD concepts and design approaches. Even during construction, the team is involved in evaluating, testing, redesign, and verification of the models used to predict building performance. The effect of change orders and modifications during construction must be monitored and evaluated by the entire team.

Performance-Based Design Flowchart

The performance-based design process involves the series of steps illustrated in Figure 17.1. These steps are outlined as follows:

Defining Project Scope. This step identifies the boundaries of the performance-based analysis or design as determined by the project stakeholders, or all who have an interest, vested or otherwise, in the project.

FIGURE 17.1
Steps in the Performance-Based Analysis and the Conceptual Design Procedure for Fire Protection.

(Source: *SFPE Engineering Guide to Performance-Based Fire Protection Analysis and Design of Buildings*, NFPA, 2000, Figure 3-2)

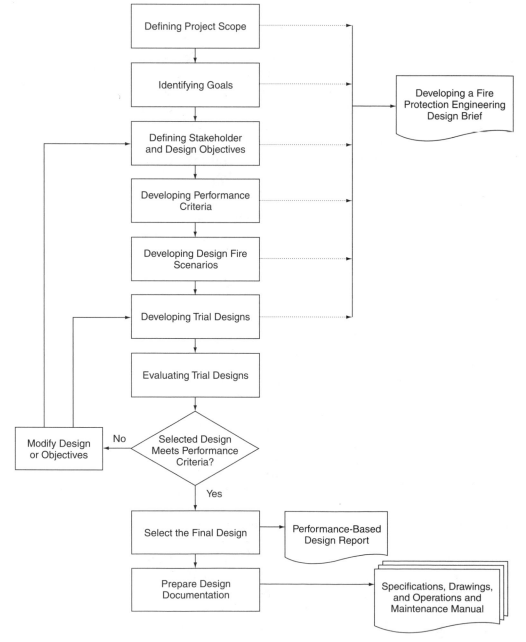

Identifying Goals. The goals for fire safety may be to protect life and property, to ensure the continuity of operations, and to limit the environmental impact of fire. Depending on the building, location, and views of the stakeholders, one or more of these goals may have to be met by the design.

Defining Stakeholder and Design Objectives. During this step, the project team defines the method used to achieve each goal, which is usually stated in terms of acceptable or sustainable loss or in terms of a desired level of risk.

Developing Performance Criteria. The process of developing performance criteria identifies the quantitative or qualitative measures that will be used to evaluate the design objective. These may include minimum code requirements, thermal effects, toxicity levels, visibility criteria or other measures.

Developing Design Fire Scenarios. At this point, the engineer predicts and analyzes possible fire situations that could occur within or in the proximity of the building. Taken into consideration are the building, occupant characteristics, and fire characteristics.

Developing Trial Design Characteristics. The trial design(s) includes proposed fire protection systems, construction features, and operations that are provided in order for a design to meet the performance criteria when evaluated using the design fire scenarios. These may include fire initiation and development, smoke management, fire detection and suppression, occupant behavior, and passive fire protection.

Evaluate Trial Designs. Using accepted methods, the assessment is performed to see if the goals and objectives are met by the proposed design method. The evaluation is performed on several levels to satisfy the objectives of all stakeholders. If the goals and objectives are not met by the trial designs, then the design must be modified. Code-established goals and objectives are deemed to be minimum and should not be reduced to a lower level.

Documentation. If the evaluation of the design is acceptable, then documentation is written to create a record of the process, with all decisions made by the design team, with supporting information provided.

Trial Designs and the Design Concepts Tree

The performance-based design process includes the development of trial designs. Using the design fire scenarios that the stakeholders have taken from the code or have independently established, the results are analyzed for achievement of the goals and objectives. When a code does not acknowledge performance-based design, a comparison evaluation might be necessary between the prescriptive-based design option and the result of the performance-based design. The different systems or features in a building combine to determine the level of fire safety in the building. Each component can be analyzed as a subsystem that is interdependent of other components or subsystems. NFPA 550, *Guide to the Fire Safety Concepts Tree*, provides a logic tree (see Figure 17.2) that can be used to work through the concepts and evaluate each subsystem. Not all the subsystems

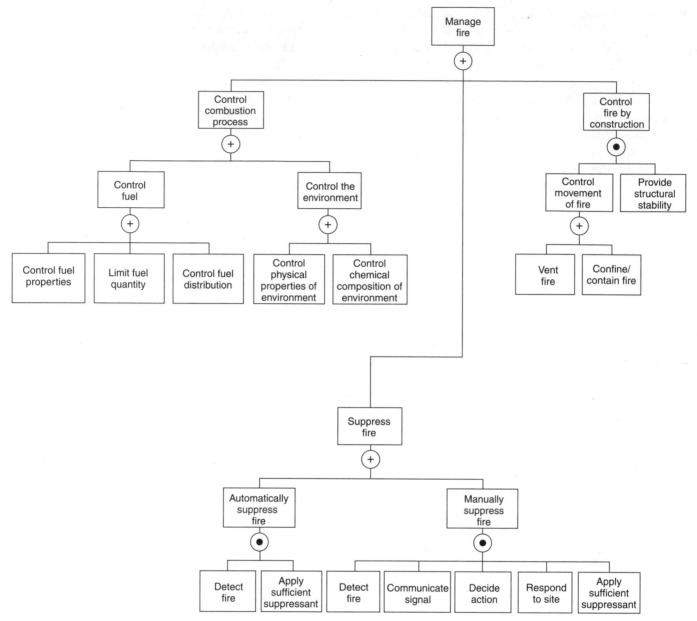

FIGURE 17.2
Manage Fire Branch of Fire Safety Concepts Tree.
(Source: NFPA 550, 2002, Figure 4.5.1)

are necessary for every design. Some PBD examples may have only a single subsystem of the analysis tree. For example, the design may rely on the presence of automatic fire sprinklers in the building, and the manual suppression option might not be used. Methods of fire detection, suppression, separation by compartmentalization, and surface flame spread and flammability control can be considered and compared by using a logic tree for analysis.

For example, to manage the trial fire design the combustion process could be controlled, the fire could be suppressed, or fire spread could be controlled by construction, or a combination of these design factors could be considered.

The first option allows the owner and designer to limit the combustibles present in the building or area of the building by selection of furnishings and interior finishes. They can remove, isolate, or select materials that meet these limitations. The third option allows the control of the spread of fire and smoke by building construction or systems that will remove smoke or otherwise limit its production. The second option allows for the suppression of the fire either by manual or automatic means. Because this concept is in keeping with the subject of this book, the following paragraph will explore this option.

Automatic fire suppression, combined with detecting the fire and applying sufficient suppressant, can quickly and reliably suppress the fire without immediate human intervention. When pre-action systems are used, the fire can be detected by either the heat-sensitive element in the fire sprinkler or by heat or smoke detection, and system operation can be predicated on one or a combination of these detection scenarios. With manual suppression several more steps are involved. The fire must be detected, then the signal communicated to the emergency responders. Someone must decide how to act on the signal and respond to the site. This may be accomplished by a local fire department or an in-house fire brigade. Following these delays, the suppressant can be applied to the fire and extinguishment will eventually result.

The fire safety concepts tree allows for the parallel comparison of the automatic and manual suppression methods. A timeline can be inserted into the boxes and a discussion can follow between the stakeholders about reliability, response times, and acceptable levels of loss in relation to the intrinsic delays of both systems.

Using Performance-Based Design in Building Design

The PBD method begins with identifying the prescriptive-based code requirements within a building, and project stakeholders may provide additional project goals. After the requirements are addressed, the engineer can investigate alternative methods or additional methods or systems to improve the fire protection features in the building. Finally, the designers will provide documentation to support how the suggested alternative methods meet the intent of the code requirements.

Frequently, the design solutions presented by this method utilize fire protection systems, such as fire alarm or fire suppression systems, to meet some of the stated objectives. As the team evaluates how it wants the building or systems to perform, it looks at several different aspects of the design. Active fire protection is often used to limit the spread of smoke or fire by automatic means, limiting the need for human intervention. Sprinklers can automatically begin suppression when heated to their activation temperature. Smoke or heat detectors can be used as an early warning system to alert occupants of the need to evacuate a building. By using active methods of fire protection, the relative risk may be more manageable to control and reduce by design. Active automatic suppression is considerably more likely to result in suppression in the early stages of the fire development than would be expected by manual suppression and would be much more likely to promote a successful result.

Fire protection engineers perform the job of evaluating and specifying the systems and methods of performance-based design with a focus on "How do we want the systems/building to perform?" This is accomplished through design guidelines and engineering specifications, drawings, and operation and maintenance manuals (O&Ms).

Sprinklers can be used to accomplish several tasks, including detection and suppression of a fire. Sprinklers can also act to limit heat exposure and smoke development, limit or control the size of a fire, limit the quantity of smoke production, and provide notification for occupant egress. It is important to note that no single PBD solution can be considered the only "correct" solution. If the performance of the systems or building govern the design, then every building will be slightly different, and the design will be expected to address the uniqueness of each structure.

A building design that has utilized PBD must be reevaluated for future building expansions, remodeling, or changes in occupancy. Even within the broad classifications of the code occupancy groups, a PBD building may have had more specific occupancy requirements that needed to be addressed. The advantage is that changes or modifications should be easy to consider if the correct procedure has been followed to evaluate possible scenarios. A new use should be compared to the same goals and objectives to see if the new use is within the limitations of the original design. A conservative approach could consider designing for increased occupancy to minimize the effect of anticipated building modifications.

Codes and Standards in Performance-Based Design

Performance-Based Codes and Standards Around the World

There are several codes and standards that include the performance-based design option, including the following:

- *Building Code of Australia*
- *Building Code of Canada*
- *Building Code of Japan*
- *Building Code of New Zealand*
- *Building Code of Norway*
- *Building Code of Spain*
- *Building Code of the United Kingdom*
- *ICC Performance Code*
- NFPA 1, *Uniform Fire Code*™, 2003 edition
- NFPA 72®, *National Fire Alarm Code*®, 2002 edition
- NFPA 101®, *The Life Safety Code*®, 2003 edition
- NFPA 914, *Code for Fire Protection of Historic Structures*, 2001 edition
- NFPA 5000®, *Building Construction and Safety Code*®, 2003 edition

NFPA 13 and Performance-Based Design

NFPA 13, *Standard for the Installation of Sprinkler Systems*, provides system and installation requirements and design approaches for fire sprinkler systems in buildings. NFPA 13 is primarily a prescriptive document that requires specific design criteria for a compliant system, including density, spacing, and calculations. Performance-based designs could result in the specification of a sprinkler system when none is otherwise required by prescriptive documents or could result in an occupancy determination that differs from NFPA 13, which could mean increased density, decreased spacing, and not permitting extended coverage sprinklers or reduced areas for quick-response sprinklers. Other possible performance-based design approaches include specifying different spacings, activation temperatures, or response time indexes than identified in NFPA 13.

There are many potential options for the use of sprinkler systems to meet performance objectives. In addition to area/density protection, sprinklers can be used to prewet unprotected structural members or to create a water curtain in place of a rated wall. Sprinkler systems can be provided above and below a ceiling to serve different performance objectives, or newer sprinkler technology can be utilized, such as ESFR or high K-factor sprinklers. The activation time of a sprinkler may be compared with the fire growth and peak heat-release rates to determine an equivalent level of safety to that afforded by the model codes for the passive fire protection of structural elements. The requirements of NFPA 13 are then used as a design guide and may be exceeded in PBD. The fire protection engineer will determine the "best" sprinkler solution for the project and present this to the stakeholders with appropriate documentation.

For example, the fire protection engineer may evaluate the ADD (actual design density) and the RDD (required design density) to determine if the density delivered to the floor is sufficient to control the actual fire load that is expected in a particular facility. This approach obviously will be more specific for a PBD building than what the occupancy classifications in NFPA 13 offer for a "typical" facility and can be used to appropriately determine if an "equal level of protection" is provided by the design.

NFPA 5000 and Performance-Based Options

Chapter 5 in the *Building Construction and Safety Code* includes a performance-based design option, which follows the same procedure presented in the *SFPE Engineering Guide to Performance-Based Fire Protection Analysis and Design of Buildings*. The method allows buildings or structures to take an alternative path to code compliance. When using this alternative path, it is not necessary to refer to the prescriptive provisions of the code.

The performance-based design option can be applied to an entire building, certain systems, or one unique portion of a project. When the entire building is not included in the analysis, it may be preferable to use the alternative means and methods section that is permitted in Section 4.3 of *NFPA 5000*. Chapter 4 of *NFPA 5000* outlines the goals and objectives of building design, the assumption of a single fire source for hazard analysis, and the options for building design and life safety compliance. *NFPA 5000* bases building design on the goals of life safety and property protection, not just prescriptive tables and numbers. The fire and life safety requirements include the goals of protecting the means of egress and limiting the spread of fire, which are fundamental fire protection concepts.

Section 5.3 of *NFPA 5000* addresses prescriptive requirements that have been retained even when a performance-based option is chosen. The means of egress, the building environment, and the mechanical and electrical systems in buildings must still meet prescribed code requirements for the building. Fire protection systems are required to comply with applicable NFPA standards for those systems and features, unless specifically addressed by the performance-based design.

Design characteristics may greatly influence the performance of systems or buildings. These are presented, along with some of the assumptions that are made concerning the fire scenarios or personnel involved. Some of these include occupant profiles, response characteristics, number of occupants and their locations in the building, and the capability of the staff to assist in the event of an emergency.

Finally, Sections 5.5 and 5.6 of *NFPA 5000* address the development of design scenarios and the evaluation of proposed designs. Guidance is provided in the *Building Construction and Safety Code Handbook* to help determine general design scenarios, fire growth, and ignition factors. The design scenarios include life safety concerns, fire development, and safety during building use.

One of the most important aspects of performance-based design is the documentation of the process, methods, and conclusions. Performance-based building design specifications require extra attention in outlining the systems, methods, and materials in relation to the more detailed process. The fire modeling used and the modeler's professional capabilities should be documented. By presenting a thorough chapter on this design option, *NFPA 5000* has provided the basis that a designer needs to work with a jurisdiction in developing, performing, and gaining acceptance and approval of a design.

Fire Modeling in Performance-Based Design

There are many calculation methods available to evaluate trial designs. Fire models simulate the response of a real fire, based on fuel load, room size, configuration, and openings. The advantage of using a fire model is that given the same input, the results will always be the same, no matter who is doing the analysis. Also, it is much less expensive to use a computer model than to have full-scale fire testing performed for a specific product or installation.

Field and Zone Models

Two types of fire models are generally available, zone models and field models. Zone models use two control volumes to describe a room—an upper layer and a lower layer, as shown in Figure 17.3.

FIGURE 17.3
Two-Layer Model with No Exchange Between Layers Except the Plume.

(Source: *The SFPE Handbook of Fire Protection Engineering*, NFPA, 2002, Figure 3-7.1)

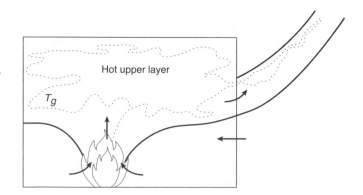

Hot upper layer

T_g

The two-layer zone model approach assumes that a fire environment can be approximated by a hot upper layer and a cool lower layer. Hot gases rise and collect at the ceiling and begin to fill the room from the ceiling downward, with the fire plume serving as a pump to inject these gases into the upper layer, drawing entrained air from the lower layer into the plume. The lower level includes the chosen fire scenario in a specific location in the room. The program calculates the time or temperature required for the smoke interface layer to reach a certain location. Common models used include FPETool, FASTLite, ASET-B, and the CFAST model used in the program FAST, available from NIST at its web site, www.nist.gov. FPETool, FASTLite, and DETACT-QS are packages of calculation tools that can perform several functions related to sprinkler design. They include a compartment fire model and several other tools that are used to calculate vent flows, radiant heating of objects, sprinkler actuation, detector responses, timed egress, and fire plume dimensions.

A field model divides the room into many more than two zones—up to thousands of grid points. Because this type of model can predict much greater variation in conditions within the layers, it takes much longer to run.

Both types of models—field and zone—are used to calculate the distribution of smoke and fire gases and the changes in temperature throughout a space during a fire. Fire models also can be used to predict sprinkler or detector response during a fire or the evacuation time from a location in a building.

For additional information on fire modeling, see *The SFPE Handbook of Fire Protection Engineering* [3] and the *Fire Protection Handbook* [4].

Plan Review of Performance-Based Design Sprinkler Systems

Plans and specifications for sprinkler systems that utilize performance-based design methods must include all criteria upon which the design is based. Criteria must be specific, including the design of the fire protection systems. Sprinkler zoning, spacing, discharge density, type of sprinklers, and many other items may be critical to the PBD. Criteria are also needed for all other systems or components that are interfaced with the PBD requirements.

A challenge involved in the enactment of a performance-based design is that future renovations of a building designed using performance-based design may be mistakenly evaluated using a prescriptive approach or a performance-based approach that is not congruent with the initial design approach. By maintaining a detailed set of record drawings, calculations, and Operations and Maintenance (O&M) manuals, the chances for establishing a sense of continuity for the building will be enhanced and any future modifications can be better addressed.

Summary

Performance-based design (PBD) is a team-based approach that bases design on the attributes of a building and analyzes the performance goals and objectives of the building in terms of fire protection. The PBD can make use of innovative technologies that can surpass the code-prescribed fire protection measures. In the case of sprinkler systems, the PBD can, for example, allow for a design that exceeds the requirements of NFPA 13 by installing new sprinkler technology, such as ESFR or high K-factor sprinklers in unique ways. Two types of fire models, zone and field, can be used when determining the need for a PBD approach to sprinkler systems. The PBD team, including architects, engineers, AHJs, the building and fire officials, insurance companies, contractors, users, and owners, must ensure the appropriateness and code-compliance of the design and must maintain detailed documentation, including the criteria upon which the design was based, to allow for future modifications to the building.

BIBLIOGRAPHY

References Cited

1. *SFPE Engineering Guide to Performance-Based Fire Protection Analysis and Design of Buildings*, NFPA, Quincy, MA, 2000.
2. *ICC Performance Code for Buildings and Facilities*, International Code Council, Inc., Country Club Hills, IL, 2003.
3. Quintiere, J. G., "Compartment Fire Modeling," in DiNenno, P. J., ed., *The SFPE Handbook of Fire Protection Engineering*, 3rd edition, NFPA, Quincy, MA, and SFPE, Bethesda, MD, 2002.
4. Beyler, C., and DiNenno, P. J., "Introduction to Fire Modeling," in Cote, A. E., ed., *Fire Protection Handbook*, 19th edition, NFPA, Quincy, MA, 2003.

NFPA Codes, Standards, and Recommended Practices

NFPA Publications National Fire Protection Association, 1 Batterymarch Park, Quincy, MA 02169-7471

The following is a list of NFPA codes, standards, and recommended practices cited in this chapter. See the latest version of the *NFPA Catalog* for availability of current editions of these documents.

NFPA 13, *Standard for the Installation of Sprinkler Systems*
NFPA 72®, *National Fire Alarm Code®*
NFPA 101®, *Life Safety Code®*
NFPA 550, *Guide to the Fire Safety Concepts Tree*
NFPA 914, *Code for Fire Protection of Historic Structures*
NFPA 5000®, *Building Construction and Safety Code®*

Additional Readings

Building Construction and Safety Code Handbook, NFPA, Quincy, MA, 2003.
Code Official's Guide to Performance-Based Design Review, Society of Fire Protection Engineers and International Code Council, 2003.
"SFPE White Paper on Ethical Peer Review," Society of Fire Protection Engineers, Bethesda, MD, 2003.
"SFPE Guidelines for Peer Review in the Fire Protection Design Process," Society of Fire Protection Engineers, Bethesda, MD, 2003.

18

Coordination with Other Professionals and Trades

Carl Anderson, PE

This chapter reviews the coordination among professionals, authorities having jurisdiction (AHJs), and owners that occurs prior to issuing construction documents for bid. A variety of potential conflicts may arise when engineers do not communicate with one another, with the AHJ, with the owner, and with the architect. These conflicts may include, but are not necessarily limited to, routing of underground mains and locations for piping entries into a building, location of the fire department connection (FDC), minimum requirements for riser room location and size, resolution of fire pump issues including power and space requirements, or special concerns specific to building design features such as atria and buildings of unusual configurations and heights.

A sprinkler system must be physically coordinated to avoid conflict with the installations of other trades, including ducts, piping, and conduits. Further, a sprinkler system must be electrically and logically coordinated with other trades to ensure proper function of electrical sprinkler components such as waterflow switches, tamper switches, air compressors, and fire pumps. In addition, a sprinkler system must be coordinated with respect to the performance objectives of equipment controlled or supervised by the fire alarm system, such as elevators, smoke control equipment, process control equipment, and other components. Physical and electrical coordination of a sprinkler system with other trades is an essential function of a fire protection engineer (FPE). Lack of coordination by the FPE may give the impression to the contractors that the FPE did little or nothing for the fee received. Also, conflicts that arise in the field during construction are generally more expensive and time-consuming to fix.

Coordination among trades to make systems integrate properly with interdependent systems is also essential. The minimum responsibility of the FPE with the respect to the coordination of a sprinkler system is to establish a requirement in the contract documents that the installing contractor attend one or more coordination meetings with other trades and to establish a framework for the proper conduct of such coordination meetings. The FPE should be in attendance at all coordination meetings and serve to resolve coordination disputes in a manner that ensures proper function of the sprinkler system and its interdependent systems.

Water Supply

Piping

Possible sources of conflict that may increase project cost or difficulty of construction related to underground supply mains are other water mains, electric lines, gas lines, wastewater lines, or any other underground utilities. Jurisdictions may enforce specific separation distance between various utilities, and the contractors' ability to do their work in the close proximity should also be considered. A thorough review of site drawings showing existing utilities and an extensive site survey are needed to adequately perform coordination of underground utilities with proposed sprinkler system piping. Consideration of the potential underground supply piping conflicts early on in the project may indicate that an alternative approach to the site may be more economical or may be more time-efficient. Discovering these problems later in the project can add cost and time to the installation of the sprinkler piping.

Current practice predominately involves the civil engineer specifying the underground portion of the sprinkler piping to a point 5 ft from the exterior face of the building and the FPE specifying the overhead portion of the system, often including underground piping from a point 5 ft from the exterior face of the building. However, it is strongly recommended that the FPE be involved in the thorough coordination of all underground piping. When this does not occur, sometimes the portion of the supply from 5 ft out to the base of the riser inside the building is missing from the specifications of both disciplines. If this seemingly small portion of work is not included in the specifications, it will not appear in anyone's contract and will lead to last-minute change orders. If this portion of the work is included in the specifications for the interior work, it may lead to confusion as to who is responsible for hydrostatically testing and flushing the entire supply main. If a utility contractor installs the underground main to a point 5 ft out and tests its portion of the work, the fire sprinkler contractor may then be forced to retest the entire underground piping system after connecting its portion of the underground pipe and running to the inside of the building. In order to ensure that the supply main is properly flushed and hydrostatically tested in its totality, the FPE must coordinate all piping and properly assign responsibilities for flushing and testing.

In cases where the civil engineer has been assigned the responsibility for coordination of underground piping, the FPE must be responsible for coordinating with the civil engineer early in the project to ensure that all portions of the sprinkler system are included in an appropriate portion of the specifications and that all testing and inspection responsibilities are clearly defined. The FPE must be responsible for ensuring that the underground sprinkler piping entries into the building are included in the appropriate portion of the specifications to facilitate testing of the entire underground supply at one time. This will necessitate careful coordination with the specifications for the interior piping to ensure that the pipe enters the building at the proper location in the riser room to allow clearance for backflow preventer and riser installation. Installation of piping to a yard FDC, which originates above the sprinkler system control valve and terminates on the exterior of the building and involves work by the sprinkler contractor and perhaps the site utilities contractor, will have similar coordination issues between trades and should be properly specified in the contract documents to correctly assign installation and testing responsibilities.

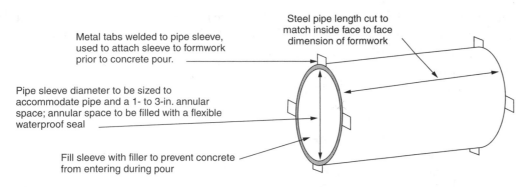

Figure 18.1
Pipe Sleeve.

Metal tabs welded to pipe sleeve, used to attach sleeve to formwork prior to concrete pour.

Steel pipe length cut to match inside face to face dimension of formwork

Pipe sleeve diameter to be sized to accommodate pipe and a 1- to 3-in. annular space; annular space to be filled with a flexible waterproof seal

Fill sleeve with filler to prevent concrete from entering during pour

Note: Sleeve should be located by fire protection engineer, and if penetrating a structural member, the location must be approved by structural engineer. Responsibility for sleeve placement must be assigned in project specifications.

Some AHJs may require that the underground piping be installed and tested by a licensed fire sprinkler contractor, and if this is the case, the FPE must be sure to verify and coordinate these features prior to the release of the contract documents and the awarding of the contract.

Other potential items associated with sprinkler water supply that may be overlooked if not properly coordinated are the locations of all pipe penetrations into the building, backflow preventor type and location, control valves, meters, and FDC location and style.

If the sprinkler supply is required to penetrate a structural member in the building foundation, coordination with the structural engineer is necessary. Also, installation of a sleeve in the foundation in advance of pouring, as shown in Figure 18.1, can be more economical than a core drilled hole after the foundation is poured. A typical budget cost for a 10-in.-diameter hole through a 12-in. concrete wall is about $200 per hole.

Some AHJs require an owner's sprinkler control valve to be installed at the property line. A post indicator valve (PIV), roadway box, or valve vault may be required, depending on the requirements of the AHJ. The underground contractor might assume that the public utility provides this item, but this is not always the case. The FPE should coordinate requirements, make a determination, and be involved in the selection of the contractor responsible for its installation. Where a PIV is required, the electrical engineer will need to make provisions for installation of wiring to a tamper switch on the PIV in most cases, while in some cases the AHJ may permit a chained and locked valve.

Metering

Some water purveyors require that a meter be installed on the fire sprinkler main. The FPE should confirm this requirement, determine the contractor responsible for installation, and determine the type and size of meter utilized. The friction loss associated with this device is often overlooked but needs to be included in the sprinkler system hydraulic calculations.

Backflow Prevention/Cross-Connection Control

The FPE should determine the type of backflow protection required by coordinating with the requirements of the local water purveyor. Some AHJs may require that the backflow preventer be installed at the property line. This requirement could necessitate the installation of a large utility vault or heated aboveground enclosure, depending on the climate and requirements of the AHJ. Other AHJs will permit the backflow preventer to be installed inside the building provided that the pipe from the public main to the backflow preventer is suitable for potable water. If the backflow preventer is installed at the property line, coordination with the electrical engineer is indicated, as the control valves on assembly may need to be equipped with tamper switches and the enclosure may need heat and light.

It is also required that provisions for testing the backflow preventer be provided. This may involve the installation of a removable cap on the system side of the backflow preventer or may require a pipe to be run to a drainage location. Where reduced pressure zone (RPZ) backflow is required, there is a significant requirement for drainage, which is often missed or only accommodated by a standard 2-in. drain.

For more information on cross connections and backflow, see Chapter 11, "Cross Connections and Backflow Prevention."

Fire Department Connection

The fire department connection (FDC), sometimes referred to as a siamese connection, is usually located on the street address side of the building. The FDC needs to be in a location accessible to the fire department and close enough to a fire hydrant so that fire crews can make their connection to bolster the fire sprinkler water supply quickly. Some AHJs have very specific location requirements, including minimum distance from buildings, maximum distance from fire hydrants, type of hose thread, and installation of quick-connect fittings. The FPE needs to determine all applicable special requirements of the local fire department. The FDC needs to be located to avoid conflicts with electric utility

FIGURE 18.2
Typical Fire Department Connection.

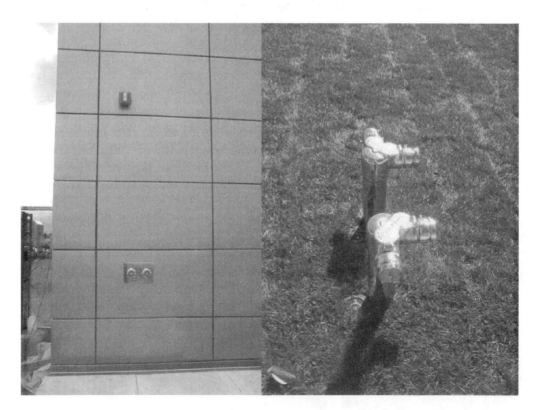

panels, dumpsters, landscaping, parking stalls, and any other obstruction to wrench rotation and hose lay from the fire department pumper.

Finally, the esthetic concerns of the architect and owner will need to be balanced with the issues just discussed in terms, for example, of exterior walls that may be unsuitable for FDC installation, such as glass walls or mirrored exterior wall panels. In cases where an FDC cannot be mounted on the exterior face of a building, underground piping may need to be supplied to facilitate its installation away from the exterior face of the building, closer to the parking lot or street. Figure 18.2 shows one typical type of FDC.

Interior Work

Sprinkler Valve Room Requirements

The transition from underground supply piping to the overhead portion of the fire sprinkler system occurs at the sprinkler control valve risers and should occur just inside the building shell, with a door to the sprinkler valve room, arranged for expeditious access from the building exterior.

Reliable heat and light are required for the sprinkler valve room as well as an adequately sized floor drain, especially when the system main drain does not discharge to the outside. It is often preferred to pipe the main drain and any inspector's test valves to the outside of the building; however, this is not always possible or permissible. For environmental reasons, some AHJs require that drains discharge to a sanitary sewer rather than directly to the outside, where the water may eventually find its way to the storm sewer.

Some AHJs require a door directly from the building exterior into the riser room; others may permit an expeditious access from the interior, such as an access from a lobby or the end of a corridor. Obviously, this requirement will limit possible locations for the riser room, and the architect will need to be made aware of agreements between the FPE and the AHJ on sprinkler valve room location early in the design process. Other times, there may be a requirement for a wall-mounted PIV that permits control of a sprinkler valve from the exterior of the building, in cases where a yard PIV has not been provided.

Coordination between the fire suppression specification and the electrical specification is necessary to ensure that all items of electrical equipment mounted on a fire suppression system are accounted for in the fire alarm design. Waterflow switches, tamper switches, and pressure switches, as well as air compressors, fire pumps, excess pressure pumps, and control panels, will be present in the sprinkler valve room, and some electrical devices may be found in locations outside the sprinkler valve room, such as would occur when providing separate control valves for each floor in a high-rise building or where providing separate valving for each tenant in an office or mercantile occupancy. The actual number and type of switches are determined by the type and number of systems to be installed.

Fire Pumps

When the fire sprinkler design calls for a fire pump, the sprinkler valve room becomes the fire pump room, and coordination considerations multiply. Fire pumps are a significant load on the floor and may need special attention from the structural engineer. The floor will need to be sloped such that water will be routed away from critical equipment, such as the pump and controller, and a floor drain of reliable capacity is required. Often the pump and controller are installed on concrete "housekeeping" pads elevated a few inches above the finished floor, to minimize damage from standing water in the pump room.

An item often missed until later in the project is the requirement that a pump room have a 1 or 2-hour fire resistance rating. Since this impacts the physical construction of the room and selection of doors and other opening protection, the architect will need this information early in the design process.

Drainage is a significant issue for fire pump rooms. An RPZ will result in very large drain sizes, often as large as the RPZ size to account for the dump valve on the RPZ. There have been significant losses associated with RPZ drainage. The flow must be addressed so that adequate drains can be provided.

The architect and electrical engineer will need to be made aware of the unique power supply requirements found in NFPA 20, *Standard for the Installation of Stationary Pumps for Fire Protection*, and NFPA 70, *National Electrical Code®*. (See Chapter 12, "The Role of Fire Pumps in Automatic Sprinkler Protection," for more detailed information on fire pumps.)

Some other coordination considerations when a fire pump is required are fuel tank location, protection for engine-driven pumps, the provision of secondary power supplies for electric motor-driven pumps, routing of exhaust pipes, location of test headers, and fire alarm interface.

Some jurisdictions require secondary containment for fuel tanks or may set volumetric limits for fuel storage, particularly within buildings, may establish requirements that specify the location of a fuel tank within a building, or prohibit fuel storage within certain buildings.

NFPA 20 requires a second source of power for electric motor-driven pumps where the height of a building exceeds the local fire department's pumping capacity. Exhaust piping for engine-driven pumps needs to be installed clear of combustibles and with due consideration for noise impacts on building occupants during frequent testing.

Test headers are required by most AHJs and NFPA 20 for fire pump testing and need to be installed in a location amenable to testing, taking into account the large volume of water to be discharged. (See Figure 18.3 for an example of test headers.) Some AHJs may permit the use of flow meters for some portions of fire pump testing, eliminating the need to discharge water as frequently. If this is the case, installation of the flow meter needs to be specified in the initial design of the system. Various aspects of pump operation or failure need to be monitored by the building's fire alarm system. The engineer should identify these monitoring points so that they will be included in the fire alarm system design.

Drain and Test Connections

A designer should make every effort to design a sprinkler system such that it can be fully drained from the sprinkler riser room. Where this is not possible, design documentation should indicate locations of drains, including where they are to discharge. Drains can easily flood a floor drain, and drains discharging to the building exterior may create a hazard where freezing could occur.

FIGURE 18.3
Typical Fire Pump Test
Connection.

Special System Requirements

Dry, preaction, deluge, and antifreeze systems are permitted by NFPA 13, *Standard for the Installation of Sprinkler Systems*. As these systems are more complex than wet pipe systems, a greater level of care in coordination with other professionals will be needed. Dry and preaction systems need power for the required air compressor, unless shop air is supplied, which may be the case in some industrial settings. Preaction systems require that a detection system be extended into all spaces served by the preaction sprinkler system in order to permit the preaction valve to open and the sprinkler system to operate (see Figure 18.4). AHJs generally require that antifreeze and foam systems be equipped with a reduced-pressure backflow prevention device. Buildings protected by deluge systems need drains sized to handle full flow during testing, and in facilities where paint or chemicals could mix with water from sprinklers, some AHJs may require containment for water discharged by the systems.

High-rise Buildings. Flow and tamper switches are required at each floor or zone, with flow switches coordinated to properly interface with the smoke control system wherever applicable. If individual floor/zone control valves are not provided, a flanged joint or mechanical coupling shall be used at the riser at each floor for connections to piping serving floor areas in excess of 5000 ft^2. Flow and tamper switch zoning needs to be coordinated with the fire alarm specification. Pressure-reducing valves (PRVs) are required where pressures exceed the allowable operating pressures of system components or where pressures at hose valves exceed the limit set forth in NFPA 14, *Standard for the Installation of Standpipe and Hose Systems*. Where PRVs are installed, a 3-in. drain riser adjacent to each standpipe may be needed to allow proper testing

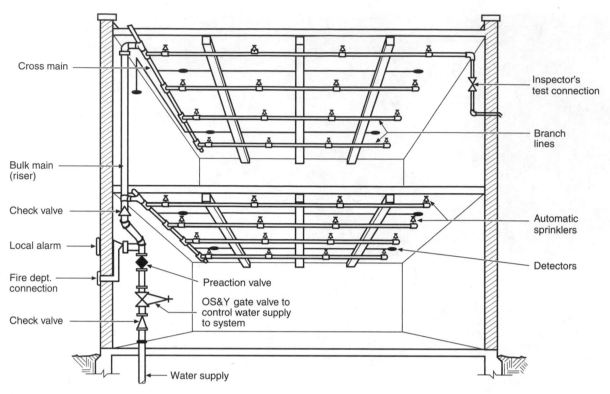

Figure 18.4
Typical Preaction Sprinkler System.
(Source: *Fire Protection Handbook*, NFPA, 2003, Figure 10.11.11)

of the PRVs. A consideration in the initial sprinkler system design might be the minimization of the need for PRVs by properly sizing the fire pump or perhaps not using the fire pump to supply the sprinkler system on lower floors of the building.

Atria. Coordination with the architect and local building and fire code officials is critical when designing a sprinkler system for a building with an atrium, as in Figure 18.5. *NFPA 5000*®, *Building Construction and Safety Code*®, requires that an atrium be separated from surrounding areas by a fire barrier with a minimum 1-hour fire resistance rating; however, sprinkler protection may be used to allow some relaxation of this requirement. *NFPA 5000* spells out very specific requirements for the placement of sprinklers associated with atrium protection and such protection needs to be specified in the architectural details of the building as well as clearly spelled out in the project specifications. Generally, glass walls and inoperable windows are permitted where sprinklers are provided on both sides of the glass within 1 ft of the glass, at six-ft spacing, and arranged to wet the entire surface of the glass. Sprinklers are not required on the atrium side of the glass if there is no walkway or other floor area on the atrium side above the main floor. Refer to subsection 8.12.3 of the 2003 edition of *NFPA 5000* for more details.

Some AHJs may require that the sprinklers in an atrium activate the atrium smoke control system, which requires that the atrium sprinklers be piped and

FIGURE 18.5
Atrium.

controlled independently of other sprinklers on the same floor that are not in the atrium.

Unenclosed Vertical Openings. Water curtains and draft stops are required around unenclosed floor openings, such as those associated with escalators or open stairs. The sprinklers must have a maximum spacing of 6 ft, while simultaneously maintaining the minimum listed distance between sprinklers, and be located between 6 and 12 in. from the draft stops on the side away from the opening. The draft stops must be located immediately adjacent to the opening, be at least 18-in. deep, and be of noncombustible or limited-combustible material arranged such that they will stay in place before and during sprinkler activation. It is important that the architect be made aware of minimum depth of the draft stops in order to comply with minimum headroom requirements and to accommodate esthetic considerations early in the design.

Elevators. The needs versus the hazards of sprinklers in elevator shafts and equipment rooms is an old debate. Fire protection professionals cite examples of fires in equipment rooms and shafts, whereas elevator professionals cite the danger of applying water to elevator electrical equipment and friction brakes.

NFPA 13 requires sidewall sprinklers at the bottom of the hoistway, except for enclosed, noncombustible elevator shafts not containing combustible hydraulic fluids, sprinklers in elevator machine rooms, and upright or pendent sprinklers at the top of hoistways, except where the hoistway is noncombustible and the elevator car enclosure meets ASME A17.1, *Safety Code for Elevators and Escalators* [1]. Note that ASME A17.1 addresses additional requirements intended to increase the safety of elevator operation should a fire actuate a sprinkler in an elevator shaft or machine room. Sprinkler protection is only permitted per ASME A17.1 where all risers and returns are located outside the machine rooms or hoistway, branchlines in the hoistway supply sprinklers at not more than one floor level, and a means is provided to automatically disconnect the mainline power supply to the affected elevator upon or prior to the application of water from sprinklers. Smoke detectors are not used to activate sprinklers in these spaces or to disconnect the mainline power supply. The requirement to disconnect power to the elevator machinery applies to sprinklers located in the machine room or in the hoistway more than 2 ft. above the pit floor. The means of disconnecting power must be independent of the elevator control and shall not be self-resetting. The activation of sprinklers outside of the hoistway or machine room shall not disconnect the mainline power supply.

Most AHJs require compliance with ASME A17.1 and many have additional requirements or policies dictating how the provisions of ASME A17.1 are to be accomplished. Methods to disconnect the main line power supply to any elevator equipment might include a shunt trip activated by heat detectors, a preaction sprinkler system that uses the heat detection system or a waterflow switch to open the shunt trip circuit breaker, or a dry pipe system that uses the waterflow switch to open the shunt trip circuit breaker. The FPE should determine requirements of the local AHJs and include the details in the specifications.

Fire Alarm–Sprinkler Interface

Monitoring and local or remote transmission of signals may be required for water flow, valve tamper, air pressure, fire pump, and other items. The FPE should specify and properly assign responsibilities for such interfaces. The sprinkler contractor typically provides and installs signal-initiating devices (flow switch, tamper switch, pressure switches, etc.), and the fire alarm contractor is typically responsible for connecting them to the fire alarm system.

Engineers should be aware of the *NFPA 72®*, *National Fire Alarm Code®*, requirement that any valve that can isolate a pressure switch used to send a waterflow alarm be supervised open with a tamper switch. Pressure switches are typically used to initiate waterflow alarm on dry pipe, preaction, and deluge systems.

Summary

The engineer tasked with the design of a fire sprinkler system needs to consider not only the prescriptive requirements found in NFPA 13 but also how the sprinkler system will fit into the overall design and function of the building. Anticipating the needs of various AHJs, considering the impact of special building features such as atria or smoke control requirements, and addressing functional and esthetic issues important to the owner and architect are an important function of the specifying FPE.

BIBLIOGRAPHY

Reference Cited

1. ASME/ANSI A17.1, *Safety Code for Elevators and Escalators*, American Society of Mechanical Engineers, New York, 2000.

NFPA Codes, Standards, and Recommended Practices

NFPA Publications National Fire Protection Association, 1 Batterymarch Park, Quincy, MA 02169-7471

The following is a list of NFPA codes, standards, and recommended practices cited in this chapter. See the latest version of the *NFPA Catalog* for availability of current editions of these documents.

NFPA 13, *Standard for the Installation of Sprinkler Systems*
NFPA 14, *Standard for the Installation of Standpipe and Hose Systems*
NFPA 20, *Standard for the Installation of Stationary Pumps for Fire Protection*
NFPA 70, *National Electrical Code*®
NFPA 72®, *National Fire Alarm Code*®
NFPA 5000®, *Building Construction and Safety Code*®

19

Professional Issues in Fire Protection Engineering Design

Robert M. Gagnon, PE, SET, FSFPE

Designers of fire protection systems are expected to possess expertise and experience commensurate with the services performed. The public demands that engineers conduct their practice in a manner befitting their ethical expectations, perform services within their area of expertise, and maintain a level of continuing education that keeps them on the crest of the wave of the technology associated with their chosen field of endeavor. This chapter provides advice and resources that assists practitioners in meeting these expectations.

Whether evaluating water supply, performing layout of sprinkler systems, hydraulically calculating sprinkler systems, or integrating sprinkler systems into a coordinated and automated building system, fire protection engineers must constantly be aware that what they do has the capability to save lives but could fall short of that goal if not done well.

Professional Ethics

Engineers enjoy a reputation that is envied by the general public. In a 1998 Harris poll, sponsored by the American Association of Engineering Societies (AAES), that evaluated the level of prestige that Americans impart to various professions, engineering ranked with doctors, scientists, teachers, ministers, and policemen as among the nation's most respected professions [1].

Of note is the fact that the public directly interacts with doctors, teachers, ministers, and police officers and is therefore more familiar with the precise duties of these professionals. The public does not generally interact directly with engineers and scientists yet still ranks them as highly respected. Further, there is confusion related to the difference between scientists and engineers (engineers actually perform many of the tasks attributed by the public to scientists). Considering the general lack of understanding among the public as to precisely what engineers do, it is significant that the public concurrently bestows respect upon them. The gap between ignorance of what engineers do versus the relatively blind respect bestowed upon them only increases the need for engineers to act competently and ethically when performing engineering services. Engineering related to the design of sprinkler systems is certainly part of the disconnect since the welfare and safety of the public are directly tied to the proper application of the current technology in fire protection to the saving of lives, and since the public has very little understanding of sprinkler systems and the manner in which they are designed.

People look up to engineers and depend upon them for their skill. Young people aspire to design and build the things with which engineers are associated. Students work toward attaining the reputation and respect that engineers enjoy. Engineering has long been considered a profession where the interests of public safety take precedence over profit or other commercial concerns. Engineers are consistently rated among the elite professions and are extended a measure of trust for their expertise, professionalism, and qualifications, especially on those occasions when the public enters buildings and crosses bridges, in a manner similar to the trust extended to a pilot of a passenger airliner. The pilot's name and qualifications are unknown, but it is expected that the pilot possesses the experience and qualifications to execute the services performed. Engineers perform a service that is no less compelling and with consequences no less immediate. The failure of a bridge, a machine, or a fire protection system has the potential for consequences that can damage the reputation of engineers for a considerable period of time, perhaps indefinitely.

The reputation of engineers can be viewed as a delicate and potentially fragile structure, built upon the combined contributions of all engineers, past and present, who have served the profession and society, from the building of the aqueducts many centuries ago, to the current development of interplanetary travel craft. The improper action of any one engineer has the potential to negatively affect the reputation of all engineers, causing significant damage to this painstakingly constructed reputation. Those practicing the profession must be vigilant to maintain the exceptional respect for the profession of engineering. The reader may be able to create a list of professions that were once held in high esteem but whose reputations suffered by a series of public disclosures of wrongdoing, or where the safety and welfare of the public was negligently compromised. This cannot be permitted to happen to our profession.

The NSPE Code of Ethics for Engineers

The National Society of Professional Engineers (NSPE) has published a document that provides advice on ethical behavior for engineers, the NSPE Code of Ethics for Engineers, which can be obtained in its entirety from the NSPE web site, http://www.NSPE.org. The Code of Ethics has been adopted with some variation by each state in the U.S. and thereby becomes a legal requirement for all licensed engineers, and each state has a board of ethical review that enforces violations of the Code of Ethics that have been reported to the state licensing board.

The Code of Ethics for Engineers, including the Fundamental Canons, Rules of Practice, and Professional Obligations, provides specific advice to assist engineers in the resolution of ethical dilemmas, including such issues as guilt by association, the application of engineering supervision to drawings and documents, the honest and timely performance of engineering service, and the protection of the profession of engineering. The Code of Ethics can be used to provide valuable assistance to young engineers as well as practicing professionals.

The NSPE Preamble. The Preamble of the Code of Ethics for Engineers emphasizes the importance of the profession and stresses the value of the performance of each engineer relative to the profession as a whole. The honesty, integrity, impartiality, fairness, and equity of all engineers are cited as traits expected of an engineer. It is further stated that engineers have a vital impact on the quality of life for all people and that it is their mission to protect the health, safety, and welfare of the public. The Preamble instills in engineers that their behavior be ethical and professional.

NSPE Fundamental Canons. The Fundamental Canons of the NSPE Code of Ethics promote six concepts under which engineers are assumed to perform: holding public safety paramount over all other design considerations, working within one's education and training, dealing with the public truthfully, being a faithful representative of their employers, avoiding deceptive acts, and promoting personal and professional conduct that honors the profession of engineering.

The Fundamental Canons are the cornerstone of the Code of Ethics. They emphasize the following points:

- That engineers are to keep the safety, health, and welfare of the public as their highest priority
- That engineers are to practice their services in areas for which they are trained and competent
- That engineers are to be truthful and objective with public statements
- That engineers are to serve their employers faithfully
- That engineers are to avoid deceptive acts
- That engineers are to conduct themselves as if the reputation of the profession of engineering singularly depended on their responsibility

NSPE Rules of Practice. The Rules of Practice provide greater detail to the Fundamental Canons so that engineers can more effectively use them as a reference when encountering ethical dilemmas.

With respect to health and safety of the public, engineers are obligated to

- Provide notification when their engineering judgment is overruled
- Provide engineering approvals to those items that conform to national standards
- Maintain the confidentiality of information obtained in the course of engineering work, unless authorized to reveal the information
- Refrain from associating with dishonest firms
- Practice engineering lawfully and associate only with those that do
- Report violations of the Code of Ethics to the appropriate authorities

Relating to practice within one's competence, the Code of Ethics specifies the following:

- Obtain training and experience in each field of engineering endeavor practiced.
- Do not sign any item in which one's competence is lacking.

- Do not sign any item that was not prepared under one's direction and control.
- Engineers are permitted to supervise an entire project, provided that each portion of that project is signed and sealed by engineers competent in the respective field of endeavor.

Relating to the truthfulness of the public statements of engineers, the Code specifies:

- Be objective and truthful and provide all relevant information.
- Make statements based on knowledge and competence.
- Avoid conflicts of interest.

Relating to faithfulness to employers, the Code directs engineers to

- Disclose conflicts of interest to employers or clients
- Not accept a fee from two parties for the same project without full disclosure
- Not take money or gifts from outside agents for work for which they are responsible
- While employed in public service, not make decisions on services solicited or provided by them to private or public engineering practice
- Not accept contracts from a governmental body on which a member of their organization serves

    With respect to deceptive acts, the Code specifies:

- Do not falsify or misrepresent one's qualifications.
- Do not participate in influence peddling.

NSPE Professional Obligations. Obligations related to honesty and integrity include

- Acknowledging errors and not distorting facts
- Advising clients when engineers believe a project will not be successful
- Working outside regular employment only with permission and when it will not negatively affect regular work
- Not recruiting engineers from competing firms under false pretenses
- Not allowing self-interest to supersede the integrity of the profession

Obligations related to serving the public interest include

- Participating in civic affairs
- Sealing documents that comply with applicable standards
- Promoting the profession of engineering to the public

Obligations related to proper professional conduct include

- Public statements that do not misrepresent or omit material fact
- Truth in advertising for engineering employment
- Proper credit when using attributed information in journal articles

Obligations related to confidentiality include

- Avoiding using information from one client to gain an advantage with another client, without permission
- Not using specialized knowledge to assist an adversary interest

Obligations related to conflict of interest include

- Not being influenced by conflicting interests
- Not accepting any commissions or allowances from contractors in association with a project that involves both parties

Obligations related to honesty in soliciting employment include

- Not accepting contingent commissions related to situations where an engineer's judgment can be compromised
- Performing part-time work ethically and consistently with the primary employer's policies
- Not using an employer's equipment for outside engineering work without permission

Obligations involving the reputation of engineers include

- Reporting engineering misconduct to the proper authorities
- Reviewing the work of another engineer only with permission and knowledge
- Reviewing the work of other engineers when the duties of governmental, industrial, or educational engineers require it
- Comparing products when an engineer serves in sales or industry

Obligations related to an engineer's personal responsibility include

- Complying with state registration laws
- Not using a nonengineer to cloak unethical acts

Obligations related to attributing credit and acknowledging proprietary interests include

- Providing names of persons who contributed to a finished product
- Not using another engineer's design without permission
- Entering into an agreement that defines ownership of design work
- Allowing work output performed for an employer to become the property of the employer
- Continuing the development of engineering skills throughout their career

The SFPE Code of Ethics

The Society of Fire Protection Engineers has published a Code of Ethics, shown as Box 19.1. This code can assist fire protection engineers in performing design of sprinkler and other fire protection systems. Some of the 14 canons of the SFPE Code of Ethics have similarities to the NSPE Code of

Ethics but are specifically tailored for the fire protection engineering profession. The Code defines a fire protection engineer, stresses the crucial nature of the work performed by fire protection engineers, and provides assistance in resolving ethical dilemmas. SFPE's web site, http://www.SFPE.org, provides additional information on professional issues for fire protection engineers.

BOX 19.1
SFPE Code of Ethics

What Is Fire Protection Engineering?

Canon of Ethics for Fire Protection Engineers

Preamble

Fire protection engineering is an important learned profession. The members of the profession recognize that their work has a direct and vital impact on the quality of life for all people. Accordingly, the services provided by fire protection engineers require honesty, impartiality, fairness and equity, and must be dedicated to the protection and enhancement of the public safety, health and welfare. In the practice of their profession, fire protection engineers must maintain and constantly improve their competence and perform under a standard of professional behavior which requires adherence to the highest principles of ethical conduct with balanced regard for the interests of the public, clients, employers, colleagues, and the profession. Fire protection engineers are expected to act in accordance with this Code and all applicable laws and actively encourage others to do so.

Fundamental Principles

Fire protection engineers uphold and advance the honor and integrity of their profession by:

   I. Using their knowledge and skill for the enhancement of human welfare.
  II. Being honest and impartial, and serving with fidelity the public, their employers, and clients;
 III. Striving to increase the competence and prestige of the fire protection engineering profession.

Canon of Ethics—Knowledge and Skill

Canon 1

Fire protection engineers shall be dedicated to the safety, health, and welfare of the public in the performance of their professional duties. If fire protection engineers become knowledgeable of hazardous conditions that threaten the present or future safety, health or welfare of the public, then they shall so advise their employers or clients. Should knowledge of such conditions not be properly acted upon, then fire protection engineers shall notify the appropriate public authority.

Canon 2

Fire protection engineers shall consider the consequences of their work and societal issues pertinent to it and shall seek to extend public understanding of those relationships.

Canon 3

Fire protection engineers shall be encouraged to contribute services for the advancement of the safety, health and welfare of the community and support worthy causes.

Canon of Ethics—Honesty and Impartiality

Canon 4

Fire protection engineers shall perform professional services only in the areas of their competence and after full disclosure of their pertinent qualifications.

Canon 5

Fire protection engineers shall be honest and truthful in presenting data and estimates, professional opinions and conclusions, and in their public statements dealing with professional matters and shall not engage in improper solicitation of professional employment or contracts.

Canon 6

Fire protection engineers shall act in professional matters for each employer or client as faithful agents or trustees and shall not disclose confidential information concerning the business affairs or technical processes of any present or former client or employer without consent.

Canon 7

Fire protection engineers' decisions shall be made and actions taken without bias because of race, religion, sex, age, national origin or physical handicaps.

Canon 8

Fire protection engineers shall make prior disclosure to all interested parties of all known or potential conflicts of interest or other circumstances which could influence or appear to influence their judgment or the quality of their service.

Canon of Ethics—Competence and Prestige

Canon 9

Fire protection engineers shall perform services and associate with others only in such a manner as to uphold and enhance the honor and integrity of the profession.

Canon 10

Fire protection engineers shall continue their professional development throughout their careers and shall provide opportunities for the professional development of those engineers under their supervision.

Canon 11

Fire protection engineers having knowledge of any alleged violation of this Code shall cooperate with the proper authorities in furnishing such information or assistance as may be required.

Canon 12

Fire protection engineers shall accept responsibility for their actions, seek, accept, and offer honest criticism of work, properly credit the contributions of others, and shall not accept credit for the work of others.

Canon 13

Fire protection engineers shall strive to advance the knowledge and skills of the profession and to make these advancements available to colleagues, clients, and the public.

Canon 14 (adopted 9/17/92)

Fire protection engineers shall perform professional services using only those engineering methods and tools for which they have an adequate understanding of the correct use and limitations.

The NICET Code of Responsibility for NICET-Certified Engineering Technicians and Technologists

The profession of engineering technology has a mission no less compelling than that of engineers and a reputation whose preservation is no less compelling. Engineering technologists, often referred to as technicians, perform layout of fire protection systems, based upon the designs provided by a fire protection engineer or other responsible engineer.

The National Institute for Certification on Engineering Technologies (NICET) has published a Code of Responsibility for use by engineering technicians in resolving ethical situations, which can be obtained in its entirety from the NICET web site, http://www.NICET.org.

The NICET Code of Responsibility recognizes the importance of work performed by engineering technicians; the impact that it has on the safety, health, property, and well-being of the public; and the value of integrity and competence in the development of fire protection design. Although its fundamental principles are similar in nature to the NSPE Code of Ethics, it is specifically tailored for engineering technicians. A summary of the key points of the NICET Code of Ethics includes the following items:

- Hold the health, safety and welfare of the public as the highest priority, and report violations that endanger that priority.
- Perform services commensurate to one's education, training, and experience.
- Perform services with efficiency, competence, fidelity, and honesty.
- Admit errors and do not distort facts.
- Avoid conflicts of interest and report violations.
- Avoid giving and receiving bribes.
- Maintain professional competency.
- Do not misrepresent one's qualifications.
- Do not divulge information obtained on a project without permission.

The SFPE Position Statement, "The Engineer and the Technician: Designing Fire Protection Systems"

The Society of Fire Protection Engineers (SFPE) has published an important position paper that details the relative responsibility of engineers and technicians, which can be obtained in its entirety from the SFPE web site, http://www.SFPE.org. Volunteers from a number of organizations representing various interests in the sprinkler design process, interested in assisting design professionals with meeting the expectations of the profession, came together to create the SFPE Position Statement. Practitioners can use the SFPE Position Statement as a guide while performing design and layout services.

Of importance in the SFPE Position Statement is the definition and scope it gives to the respective roles of engineers and technicians. It was considered important by the profession to establish roles and guidelines for engineers and technicians so that they may work within their respective roles and not practice outside them, with public safety being the primary measure.

Fire protection engineers are in the best position to view a building project in its totality, as opposed to viewing it as simply a sprinklered risk. Two churches that are seemingly identical can be considered to illustrate this principle. One congregation was in a position to fabricate its church entirely of solid marble. The other congregation could not afford the marble, but commissioned a wood-frame church lined with polymethylmethacrylate (PMMA). Both churches looked identical in size and appearance. Which one was light hazard? NFPA 13, *Standard for the Installation of Sprinkler Systems*, states that light hazard occupancy is defined as "occupancies or portions of other occupancies where the quantity and/or combustibility of the contents is low and fires with relatively low rates of heat release are expected." Note that definitions for "low" and "relatively" are unstated and that there is no direct statement that a church is a light hazard occupancy. Examples of typical light hazard occupancies are given in the annex of NFPA 13, which is helpful commentary but is not code language. Further, there no direct mention of PMMA (a plastic with extraordinarily high flame spread that produces a potentially deadly toxic gas when undergoing combustion). Most people would agree that the marble church is light hazard, but it is hoped that there would be agreement the PMMA church should be evaluated by a fire protection engineer for egress, containment, detection, and suppression.

However, the contents of the building are the focus of the NFPA 13 definition, not the construction of the building. What if the interior of the marble church was entirely transformed into a manger scene diorama, with straw, old weathered wood structures, and gas-fed torches to provide lighting for the scene. Would it still be light hazard? What if the diorama were fabricated with PMMA instead of straw and wood? Would it still be light hazard? It is clear that a competent fire protection engineer should evaluate life safety, detection, and suppression for these scenarios as well.

How about a common hardware store? In the 1950s, these were simple buildings to evaluate. Today, they contain pool chemicals, fertilizers, paints, and plastic items stored on pallets or in flammable or encapsulated containers to heights of 25 ft or more. A competent engineering evaluation is in order for this scenario as well. The way things are stored has changed, and people are at risk without an engineering evaluation of the totality of the risk in a building.

Liability and Responsibility

Engineers performing sprinkler system design may encounter cases where liability is a concern, whether it be liability for errors and omissions by the engineer; liability for damage to personal vehicles, fire protection equipment, or other equipment while on a job site; or liability associated with being a "good Samaritan" in a time of emergency or crisis. In some cases, it is a contractual requirement to have some level of liability coverage for an engineer performing sprinkler or fire protection system design, supervision, or testing. In other cases, it may be prudent to carry such coverage even though not required.

The National Society of Professional Engineers does important work to protect the reputation of engineers, and the NSPE web site, http://www.NSPE.org, has a wealth of information to assist practitioners. Engineers enjoy numerous benefits but have liabilities and responsibilities that must be considered. NSPE, in conjunction with the American Institute of Architects (AIA), the American Consulting Engineers Council (ACEC), the American Society of Civil Engineers (ASCE), and their affiliated state organizations, has created a model that can be adopted by states or local jurisdictions. The model addresses proposals for limitation of liability to engineers in certain situations, such as "good Samaritan" acts, where engineers intercede in emergency situations to apply engineering knowledge on a voluntary, nonpaid basis in times of crisis, such as a catastrophic fire or earthquake. The concern is that unless liability is limited for such acts, some engineers may opt not to act, even when the loss of lives is imminent. The model provides immunity for certain properly authorized acts by engineers, in hopes that engineers will do the right thing when appropriate protections are provided. Other areas covered by the model are certificates of merit and workers compensation.

NSPE has been active in promoting state and local liability reforms, including

- Statutes of repose
- Sole source worker's compensation statutes
- Certificate of merit statutes
- Good Samaritan statutes

Some states have been more successful than others in adopting such reforms. Examples where states have been successful in promoting liability reform include the following:

- Establishment of architect and engineer liability coalitions, representing all states
- Development of legislative language approved and supported by all of the state A/E organizations
- Working through industry-wide coalitions

State legislators are much more responsive to considering proposed legislation that has gained the endorsement of most if not all of the affected architectural and engineering groups in the state. NSPE's goal is to encourage architects and engineers and their state AIA, ACEC, ASCE and NSPE groups to enact legislation to reduce the number of lawsuits brought against architects and engineers.

Aside from the legal initiatives by NSPE relative to liability limitation, engineers still retain a level of liability on a given project. They are expected, and sometimes required, to hold some level of insurance that covers them with respect to errors and omissions. This insurance provides relief or correction where errors or omissions result in damage, loss, or improper design of a project. In addition, it is often required that engineers hold some level of job site liability coverage. Such coverage may include protection from damage to or from automobiles while on a construction site, damage from improper or incorrect operation of machinery or devices while on a job site, or other accidental damage caused by an engineer while on a construction site.

Continuing Education

The previous discussion of ethics had a recurring theme relating to competency for a given project, which may require occasional reeducation and training. The previous discussion on liability may be enlightened by the realization that one of the better ways to reduce liability on a project is to ensure that one's knowledge is in keeping with the complexity of a project. For example, although engineers are expected to be knowledgeable about current sprinkler technology, some may not be capable of integrating a sprinkler system with a fire alarm and detection system, smoke control system, elevator control system, HVAC control system, security system, and building automation system. Further, some engineers may require additional training in order to perform performance-based design of sprinkler and other fire protection systems, fire modeling, design of fire-rated assemblies, or egress modeling. Even prescriptive sprinkler system design by itself is growing and changing so rapidly that engineers require occasional updating in the newest technology.

Fire protection engineering as a discipline is continually expanding in scope and increasing in complexity. Fire protection engineers must keep up with the current technologies in fire protection engineering by participating in professional activities and by attending seminars and training sessions.

Continuing Education from the Society of Fire Protection Engineers

The Society of Fire Protection Engineers and its local chapters are an excellent source for the training and the continuing education of the fire protection engineering professional. Refer to the SFPE web site, http://www.SFPE.org, for information on current course offerings.

Continuing Education from the National Fire Protection Association

The National Fire Protection Association accomplishes a number of missions, which include fire protection education. The NFPA provides training in a number of topics that will assist practitioners in advancing and sharpening their skills. These courses are offered regionally and at the NFPA World Safety Congress and Exposition, held yearly. Complete information on NFPA training opportunities may be found on the NFPA web site, http://www.NFPA.org.

Continuing Education from Sprinkler Associations

The American Fire Sprinkler Association (AFSA) and the National Fire Sprinkler Association (NFSA) both offer a variety of courses. For information on their current courses, see the AFSA web site, http://www.firesprinkler.org, and the NFSA web site, http://www.NFSA.org.

Colleges and Universities

There are resources for further training in fire protection engineering, and degrees in fire protection engineering are available. Some of the schools offering training in fire protection engineering include the following.

Lund University Department of Fire Safety Engineering. The department has a research program in the fields of fire safety and risk analysis. Research projects are concerned with risk analysis, risk-based fire safety design, modeling of fires, extinguishing media and extinguishing processes, industrial risks, and rescue services. The Department of Fire Safety Engineering is a part of the Lund Combustion Center. The divisions within the Combustion Center are involved in several networks, nationally as well as internationally, and they have connections to national and international industries. Since a graduate school, the Center of Excellence in Combustion Science and Technology (CECOST) is also based at Lund Institute of Technology, there are possibilities to pursue advanced courses in the field. For additional information, contact Department of Fire Safety Engineering, Lund University, P.O. Box 118, SE-221 00 Lund, Sweden.

Oklahoma State University. The Oklahoma State University (OSU) School of Fire Protection and Safety Technology (FPST) is located on the Stillwater campus of OSU. The FPST Program is a 4-year degree program that concludes with a Bachelor of Science in Engineering Technology. The FPST curriculum is designed to prepare graduates to assess and reduce the loss potential existing in an operation with respect to fire, safety, industrial hygiene, and hazardous material accidents.

The School of Fire Protection and Safety Technology is under the College of Engineering, Architecture, and Technology (CEAT). Note that OSU also offers nondegree programs such as seminars. For additional information, contact Department of Fire Safety and Technology, Oklahoma State University, 303 Campus Fire Station, Stillwater, OK 74078-4082, Phone: (405) 744-5721, Fax: (405) 744-6758.

University of Maryland at College Park, Clark School of Engineering. The University of Maryland Department of Fire Protection Engineering offers undergraduate and graduate degrees in fire protection engineering. Candidates for the master's degree may choose either a thesis or nonthesis program. In addition, a Professional Master's or Master of Engineering degree is offered for practicing engineers. Distance learning opportunities are available. For additional information, contact Department of Fire Protection Engineering, 0151 Engineering Classroom Building, University of Maryland, College Park, MD 20742-3031, Phone: (301) 405-3992, http://www.enfp.umd.edu.

University of New Haven Department of Fire Science. The University of New Haven offers three undergraduate degrees and four certificate programs designed for individuals entering the field of fire science. Classroom lectures, laboratory sessions, case studies, and field trips are utilized to give the student the exposure in this area of study. Internships are used to allow the student to obtain real-life work experience in this specialized field. The university also offers graduate certificate programs and a master's degree in fire science for those completing their bachelor's degrees. For more information contact University of New Haven, 300 Orange Avenue, West Haven, Connecticut 06516.

Worcester Polytechnic Institute. Worcester Polytechnic Institute offers a dual-degree 5-year program for high school graduates as well as an MS and Ph.D. in fire protection engineering. Thesis and nonthesis options are both available. The master's degree, graduate certificate, and advanced certificate are available to practicing engineers both on campus and off campus via distance learning. Worcester Polytechnic Institute offers fully paid internships and undergraduate co-cop jobs for full-time students. For additional information, contact Center for Firesafety Studies, Worcester Polytechnic Institute, 100 Institute Road, Worcester, MA 01609, Phone: (508) 831-5593, http://www.wpi.edu/+FPE.

Other Institutions. Other schools that offer opportunities in fire protection engineering education include

- Carleton University, Canada
- Fire Safety Engineering College (Muscat and Oman)
- Hong Kong Polytechnic University (Hong Kong and China)
- South Bank University (England)
- Science University of Tokyo (Japan)
- Swiss Federal Institute of Technology (Switzerland)
- University of California at Davis—certificate in fire protection
- University of Canterbury (New Zealand)
- University of Central Lancashire (England)
- University of Edinburgh (Scotland)
- University of Greenwich (England)
- University of Leeds (England)
- University of New Brunswick (Canada)
- University of Ulster (Northern Ireland)
- Victoria University of Technology (Australia)

Summary

Sprinkler system design is an important function of a fire protection engineer. The future of the profession depends not only on competent design but on ethical design. This chapter has summarized ways in which engineers can remain current relative to competency and has reviewed sources that may be used by engineers in meeting the ethical expectations of the profession.

BIBLIOGRAPHY

Reference Cited

1. *Raising Public Awareness of Engineering*, National Academy of Engineering, The National Academies Press, Washington DC, 2002, Table 2-1, p. 11.

NFPA Codes, Standards, and Recommended Practices

NFPA Publications National Fire Protection Association, 1 Batterymarch Park, Quincy, MA 02169-7471

The following is a list of NFPA codes, standards, and recommended practices cited in this chapter. See the latest version of the *NFPA Catalog* for availability of current editions of these documents.

NFPA 13, *Standard for the Installation of Sprinkler Systems*

Index

Note: page numbers followed by *f* have figures; page numbers followed by *t* have tables.